Modeling & Simulation-Based Data Engineering

Modeling & Simulation-Based Data Engineering:

Introducing Pragmatics into Ontologies for Net-Centric Information Exchange

Bernard P. Zeigler
Phillip E. Hammonds

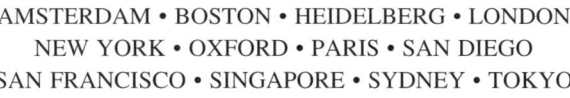

AMSTERDAM • BOSTON • HEIDELBERG • LONDON
NEW YORK • OXFORD • PARIS • SAN DIEGO
SAN FRANCISCO • SINGAPORE • SYDNEY • TOKYO

Academic Press is an imprint of Elsevier

Elsevier Academic Press
30 Corporate Drive, Suite 400, Burlington, MA 01803, USA
525 B Street, Suite 1900, San Diego, California 92101-4495, USA
84 Theobald's Road, London WC1X 8RR, UK

This book is printed on acid-free paper. ∞

Copyright © 2007, Elsevier Inc. All rights reserved.

No part of this publication may be reproduced or transmitted in any form or
by any means, electronic or mechanical, including photocopy, recording, or any
information storage and retrieval system, without permission in writing from
the publisher.

Permissions may be sought directly from Elsevier's Science & Technology
Rights Department in Oxford, UK: phone: (+44) 1865 843830, fax:
(+44) 1865 853333, E-mail: permissions@elsevier.co.uk. You may also complete
your request on-line via the Elsevier homepage (http://elsevier.com), by
selecting "Customer Support" and then "Obtaining Permissions."

Library of Congress Cataloging-in-Publication Data
APPLICATION SUBMITTED

British Library Cataloguing in Publication Data
A catalogue record for this book is available from the British Library

ISBN 13: 978-0-12-372515-8
ISBN 10: 0-12-372515-1

For all information on all Elsevier Academic Press publications
visit our Web site at *www.books.elsevier.com*

Printed in the United States of America
07 08 09 10   9 8 7 6 5 4 3 2 1

**Working together to grow
libraries in developing countries**

www.elsevier.com | www.bookaid.org | www.sabre.org

ELSEVIER   BOOK AID International   Sabre Foundation

This book owes much to a multitude of colleagues whose ideas and hard work contributed to its development. Among those that come to mind are: Dasia Benson, Steve Bridges, Erika Carswell, Saehoon Cheon, Rich Clarke, Jerry Couretas, Robert Flasher, Dale Fulton, Anthony Galassi, Steve Gayer, Xiaolin Hu, Moon Hwang, Steve Kerr, Doowhan Kim, Ken Kingston, Rodney Leist, Stephen Madden, Eddie Mak, Saurabh Mittal, Robin Moore, Kimberly Nunn, James Nutaro, Greg Plum, Allan Reifer, Hessam Sarjoughian, Chad Schulenberg, Chungman Seo, and Ming Zhang. To them, and all others we may have failed to mention, we offer our heartfelt appreciation.

Bernard P. Zeigler
Phillip E. Hammonds
Arizona, 2007

Contents

FOREWORD xi

PART I FOUNDATIONS

1 ONTOLOGIES AND INFORMATION EXCHANGE: A PRAGMATIC FRAMEWORK 3

How Ontologies generate world state descriptions for use in application contexts called pragmatic frames

2 BACKGROUND: SYNTAX, SEMANTICS, AND ONTOLOGY FRAMEWORKS 13

Introduction to XML and related ontology environments

3 FORMULATING PRAGMATIC FRAMES AND ONTOLOGIES: GEOSPATIAL SENSOR DATA 35

Illustrating pragmatic frame concepts in the geospatial sensing context

PART II SYSTEM ENTITY STRUCTURE CONCEPTS AND OPERATIONS

4 INTRODUCTION TO THE SYSTEM ENTITY STRUCTURE 53
Basic concepts of system entity structure (SES) with examples of how to apply them

5 SYSTEM ENTITY STRUCTURE AXIOMS: INTERPRETATIONS AND APPLICATIONS 71
Illustrated presentation of the set theoretic formulation of SES and how the axioms are interpreted and how they impact application

6 SYSTEM ENTITY STRUCTURE: COMPUTATIONAL REPRESENTATIONS 90
Representing SES in XML, DOM, JAVA, and Natural Language and their transformations

7 MAPPINGS: TRANSFORMATIONS AND RESTRUCTURINGS 113
Restructurings of the SES that support development and harmonization

8 PRUNED ENTITY STRUCTURES AND XML SCHEMA INSTANCES 128
Pruning as a process for generating world state descriptions and their encodings in XML

9 CONSTRAINED PRUNING 146
Using rules and relations to constrain pruned entity structures (PES) to better describe the application domain

10 PRUNED ENTITY STRUCTURES: DATA EXTRACTION AND CHANGE-BASED INFORMATION EXCHANGE 176
Applying the SES and PES structures to implement efficient data extraction and information exchange

PART III MODELING AND SIMULATION AND DATA ENGINEERING

11 HIERARCHICAL SYSTEMS, MODELS, AND SIMULATIONS: THE SES ONTOLOGY 195

Connects the SES with the domain of simulation modeling using application examples

12 MANAGING SYSTEM ENTITY STRUCTURES: COMPOSING LARGE SYSTEMS 210

Merging SESs and PESs to compose large hierarchical structures from components

13 HARMONIZING DATA REPRESENTATIONS AND ONTOLOGIES WITHIN PRAGMATIC FRAMES 219

How to use Pragmatic Frames and SES concepts to provide a sound basis for information exchange

14 GEOSPATIAL SENSOR DATA: THE UNIVERSAL PHASE HISTORY DATA (UPHD) STANDARD 242

An example of the application of SES-based methodology to a significant real-world problem domain

15 PROCESSING NETWORKS AND PRAGMATIC FRAMES 273

Formulates networks of information processing nodes in terms of metadata operations and pragmatic frames

16 DYNAMIC PRAGMATICS: ISSUES AND METHODOLOGY 288

Discusses how to evaluate metadata in existing data exchange standards and to manage evolution over time

PART IV TESTING IN NET-CENTRIC ENVIRONMENTS

17 TESTING IN A NET-CENTRIC ENVIRONMENT: TECHNOLOGY BASIS 323

Automation based on the book's concepts achieves rigor and faster test development in a net-centric environment

18 TESTING IN A NET-CENTRIC ENVIRONMENT: MULTIPLE LEVELS 361

How to use the framework to test at syntactic, semantic, and pragmatic levels

19 BRINGING IT ALL TOGETHER: MODELING AND SIMULATION-BASED DATA ENGINEERING 388

In-depth examination of the book's contribution and its relation to the state-of-the-art

GLOSSARY 418

INDEX 423

Online companion site:
http://books.elsevier.com/companions/9780123725158

Foreword

This book was born out of necessity of our work with various government agencies, testing facilities, standards organizations, and developers who are concerned with interoperability. While interoperability used to be a once-in-a-while issue, it is now becoming an institutional necessity with the planned transition of all Department of Defense operations to the Global Information Grid (GIG), a high-speed version of the World Wide Web. There are many compatibility issues because organizations have tended to develop their own unique approaches to representing data over the years. Now they will have to reconcile these differences so they can share data needed for effective collaboration. Unfortunately, there are few well-established methods for harmonizing data representations and testing their validity in real-world contexts. This book aims to fill some of the void.

The GIG is to be organized as a marketplace of open and discoverable web services called the Service Oriented Architecture (SOA), incorporating Semantic Web technologies as they mature. Indeed, the transition to the GIG/SOA is part of an accelerating trend to greatly increased complexity of information technology-based systems. And this complexity requires that interoperability testing become more rigorous, in-depth, and thorough. At the same time, to keep up with the rapid change and short development life cycles expected of web services, tests have to be ready to be conducted in much shorter time scales. This brings out the second motivation of the book: introducing model-based dynamic data engineering and simulation to automate testing methodology, thereby increasing rigor and speeding up test development in the emerging net-centric environment.

Where does the problem just outlined fit into the current array of concepts, methodologies, and technologies? Data must be encoded into a format to send it from a producer to a consumer. So there is a purely syntactic dimension to the problem: assure that the message formats are standardized and develop tests to see that the standards are adhered to. On the web, the eXtensible Markup Language (XML) is the standard format, and the problem can be viewed as assuring that all data sets are encoded into XML and that the standards for sending and receiving XML are adhered to. Unfortunately, the problem just starts at this level. There are myriad ways, or schemata, to encode data into XML and a good number of such schemata have already been developed. More often than not, they are different in detail when applied to the same domains. What explains this incompatibility?

There are various ways of breaking down a real situation and expressing it using a collection of terms. Two decompositions might well refer to slightly, or even markedly, different aspects of the same thing using the same terms. In other words, to interoperate there must be a common understanding of the semantics — what the data refer to. A similar problem in semantic web research has led to a focus on ontologies. These are logical languages that provide a common vocabulary of terms and axiomatic relations among them for a subject area. However, semantic web researchers typically seek to develop intelligent agents that can draw logical inferences from diverse, possibly contradictory, ontologies such as a web search might discover. In contrast, the newly emerging area of ontology integration assumes that human understanding and collaboration will not be replaced by intelligent agents. Therefore the goal is to create concepts and tools to help people develop practical solutions to incompatibility problems that impede "effective" exchange of data and ways of testing that such solutions have been correctly implemented.

But what does "effective" mean in the last sentence? A command is effective if it has the effect that the general intended it to have on his subordinates, which is to say, they carry out the given orders. So effective data exchange requires not only shared agreement on syntax and on semantics, but also on pragmatics, the use to which data will be put.

Although the approach is generic, much of the application context of this book relates to geographic information as derived from intensive processing of terabytes of data collected from satellite and airborne sensors. This context is not only important on its own, but provides a microcosm for the larger problem domain of GIG/SOA interoperability. Geospatial sensors are sophisticated physics-based dynamic systems that are expensive to build and highly dependent on advanced engineering knowledge for their operation. This makes it nontrivial to retrofit them to the SOA concept of data-centered, interface-driven, loose coupling between producers and consumers. The SOA concept requires the development of platform-independent, community-accepted standards that allow raw data to be syntactically packaged into XML and accompanied by metadata that describes the semantic and pragmatic information needed to effectively process the data into increasingly higher-value products downstream.

This book presents an approach that was developed, in part, to meet the demands of developing interoperability standards and testing for their compliance in the geospatial sensor context. Given the physics-based nature of sensors it was natural to turn to a generic framework that had been developed to model and simulate such dynamic, continuous/discrete systems. The book presents a rigorous, yet intuitive, development of the System Entity Structure (SES) and its role as an ontology framework for static and dynamic world state descriptions, the latter expressed within the Discrete Event Systems Specification (DEVS) formalism.

From a static point of view, the SES is an ontology framework that is much closer to XML than that of semantic web ontology. At this level, the SES automates the creation of XML schemata using a data model that reflects system engineering concepts of hierarchical decomposition and specialization. In this sense, it occupies a space in the modeling pantheon similar to a commonly employed software engineering scheme, the Unified Modeling Language (UML). Among the advantages that it offers in this regard are a conceptual framework that is more expressive for the domain of real-world engineered systems and supporting set of tools for Schema development and testing in a virtual workspace at www.devsworld.org.

From a dynamic point of view, the SES together with the DEVS formalism offers a powerful system-theoretic framework for specifying families of dynamic services that can execute in simulated or real-time and interact with other services in a net-centric environment. In this guise, the book provides an in-depth discussion of the automated development of test agent federations. These agents can validate the behavior, and evaluate the performance and effectiveness, of web-service collaborations through observation of their external message exchanges. As an example, we discuss end-to-end and multilevel testing of collaborating services on the GIG/SOA to achieve prespecified mission goals.

Part I

Foundations

1

Ontologies and Information Exchange: A Pragmatic Framework

This chapter sets out the foundation for the rest of the book. It introduces the concept of an ontology to describe a state of the world and its changes over time. Information exchange occurs when a person or information system reports a new state of the world, or changes in previous states, to another person or information system. The person or system that generates and reports the information is called a *producer* while the user is called a *consumer*. The principle that we want to emphasize in this book is that the consumer's *use* of the information should determine the description mechanism, or ontology, used by the producer. We'll formulate a means of characterizing the consumer's use of the information and call it a *pragmatic frame*. The developer of the ontology, also called the data engineer, has the task of tuning the ontology to the pragmatic frame. We'll discuss the nature of this task once the concepts of ontology and pragmatic frame have been clarified. Later in the book, we will look at the exchange as a conversation, a metaphor that suggests further additions to the pragmatic frame concept and its applications.

Before proceeding further, we will introduce the concept of ontology with an example that tries to describe the social relations of four people marooned on an island.

Ontology for Social Relations on an Island

There are four people stuck on an island: a, b, c, and d (you are free to associate your favorite movie or sitcom characters with these letters). Because people in such a situation frequently change their affection for each other, we want to represent in some way their daily likes and dislikes of one another. We begin with the following matrices

Likes

	a	b	c	d
a		1	1	
b	1			
c	1			1
d		1		

Dislikes

	a	b	c	d
a				1
b				
c		1		1
d	1			

Using this framework, there is a large but easily computable number of possible states of affection that characterize this small number of people. Indeed, there are $4 \times 4 = 16$ cells in the *Likes* matrix, each of which can independently have a 1 or a blank. Therefore, there are 2^{16} states of the *Likes* matrix and, again assuming independence, the same number for the *Dislikes* matrix, for a total of 2^{32} that is the product of the two groups.

In reality, the number of states is much smaller because the following constraints, which we will examine in more detail later in this chapter, might apply:

1. No one likes or dislikes himself or herself;
2. Affection is mutual and so is disaffection (for example, you like someone who likes you, and the same applies to disliking someone);
3. Liking and disliking are mutually exclusive — you can't like and dislike someone at the same time; and
4. To quote an ancient proverb, "The enemy of my enemy is my friend."

Definition of Ontology

Numerous definitions of the concept of ontology can be found in the literature.[1-5] We'll employ one that is particularly appropriate to the point of view taken in this book. The Federation of Intelligence Physical Agents (FIPA) provides a discussion of ontology that can be found on the web.[6] Unfortunately, like most frameworks that attempt to capture a concept with some fair degree of formal rigor, it may appear forbidding to some readers. So instead we'll leave examining the full definition to the interested reader and instead illustrate its essence with our small "island of affection" example.

In the following, we try to explain the underlying motivation for the terms used in the FIPA definition in the context of the just-mentioned example.

First, set theory is used to describe the ontology developer's conceptualization of the domain — also called the *universe of discourse*:

- *domain space*, $\langle D, W \rangle$, where D = the set of individuals, $\{a, b, c, d\}$ and W = the 2^{32} possible world structures — these are the states of affection of the people on the island before any restrictions are placed on the possible combinations.
- *conceptualization*, $\mathbf{C} = \langle D, W, \mathfrak{R} \rangle$ where $\mathfrak{R}$ consists of the two conceptual relations, *Likes* and *Dislikes* both having two arguments (arity = 2). Here, let us say, that we require the first two constraints above to hold (the relations are irreflexive and symmetrical). From this point of view, there are many ways to conceptualize the same domain — for example, by accepting some subset of the above constraints.
- *intended world structures according to* $\mathbf{C}$, $\mathbf{S_C}$ is the subset of structures that are compatible with the conceptual relations. Note that the number is greatly reduced because under the first two constraints the only possible assignments to the matrices are to their upper-right triangles. In this case, there are six independent elements; so there are $2^6 \times 2^6 = 2^{12}$ world structures **intended** or allowed by the conceptualization C.

Then the concept of a logical language is used to capture the difference between a symbolic way of describing the universe of discourse and the possible descriptions that it allows:

- *language L* has a vocabulary, V, with constant symbols and predicate symbols. In our example, let the constant symbols be

a, b, c, d and the predicate symbols be *likes* and *dislikes* which can be read, e.g., *likes(a, b)*.
- *intensional* interpretation of L is a structure $\langle \mathbf{C}, \mathfrak{F} \rangle$, where $\mathfrak{F}$: $V \rightarrow D \cup \mathfrak{R}$ is a function assigning elements of D to constant symbols of V, and elements of $\mathfrak{R}$ to predicate symbols of V. Here the mapping is obvious — constants map to names (e.g., constant a to person a), and predicates map to corresponding relations (e.g., *likes* to *Likes*).
- The set of *models* of L is (in one-to-one correspondence with) the set of intended world structures $\mathbf{S}_C$. In other words, L provides a way of writing down the domain and its relations; its interpretations, called its models, are just the possible states allowed by the relations. For example, the statement:

likes(a,b),likes(b,a), dislikes(c,d),dislikes(d,c)

represents the state of affection:

Likes	a	b	c	d
a		1		
b	1			
c				
d				

Dislikes	a	b	c	d
a				
b				
c				1
d			1	

Now, let's examine in more detail how a logical language can express the constraints that might capture our conceptualization of a particular domain.

In our example, having decided on two of the four restrictions, the ontology will have the *axioms* (statements that are accepted as true without proof):

$$\neg likes(x,x)$$
$$\neg dislikes(x,x)$$

(for irreflexivity) and

$$likes(x,y) \Rightarrow likes(y,x)$$
$$dislikes(x,y) \Rightarrow dislikes(y,x)$$

(for symmetry).

Now, suppose our conceptualization wants to capture the states of affection that are constrained by all four restrictions above. Then the ontology developed so far would include many other states. For example, it allows the state:

Definition of Ontology

	Likes					Dislikes			
	a	b	c	d		a	b	c	d
a		1			a		1		
b	1				b	1			
c					c				
d					d				

in which two people both like and dislike each other.

Suppose then we add the axiom:

$$likes(x,y) \Rightarrow \neg dislikes(x,y)$$
$$dislikes(x,y) \Rightarrow \neg likes(y,x)$$

This reduces the possible world structures to 2^6 because we can only choose one of the matrices freely — the other is completely determined by the first.

However, this ontology would still fail to restrict the possible states far enough. We would have to add the axiom:

$$dislikes(x,y) \land dislikes(y,z) \Rightarrow likes(x,z)$$

to enforce the dictum, "the enemy of my enemy is my friend."

For example, if a dislikes b and b dislikes c, we have:

	Dislikes			
	a	b	c	d
a		1		
b	1		1	
c		1		
d				

Then we must have a likes c as in:

	Likes			
	a	b	c	d
a			1	
b				
c	1			
d				

Exercise

What is the size of the resulting set of possible world states?

Disclaimer: The example we used to illustrate the concepts underlying ontologies and their attempts to capture some aspect of reality was somewhat deceptive. We laid out the axioms initially that would be considered later as the development proceeded. This may give a false impression that such axioms are always known in advance. However, in practice, the development of an ontology may be a series of approximations to the desired conceptualization, with probable wrong turns along the way. As in all modeling of reality, it is important to have clear objectives and to let them guide your focus on discovering the critical relationships, or axioms, that best structure the domain of interest. We'll introduce the concept of pragmatic frame to help develop these objectives in the next section.

Ontology Summary: Prelude to the Framework in This Book

FIPA's concept of an ontology is a logical formalism that tries to express a set of possible world structures through its family of models. The definition itself tries to make clear that the language with its axioms usually cannot exactly capture the set of world structures that the definer intends, so it is an approximation to this intended conceptualization.

In our approach, we will not necessarily restrict the defining mechanism to a logical formalism. For example, eXtensible Markup Language (XML) might be used to capture a set of possible world states while itself not being explicitly formulated within the approach of classical logic. We will explicitly incorporate a concept of *pragmatic frame*, which delineates a data engineer's domain of interest and relates an ontology as being adequate or not to this domain. It will turn out, for example, that without further validity-checking tools, XML cannot represent the worlds in which enemies of enemies are always friends. Further, we will also want to explicitly consider how world states can change over time. For example, it might be that our islanders can't get along with each other for more than one day at a time, therefore, if x likes y today, then x will dislike y tomorrow. Further, suppose that it takes a whole week for people who have a falling out to reconcile. So, if x dislikes y today, then only this time next week will the corresponding "1" in the *Dislike* matrix disappear. Thus, we'll be interested not only in static data engineering but also in dynamic data engineering, and this will lead us toward including the full capability of modeling and simulation in our tool kit.

Pragmatics: The Information Exchange Framework

A producer observes the state of the world at its location and encapsulates it in a message to a consumer that stores it for later use. As we noted previously, the world state is a member of a set delineated by an ontology that reflects the designer's conceptualization. But what could that conceptualization be? By taking into account the *pragmatics* of the situation, we get a way to answer this question. By pragmatics we mean the aspect of language that has to do with how the information transmitted by a message will be used. In Figure 1.1, we can point out how the stored data will be used later.

Consider two examples.

EXAMPLE: *Information Sent by a Dealer to Department of Motor Vehicles*

Whenever you buy a new car, some information is sent from the dealer to the Department of Motor Vehicles where you live, which stores it in its database. The pragmatics of the use here is the subsequent processing of this data — it will be used when you register the vehicle and pick up your plates, and to compute taxes and annual renewals.

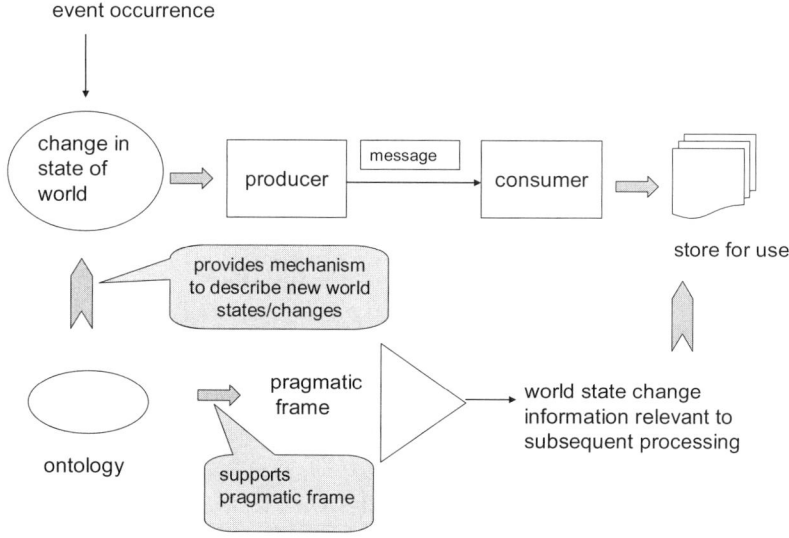

Figure 1.1

Information Exchange Framework

EXAMPLE: *Information Sent by a Dealer to Manufacturer's Headquarters*

Every month a car dealership sends a sales report to the manufacturer's headquarters to allow it to assess inventory levels and plan its production schedule for the next month. Table 1.1 illustrates the framework.

In our framework, an event occurred when you bought the car. This changed the world because the car left the dealer's lot and it is now yours to drive. The producer is the same — the car dealer — but the pragmatics are different in the examples just given: the consumers are different (Department of Motor Vehicles and manufacturer) and their subsequent use of the received information will be different. **The point is this: notice how the pragmatics determine the nature of the data to be sent, namely, the contents of the message.** In the first case, the dealer needs to provide specific information about the buyer (name, address, etc.) and vehicle (identification number, make, model, etc.). In the second case, this information is not relevant. Instead, the dealer needs to inform the manufacturer of how many cars of each make and model he sold during the month.

We'll say that an ontology *supports* (or *is applicable to*) a *pragmatic frame* if the world states (or state changes) that it can describe include those that are needed by the frame. In other words, the message contents encode the right information for the intended use. Notice that an ontology can be designed to describe a set of world states for some domain. This might be feasible for

Table 1.1. The Purchase of the Car Caused a World State Change

Event causing world state change	Producer	Consumer	Pragmatics: subsequent use	Message contents
New car purchase — car ownership is transferred from dealer to buyer	Dealer	Department of Motor Vehicles	Register owner's vehicle and give out plates	Buyer and vehicle identification
Same as above	Same as above	Manufacturer Headquarters	Assess inventory levels and production schedule	Number of cars of each make and model that were sold during the month

a limited situation such as the states of affection of four people on an island. However, in many situations, it would be much more efficient to describe the change in state caused by an event in contrast to the new state that was engendered. So we allow for ontologies that describe state changes rather than states. For example, in the case of an island with one million people, we might explicitly convey the few pairs of people that started, or ceased, liking each other since the last update. Indeed, we'll later discuss such change-based updating in depth. For the moment, we note that the producer and consumer must have agreed initially that entries in the database remain valid unless explicitly updated.

An ontology is *minimal* for a frame if it supports only that frame, not a larger one. For example, we could have minimal ontologies for the vehicle registration and inventory update frames, respectively. As illustrated in Figure 1.2, the first ontology would provide a means to describe the buyer's name, address, and Social Security number and the vehicle's make, model, and identification number. The second would provide a means to describe the number of cars of each make and model sold during the last month. On the other hand, we could have one larger, more general ontology to cover both needs (and possibly others). The minimalist approach offers conciseness and efficiency, whereas the generalist approach potentially offers better

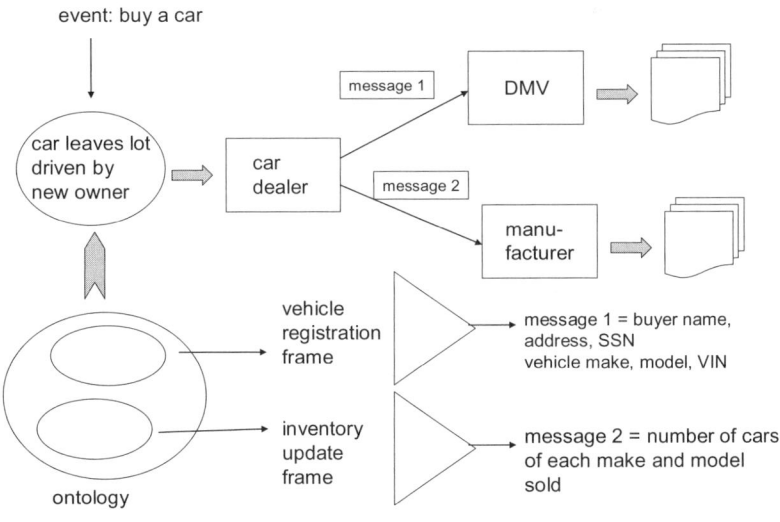

Figure 1.2

Car Purchase Example of the Information Exchange Framework

integration and, if designed with forethought, potential extensibility to meet future demands. There will always be a tradeoff between these approaches, but the *scope* of the ontology — the part of the world that it can help to describe — must ultimately be limited by a deliberate choice of the developer.

Summary

An ontology should be designed to support one or more pragmatic (or subsequent use) frames. Such an ontology provides the symbolic means (e.g., logical language or XML) to describe world state changes that a producer sends in the contents of a message to a consumer. The contents of the message encode state changes in a way that can be subsequently processed by a consumer for its use in one of the supported frames. The pragmatic frames of vehicle registration and inventory update are quite different and might be supported by two distinct minimalist ontologies even though the triggering event and producer are the same. On the other hand, we could have a more generic ontology that supports both frames. The data engineer must decide on the proper balance between conciseness and extensibility in such ontology design.

References

1. Grigoris, Antoniou, and Frank van Harmelen, "A Semantic Web Primer," Cambridge: MIT Press, 2004
2. Asuncion, Gomez-Perez, Oscar Corcho, and Mariano Fernandez-Lopez, "Ontological Engineering," New York: Springer, 2004
3. Dieter Fensel, "Ontologies: A Silver Bullet for Knowledge Management and Electronic Commerce," New York: Springer, 2005
4. Alexiev, V., M. Breu, J. de Bruijn, D. Fensel, R. Lara, and H. Lausen, "Information Integration with Ontologies," Hoboken, NJ: John Wiley, 2005
5. Nicola Guarino, "Understanding, building and using ontologies," International Journal of Human Computer Studies, Incorporating Knowledge Acquisition, Vol. 46, Number 2/3, February/March 1997
6. Ontology Service Specification, Foundation For Intelligent Physical Agents, http://fipa.org/specs/fipa00086/XC00086D.html#_Toc 505571321 (accessed Nov. 2006)

2

Background: Syntax, Semantics, and Ontology Environments

This chapter examines the mechanics of information exchange, semantics, and syntax, and their role in supporting the specification of ontologies. We review the strengths of commonly used data engineering and knowledge representation schemes to support ontology specification.

Returning to Figure 1.1, we have seen that pragmatics determine the state change information that is of interest to the consumer. *Semantics* and *syntax* support the content and form of the messages in which the information is packaged. For a natural language, such as English, *semantics* relates to the meaning of a word, sentence, or larger portion of text. The meaning might be conveyed by a dictionary whose entries try to explain the meaning of a word by offering one or more phrases that might be used in place of it. Alternatively, the meaning might be conveyed by providing word equivalents in another language, e.g., an English to French dictionary. Semantics also concerns how meanings of composites such as sentences are built up from the meanings of their constituents. *Syntax* refers to the rules governing the structure of sentences and their constituents. Syntax thus underlies semantics: when producer and consumer adhere to a mutually agreed to syntax, the consumer can more easily understand the intended meaning of a sentence. For example, if nouns always precede verbs, then we would interpret "barks the dog" as asserting that the name of a dog is "barks"; whereas if both orders are allowed, we could also interpret this as asserting that a dog is barking (the more likely meaning!).

Semantics: Testing Web Services

For ontologies, semantics and syntax refer to similar concepts as for natural language; however, they now apply to computer programs with varying degrees of intelligence. The most fundamental way to deal with semantics in this context is to relate the meaning of a message to the eventual outcome of the processing that it supports. For example, we would say that the meaning of a message in the vehicle registration frame is the manner in which a correct processing algorithm would handle the buyer's and vehicle's identification data fields. For correct processing, the make, model and vehicle identification number (VIN) must be filed under the buyer's name. Later when the buyer appears at the Department of Motor Vehicles (DMV), the clerk can retrieve the vehicle information by querying the buyer's name and proceed to compute the correct tax. If the system treats the VIN as if it were the buyer's name and misfiles the information, the intended transaction will not complete and the buyer will be most unhappy. Thus, as a test of semantic comprehension of the registration frame ontology, we could simulate the purchases of several customers and test whether they are successfully registered at the DMV. If all transactions are on the web, we could launch test agents that might execute the state transitions illustrated in Figure 2.1. Later we'll discuss how such test agents can be automatically generated from the ontology and the rules that the downstream processor is expected to live by (Chapter 16). Note that our test of comprehension (used only in a metaphoric sense) is a pragmatic test of behavior — does the system under test respond in the right way when exchanging messages with a test agent?

Exercise

Develop a test for correct comprehension of the inventory update ontology information. Provide your own rules for the downstream processing.

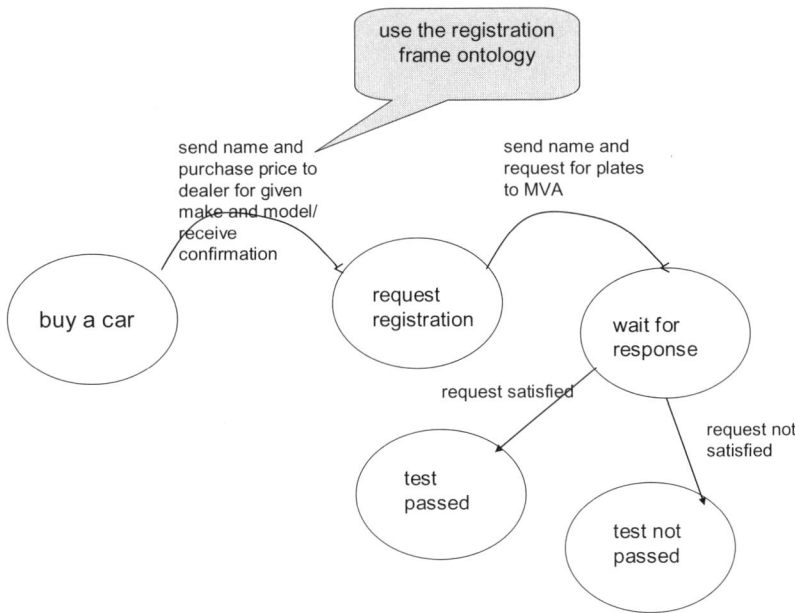

Figure 2.1

State Transition Diagram for an Agent to Test for Correct Comprehension of Registration Ontology

Syntax and XML

Because syntax is a well-developed topic in computer science, we refer to basic texts on the subject.[1–3] The precursor to eXtensible Markup Language (XML[4,5]), known as Hypertext Markup Language (HTML), has a well-defined syntax in which a small set of tags may be nested within each other to delineate portions of text in a document. For example, open your browser to a favorite page and click on "view source." You'll see <html> <head> <title> some text </title> </head> </html>. This allows your browser to treat "some text" as a title and place this text in an appropriate area for display. A parser can traverse the recursive structures defined by such context-free syntax and extract the marked text fragments. But basically this all has to do with layout and links to other Uniform Resource Locators (URLs), not with the meaning of the document. XML goes one major step further. It allows documents to carry or point to definitions of the tags used within them, thereby setting up the basis for their semantic interpretation. For example, a notice for a buyer and vehicle might be defined by a document type definition (DTD) as follows:

```
<?xml version="1.0" encoding="us-ascii"?>
<!ELEMENT notice (buyer,vehicle)>
<!ELEMENT buyer (#PCDATA)>
<!ATTLIST buyer
  firstName CDATA #REQUIRED
  lastName CDATA #REQUIRED
  SSN CDATA #REQUIRED
>
<!ELEMENT vehicle (#PCDATA)>
<!ATTLIST vehicle
  make CDATA #REQUIRED
  model CDATA #REQUIRED
  VIN CDATA #REQUIRED
>
```

Then a notice for my purchase of a Toyota Prius might be sent over the web as follows:

```
<?xml version="1.0" encoding="UTF-8"?>
<!DOCTYPE notice DTDForRegistration.dtd">
<notice>
        <buyer firstName="Bernard"
        lastName="Zeigler"
        SSN="000-00-0000"/>
        <vehicle make="Toyota" model="Prius"
        VIN="00000-00000-00000-00000"/>
</notice>
```

When a consumer receives this *instance* of notice, it has to have 1) access to the proper DTD (DTDForRegistration.dtd) and 2) an XML parser that can tear apart the text and isolate the pieces delineated by the tags — these are the ELEMENTs defined in the DTD. What's different from HTML is that the tags now have some semantic content: the buyer tag refers to its namesake and has attributes that provide the referent's first and last names and Social Security number. The same applies to the vehicle I purchased. Indeed, this looks very much like the definition of a class for notice in an object-oriented programming language referencing two classes — buyer and vehicle — each with some instance variables. The difference is that the whole document is an ASCII text file that can be created/interpreted by any producer/consumer pair on the web using any language with string handling capability. For example, the producer might employ C++ to generate an instance document and the consumer might employ Java to parse it and convert it into a Java object. For this reason, XML has become the standard for information exchange over the web and is employed as the support layer within all proposed architectures for web services.

Semantics — Ontology's Ability to Match Its Set of World Structures to the Pragmatic Frames It Supports

In our framework, the measure of success of a producer/consumer information exchange is whether the consumer correctly stores the world state/change and that downstream processing of this information correctly executes the intended pragmatic frame requirements. As we have seen, syntax provides the means by which world states or changes thereof are cast into a form that can be broken down into their meaningful pieces. Semantics provides the means by which meaning is actually assigned to the pieces and to the whole. *An ontology uses syntactical and semantical constructs to match the set of world states/changes it can describe to those required by the one or more pragmatic frames it supports.*

In practice, ontologies are developed using specific languages and associated environments that support defining the necessary syntax and semantics. These languages and environments also provide tools for verifying whether the world states/changes so described match those intended by the data engineer's pragmatic frames. However, such languages/environments tend to differ in the nature of the syntactical/semantical expressive power and verification support they offer. One of the objectives of this book is to provide you, as a data engineer, with an appreciation of the strengths and limitations of commonly employed approaches so that you make the most of their capabilities and avoid some pitfalls that may otherwise intervene. Another objective is to provide you with a more encompassing approach with associated tools that might help to solve some of the problems current approaches don't address. For now, we'll review some common approaches whose relative capabilities are illustrated in Figure 2.8 and summarized in Table 2.1.

XML Schema: Strengths and Limitations

We have already discussed XML's capability to define tags and attributes using DTDs that provide an analog to object classes

Table 2.1. Some Commonly Employed Languages/Environments for Ontology Development

Language/environment for developing ontology	Features	Tools
XML (DTD)	Tags and attributes relate to object classes and instance variables	Syntax — well-formedness checker — validity checker
XML (schema)	— Simple data types (strings, integers, etc.) — Complex data types (composites of simple ones) with regular expression restrictions — Element multiplicities	Validity checker extended to check — whether data conforms to type specification — if all multiplicities are honored
UML 1.4	Objects, classes, attributes, methods, generalization, associations with multiplicities, interactions, state charts	— Methods check for consistency among various diagrams (class, interaction, collaboration, etc) — XMI provides mapping into XML (See Chapter 19 for consideration of UML 2.0)
OKBC (Open Knowledge Base Connectivity) Knowledge Model	— Object-oriented representation of knowledge — Constants, frames, slots, facets, classes, individuals, and knowledge bases — Specified within Knowledge Interchange Format (KIF), a first-order predicate logic language with set theory	Protégé, Ontolingua
OWL	— Classes, attributes, class hierarchies — Properties (stand-alone binary relations)	Protégé, Ontolingua (See Chapters 13 and 19 for further consideration of OWL)

and instance variables. However, the values that can be placed in a document's slots (i.e., within tags and assigned to attributes) can only be strings. The XML *schema* was introduced to allow you to specify that one of a set of simple data types (strings, integers, etc.) must be used in such slots. Complex data types,

XML Schema: Strengths and Limitations

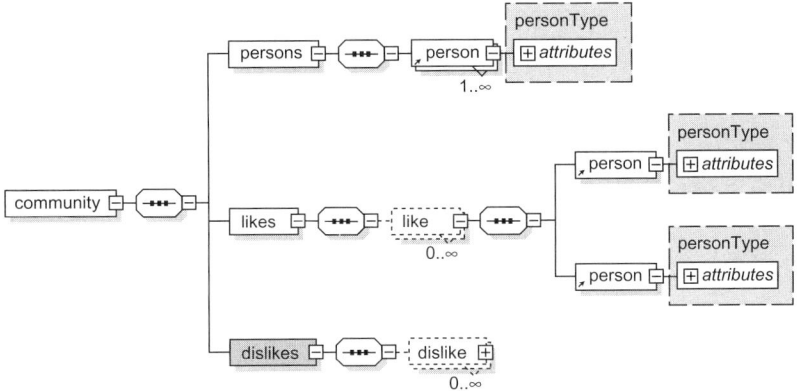

Figure 2.2
XML Schema for the Island Social Relations Ontology (illustrated using XMLSpy®[6,7])

which are extensions and composites of simple ones, can also be defined, and restrictions on values can be specified using regular expressions. Multiplicities can be specified for occurrences of elements. Further, a tool can verify that a document purported to be an instance of a schema adheres to such specifications.

As an example, consider a formulation of the island social relations ontology shown in Figure 2.2.

A sample instance of this schema is shown below. We see that a community breaks out into a sequence of three elements: *persons*, *likes*, and *dislikes*. Element *persons* can have any number of elements of type *personType*, defined as a complexType having attribute *name*, a string. Element *likes* can have any number of elements of *like*, a complexType having a pair of persons. Dislikes is similarly structured.

```
<?xml version="1.0" encoding="UTF-8"?>
<community xmlns:xsi="http://www.w3.org/2001/
XMLSchema-instance" xsi:noNamespaceSchemaLocation=
"community.xsd">

<community>

<persons>
<person name="a"></person>
<person name="b"></person>
```

```
<person name="c"></person>
</persons>

<likes>
<like>
<person name="a"></person>
<person name="b"></person>
</like>
<like>
<person name="a"></person>
<person name="c"></person>
</like>
</likes>

<dislikes>
<dislike>
<person name="b"></person>
<person name="c"></person>
</dislike>
</dislikes>

</community>
```

Let's see how the schema in Figure 2.2 can be considered an ontology. The following table provides a mapping that shows how to interpret any instance of the schema as one set of world state descriptions.

XML segment	Maps to

`<persons>`
`<person name="x1"/>`
`<person name="x2"/>`
`...`
`<person name="xn"/></persons>`
	community members =
	
	 {x1,x2...,xn}

`<likes>`
`  <like><person name="x1"/>`
`  <person name="x2"/></like>`
`...`
`<like><person name="xn"/>`
`<person name="xm"/></like>`
`</likes>`
ditto for dislikes
	Likes=
	 {(x1,x2)..., (xn,xm)}

For example, the instance above is interpreted as stating that the community members are persons with names, a, b, and c, where a likes b and c, while b dislikes c.

XML Schema: Strengths and Limitations

You'll notice that XML is quite liberal with pairs that it accepts in the *likes* and *dislikes* relations. True, XML does allow us to restrict the elements of a *like (dislike)* pair to be *personTypes*. So the ontology we have defined does place a strong restriction limiting affections described to be between persons (not involving pets, for example). *However, none of the restrictions that were earlier expressed as axioms can be imposed or checked for with standard XML validation tools.* For example, we can easily write

```
<like>
<person name="a"></person>
<person name="a"></person>
</like>
```

with impunity.

Worse, a document can include a person in a relation such as likes who is not currently in the community. For example, we can easily write

```
<like>
<person name="a"></person>
<person name="d"></person>
</like>
```

where d is not a person name included under the *persons* element.

In general, an XML schema has only a limited ability to correlate data placed inside the slots of its instances. That is, while restrictions can be placed on values going into individual slots, there is no ability to place constraints on values that simultaneously go into any pair of slots. Thus we conclude that XML has very limited capability to support data engineering that requires specifying ontologies in which relations play a major role such as the *likes* and *dislikes* relations of our example.

In contrast, we can easily expand the logic formulation to include the constraint that anything involved in a *likes* or *dislikes* relation must be among the persons belonging to the community.

Exercise

Write out the set of axioms for the expanded ontology for states of affection in a community. Provide some examples of world states that it describes. *Hint*: Introduce a predicate:

$$inComunity(x)$$

Then add another axiom that requires that for any pair of persons in the *likes* relation, both are also in the community:

$$likes(x,y) \Rightarrow inComunity(x) \wedge inComunity(y)$$

UML

Unified Modeling Language (UML[10]) is a software development language and environment that has found some application in data engineering.[9] UML targets design using object-oriented modeling and therefore supports object-oriented elements such as objects, classes, attributes, methods, generalization, associations with multiplicities, interactions, and state charts. To support data modeling, a tool, XMI,[11] can be used that maps UML specifications into XML with the help of encoding rules.

In this introduction we review features of UML 1.4,[12] a widely employed version that has considerable vendor support. A more advanced version, UML 2.0,[10] has recently been introduced that addresses some of the issues that have been raised with the earlier version. Later we return to consider the new features of UML 2.0 (Chapter 18). To illustrate, Figure 2.3 states that there are two classes — *community* and *person*: a *community* is a collection of instances of class *person*, and there are two relations, *likes* and *dislikes*, whose left and right members are instances of class person. The rhombic arrow, denoting aggregation, expresses the *community-person* relation, while the other lines, called associations, express the *person-person* relations. Multiplicities can be specified at both ends of the arrows or lines to characterize the type of relation such as one-to-one, one-to-many, etc. Beyond this, UML 1.4 is like XML in that it does not provide any means for restricting the contents of relations, similar to what is needed for the states of affection example. Such constraints must be posted as informal notes to the reader.

UML 1.4 provides for attributes (or instance variables in object orientation terms) of classes and generalization/specialization relations through which attributes are inherited. In Figure 2.4, *Payment* is an abstract class meaning that it cannot be used by itself but serves to organize specialized classes, such as *Credit*, *Cash*, and *Check*. *Payment* contains the key attribute, *amount*, which is common to each form of payment. However, different

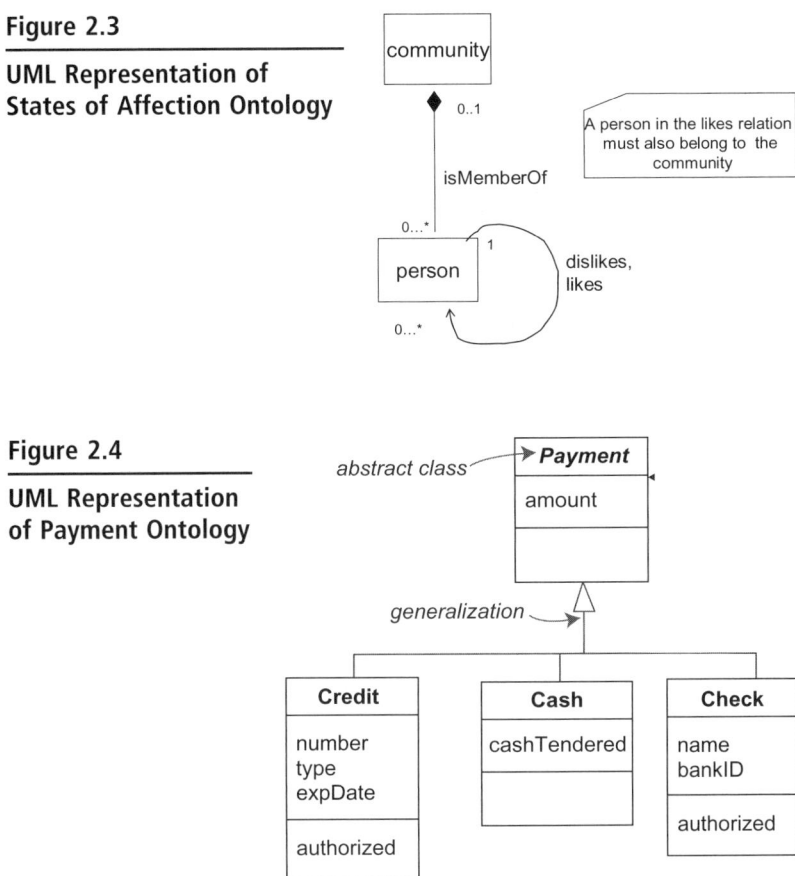

Figure 2.3

UML Representation of States of Affection Ontology

Figure 2.4

UML Representation of Payment Ontology

payment forms require their own attributes to provide the identification needed for processing.

Generalization can be combined with other relations to allow specializations to have additional substructures in addition to attributes. In Figure 2.5, a math model has two forms: a math location model and a normalized form. The normalized form has a math location model (its normalized form) but also has functions for inverting the form back to the original.

To summarize, UML 1.4, as an environment for developing ontologies, has quite powerful mechanisms for specifying taxonomic relationships (such as forms of payment) which may be many layers deep since generalization can be recursively applied. Within such taxonomic relationships, specialized classes inherit structures from their ancestors as well as adding more specific structures of their own.

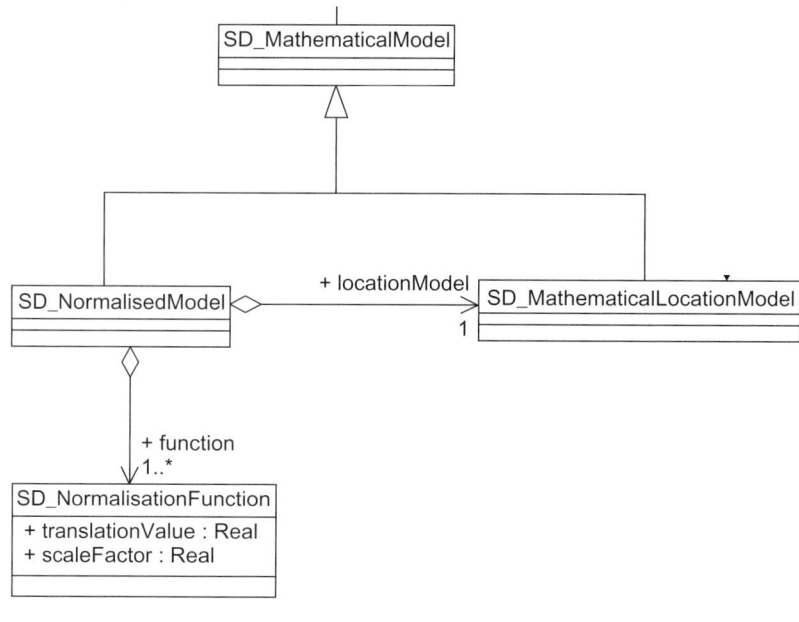

Figure 2.5
UML Representation of Math Model Ontology

Exercise

Develop a taxonomy for vehicles in UML, starting with general forms of land, sea, and air classes, and proceeding to more special forms such as cars and trucks, and to specific makes and models. Represent the fact that land vehicles have wheels, a substructure that is not present in other types of vehicles.

UML and XML Combined: Strengths and Limitations

The combination of UML 1.4 and XML offers an attractive environment for ontology development. For example, the International Standards Organization is using UML 1.4 to develop its suite of geographic information data standards. Mapping of such UML specifications into XML using a tool such as xli provides a usable product for mediating producer/consumer information exchanges. However, there are limitations to an approach based on this environment. These include limitations of expressive power and limitations related to methodology.

Limitations of Expressive Power: While UML 1.4 supports definition of relations — the heart of any ontology — it has very limited capability to control the nature and interdependence of such relations. We have seen that certain fundamental properties of relations, such as symmetry, cannot be specified. More generally, there is no ability to place constraints on concurrent value assignments (e.g., to assure that a person in a *likes* relation is also a member of the community). Moreover, only binary relations are supported so that relations of higher arity cannot be specified. For example, consider the ternary relation:

$$customerPaidFor(x,y,z)$$

meaning customer x has paid y dollars for item z.

Then we can add the axiom

$$customerPaidFor(x,y,z) \Rightarrow LessThan(y, Price(z))$$

which assures that the customer was not overcharged.

Limitations Related to Methodology: UML is a software engineering paradigm and does not provide a set of concepts or a framework dedicated to data engineering and development of ontologies. Such a set of concepts (or meta-ontology) must define entities and relationships that help to do the task. For example, our pragmatically based framework, as introduced in the first chapter, identifies producer and consumer entities, pragmatic frames, and messages as elements of interest in the development of ontologies for dynamic data engineering. In the subsequent portions of this book, we will elaborate on this framework and introduce concepts such as the System Entity Structure that provide direct support for doing data engineering more effectively.

Open Knowledge Base Connectivity Ontology

The Foundation For Intelligent Physical Agents (FIPA) employs the Open Knowledge Base Connectivity (OKBC) Knowledge Model as the meta-language for specifying interoperable ontologies. The OKBC meta-ontology supports an object-oriented representation of knowledge using a frame-based approach and is specified in Knowledge Interchange Format (KIF), a first-order predicate logic language with set theory.[13] Appendix A provides a flavor of the specification as formulated in outfix notation in contrast to the linear prefix notation employed in the FIPA specification.[1]

The OKBC Knowledge Model includes the usual *constants* of basic types such as string, integer, etc., and *frames* that play the role of objects in object-oriented programming languages. Frames have *slots* that hold *values* much like objects have attributes, with the major distinction that values to such slots are assigned under constraints called *facets*.

Using the flexibility of its facets, OKBC can impose constraints that are not specifiable in UML and XML. The following ontology restricts the members of a community to be persons, and the *likes* and dislikes relations to relate to persons who are currently members of the community. The description below is informal but could be defined within the formal specification reviewed in Appendix A.

As illustrated in Figure 2.6,[2] *Community, Person,* and *CommPair* class frames with the following slots and facets:

- *Community* has three slots — *members, likes,* and *dislikes*. The *members* slot has two facets: *CollectionType* restricts its value to sets, while *ValueType* restricts the elements of the current set to be of type *Person*. The *likes* and *dislikes* slots each have two facets: *CollectionType* (not shown in the figure) restricts its value to sets, while *ValueType* restricts the elements of the current set to be of type *CommPair*.
- *Person* has a name slot with facet *ValueType*, that restricts its values to strings.
- *CommPair* (short for community pair) has two slots, *Left* and *Right*, each of which has a facet *AllowableValues*, which restricts the slot to a single value that must be contained in the value of the current *Community* slot, *members*.

The ontology of Figure 2.6 restricts membership in social relations to current community members, but it does not further constrain the relations and their interactions. Unfortunately, to enforce these constraints is somewhat unwieldy. The problem is that OKBC requires that constraints be specified as facets of slots; if appropriate slots are not available for such facets, then they must be created. For example, in a CommPair

---

[1] Outfix is the familiar fn(x,y,z) form; linear prefix uses (fn x y z), a legacy of the LISP language, a standard of Artificial Intelligence.
[2] To avoid clutter, not all items mentioned in the text are shown in the figure (this will be true of many figures in the text).

Open Knowledge Base Connectivity Ontology

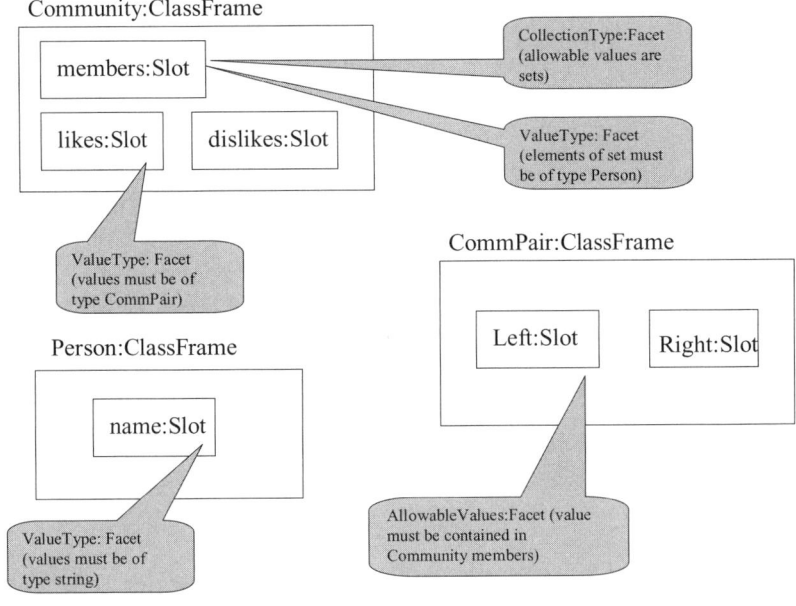

Figure 2.6

OKBC Ontology for Social Relations

frame we cannot directly require that the values assigned to the Left and Right slots be different (as required by irreflexivity). The reason is that such a constraint would apply to both slots *jointly*, and it could not be expressed as a facet for either one. As shown in Figure 2.7, a solution is to group these slots in a composite frame, called *likesDislikesPair*. We then replace the individual slots in *CommunityClassFrame* with a single slot, *pairedRel*, that takes on instances of *likesDislikesPair* as values (as specified by a *ValueType* facet). We now can associate facets with the *pairedRel* slot that place the desired constraints on interaction between its encapsulated relations, *likes* and *dislikes*.

Exercise

Suppose that you only want to enforce irreflexivity on the *likes* relation. How would the design of Figure 2.7 change?

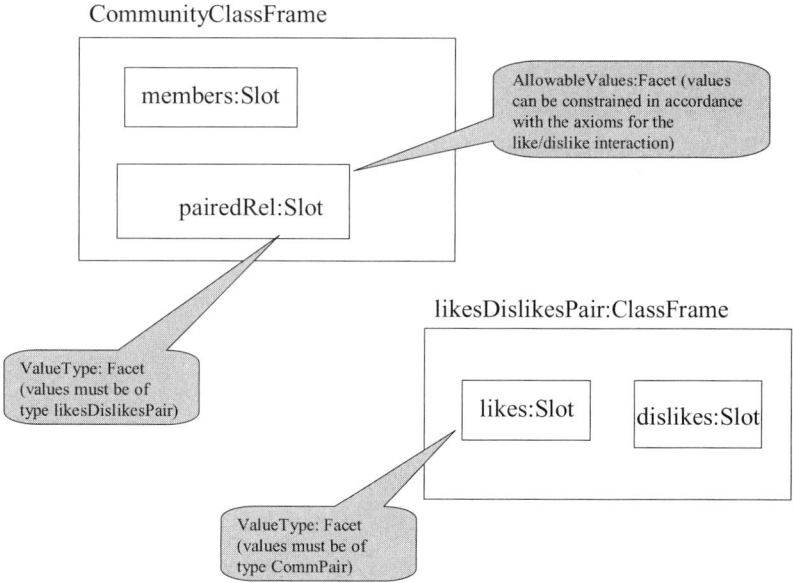

Figure 2.7

Introducing Slots to Hold Facets for Constraining Likes/Dislikes Interaction

Exercise

Develop an OKBC ontology for the customer/order/payment example — first, without enforcing the ternary "don't overcharge the customer" axiom, and then with such enforcement.

The OKBC Knowledge Model enjoys the capabilities of its underlying first order logic. However, it uses facets on individual slots to represent such logical assertions. This may lead to needless introduction of composite classes and container classes. Perhaps, allowing facets to act on frame contents rather than slots, would lead to a more friendly formulation.

Semantic Web Ontologies and Environments

The Semantic Web represents a conception for a future World Wide Web in which web services and intelligent agents are endowed with advanced intelligence to access web resources in response to sophisticated searches.[15] The Web Ontology Language (OWL) is a W3C recommended ontology standard under development to support such intelligent queries.[14] The standard example to explain the goals and capabilities of OWL is the wine agent that is to be capable of searching the web for answers to such queries as what wine would go with a particular dinner course.[16] The basic ontology framework employs classes, class hierarchies, and, notably, properties, which are stand-alone binary relations. The primary objective of OWL is to support automated logical reasoners that can derive new inferences when applied to combined ontologies from multiple web sites. OWL is intended to operate in the open web environment where ontologies are not developed under central control. Since information cannot be retracted, reasoners must be prepared to cope with contradictions that can arise from merging diverse independently developed ontologies. This represents a departure from earlier applications of formal logic that employed belief revision mechanisms to maintain logical consistency under the addition of new knowledge.

Two research projects at Stanford University implement an OKBC-compatible knowledge model and OWL. Protégé is a free, open source ontology editor and knowledge-base framework.[17] Ontolingua is a distributed collaborative environment to browse, create, edit, modify, and use ontologies.[18]

Summary

We have briefly reviewed some commonly employed languages/environments for ontology development. Their relative capabilities are illustrated in Figure 2.8. The inclusion relationships reflect the distinction between lightweight and heavyweight ontologies as defined by,[8] where the latter go beyond the former in that they include axioms and constraints to restrict the ontology's set of models.

Although HTML is not truly an ontology development language, we include it to note that the technology we are discussing ultimately is related to message exchanges over the web. The Venn diagram also tries to capture the capability relationships of the languages/environments as discussed in Table 2.1.

Note that in Figure 2.8 by not including the UML 1.4 set entirely within the OKBC set we allow for the fact that the UML has capabilities of expressing dynamic relationships such as timing and interaction diagrams that are not readily expressed

Figure 2.8

Some Commonly Employed Languages/ Environments for Ontology Development

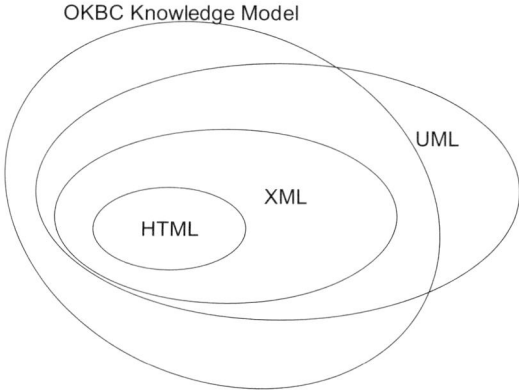

in first order logic. We'll return to the consideration of dynamics later in the book (Chapter 11).

We have only briefly discussed the points in this table. You may wish to pursue more detail through the sources that have been mentioned on these topics. We return in Chapter 19 to further consider ontology frameworks, including UML 2.0 and OWL, after presenting the dynamic data engineering framework developed in this book.

References

1. Linguistics, Syntax, Semantics, http://en.wikipedia.org/wiki/Syntax (accessed Nov. 2006)
2. Hopcroft, J., and Ullman, J. "Introduction to Automata Theory, Languages, and Computation," Reading, MA: Addison Wesley, 1979
3. Noam Chomsky, "Aspects of the Theory of Syntax," Cambridge, M.I.T. Press, 1965
4. Extensible Markup Language (XML) 1.0, http://www.w3.org/TR/2004/REC-xml-20040204 (accessed Nov. 2006)
5. Jacobs, Sas, "Beginning XML with DOM and Ajax: From Novice to Professional," Springer, New York: Apress, 2006
6. Altova XMLSpy® Enterprise Edition, http://altova.com/ (accessed Nov. 2006)
7. Kim, Larry, "Official XMLSPY Handbook," Indianapolis, IN: Wiley, 2003
8. Alexiev, V., M. Breu, J. de Bruijn, D. Fensel, R. Lara, and H. Lausen, "Information Integration with Ontologies," Hoboken, NJ: John Wiley, 2005
9. ISO/TS 19103:2005 Geographic information — Conceptual schema language, http://www.iso.org/iso/en/CatalogueDetailPage.CatalogueDetail?CSNUMBER=37800 (accessed Nov. 2006)

10. Unified Modeling Language (UML), version 2.0, http://www.omg.org/technology/documents/formal/uml.htm (accessed Nov. 2006)
11. MOF 2.0/XMI Mapping Specification, v2.1, http://www.omg.org/technology/documents/formal/xmi.htm/ (accessed Nov. 2006)
12. Fowler, Martin, "UML Distilled: A Brief Guide to the Standard Object Modeling Language," Reading, MA: Addison-Wesley, 2000
13. Ontology Service Specification, Foundation For Intelligent Physical Agents, http://fipa.org/specs/fipa00086/XC00086D.html#_Toc 505571321 (accessed Nov. 2006)
14. http://www.w3.org/TR/2004/REC-owl-guide-20040210/ (accessed Nov. 2006)
15. Grigoris, Antoniou, and Frank van Harmelen, "A Semantic Web Primer," MIT Press, 2004
16. http://www-ksl.stanford.edu/projects/wine/explanation.html/ (accessed Nov. 2006)
17. Protégé ontology environment, http://protege.stanford.edu/ (accessed Nov. 2006)
18. Ontolingua, http://www.ksl.stanford.edu/software/ontolingua/ (accessed Nov. 2006)

Appendix A: Review of OKBC Knowledge Specification (based on[13])

Frames, Own Slots, and Own Facets

There are class frames and individual frames.

EXAMPLE: Person is a class frame and Fred is an individual frame.

A frame has associated with it a set of *own slots*, and each own slot of a frame has associated with it a set of entities called *slot values*.

holds(Slot,Frame,Value) asserts that an own slot of a frame has a value description

EXAMPLE: *holds(FavoriteFood,Fred,"iceCream")* asserts that the favorite food slot of the Fred frame has value "iceCream."

An own slot of a frame has associated with it a set of *own facets*, and each own facet of a slot of a frame has associated with it a set of entities called *facet values*.

holds(Facet,Slot,Frame,Value) asserts that a facet of an own slot of a frame has a value description.

EXAMPLE: *holds(ValueType,FavoriteFood,Fred,EdibleFoods)* asserts that the ValueType facet of the favorite food slot of the Fred frame is EdibleFoods.

Classes and Individuals

Axiom: Classes and Individuals are Disjoint Frames

$$class(x) \Leftrightarrow \neg individual(x)$$

Axiom: *holds* and *instanceOf* predicates state the same relationship

$$holds(Class,Individual) \Leftrightarrow instanceOf(Individual,Class)$$

Axiom: instances of subclasses are instances of their superclasses

$$subclassOf\ (Csub,Csuper)$$
$$\Leftrightarrow \forall x(instanceOf(x,Csub) \Rightarrow instanceOf(x,Csuper))$$

Class Frames, Template Slots, and Template Facets

A class frame has associated with it a collection of *template slots* that describe own slot values considered to hold for each instance of the class represented by the frame.

A *templateSlotValue (Slot,Class,Value)* asserts that a *Slot* for a *Class* has a *Value*.

Axiom: slot values are inherited from classes to their individual elements and their subclasses

$$templateSlotValue\ (Slot,Class,Value)$$
$$\Rightarrow instanceOf\ (Individual,\ Class)$$
$$\wedge holds(Slot,Individual,Value)$$

$$templateSlotValue\ (Slot,Class,Value)$$
$$\Rightarrow subclassOf\ (subClass,\ Class)$$
$$\wedge templateSlotValue\ (Slot,subClass,Value)$$

EXAMPLE: *templateSlotValue (Gender,FemalePerson,Female)* asserts that the gender of class FemalePerson is female;

$$instanceOf\ (Mary,\ FemalePerson)$$
$$\Rightarrow holds(Gender,Mary,Female)$$

concludes that an instance of that class, Mary, will have gender = female.

A template slot of a class frame has associated with it a collection of *template facets* that describe own facet values considered to hold for the corresponding own slot of each instance of the class represented by the class frame.

Axiom: the values of a template facet are inherited to instances as values of the corresponding own facet and to subclasses as values of the same template facet.

$$templateFacetValue\ (Frame,Slot,Class,Value)$$
$$\Rightarrow instanceOf\ (Individual,\ Class)$$
$$\wedge\ holds(Frame,Slot,Individual,Value)$$

$$templateFacetValue\ (Frame,Slot,Class,Value)$$
$$\Rightarrow subclassOf\ (subClass,\ Class)$$
$$\wedge\ templateFacetValue(Frame,Slot,subClass,Value)$$

Associating Slots and Facets with Frames

Each frame has associated with it a collection of slots, and each frame-slot pair has associated with it a collection of facets. These associations are derived from the basic relationships among all elements defined above.

Derive definitions for:

- *facetOf(Facet,Slot,Frame)*
- *templateFacetOf(Facet,Slot,Class)*
- *slotOf(Slot,Frame)*
- *templateSlotOf(Slot,Class)*

as follows:

$$\exists Value{:}holds(Facet,Slot,Frame,Value)$$
$$\Rightarrow facetOf(Facet,Slot,Frame)$$

$$\exists Value{:}templateFacetValue(Facet,Slot,Frame,Class,Value)$$
$$\Rightarrow templateFaceOf(Facet,Slot,Class)$$

$\exists Value{:}holds(Slot,Frame,Value)$
$\vee$
$\exists Facet{:}facetOf(Facet,Slot,Frame)$
$\Rightarrow slotOf(Slot,Frame)$

$\exists Value{:}templateSlotValue(Slot,Class,Value)$
$\vee$
$\exists Facet{:}templateFacetOf(Facet,Slot,Class)$
$\Rightarrow templateSlotOf(Slot,Class)$

EXAMPLE: *templateSlotOf(Age,Person)* asserts that Age is a templateSlotOf class Person, so that every person has an Age slot.

EXAMPLE: *templateFacetOf(NumericMinimum,Age,Person)* asserts that NumericMinimum is a templateFacetOf Age for class Person, so that every person has a minimum age.

Slots on Slot Frames

:DOMAIN specifies the class of frames to which a slot applies. A slot frame S having a value C for own slot :DOMAIN means that every frame that has a value for own slot S must be an instance of C, and every frame that has a value for template slot S must be C or a subclass of C.

$:DOMAIN(Slot,Class)$
$\Rightarrow$
$:SLOT(Slot)$
$\wedge :CLASS(Class)$
$\wedge holds(Slot,Frame,Value) \Rightarrow instanceOf(Frame,Class)$
$\wedge templateSlotValue(Slot,Frame,Value) \Rightarrow$
$Frame = Class \vee subClassOf(Frame,Class)$

Other such slots include, e.g., :SLOT-VALUE-TYPE, :SLOT-INVERSE, :SLOT-CARDINALITY and :SLOT-COLLECTION-TYPE.

Formulating Pragmatic Frames and Ontologies: Geospatial Sensor Data

We have presented a framework in which an ontology provides the symbolic means to describe world state changes that a producer sends in the contents of a message to a consumer. Pragmatic frames are developed to capture the end-use of the information. Each frame specifies how the contents of the message will subsequently be processed by a consumer. In this chapter, we discuss the framework concepts within the context of geospatial sensing and interpretation. After illustrating the concepts in this important application domain, we close with a set of principles to guide pragmatic frame and ontology/data model development.

Figure 3.1 shows how our framework applies to sensors that gather real-world data that is interpreted, perhaps in stages, by downstream processes. By *sensor* we mean any device that can sense or detect some aspect of the world or some change in such an aspect. Such sensors might range from very simple, such as a thermometer, to very sophisticated, such as a video camera on board a satellite. Sensors typically have limited computational capability, so the data must be sent to remote locations for

36 Chapter 3 **Formulating Pragmatic Frames and Ontologies**

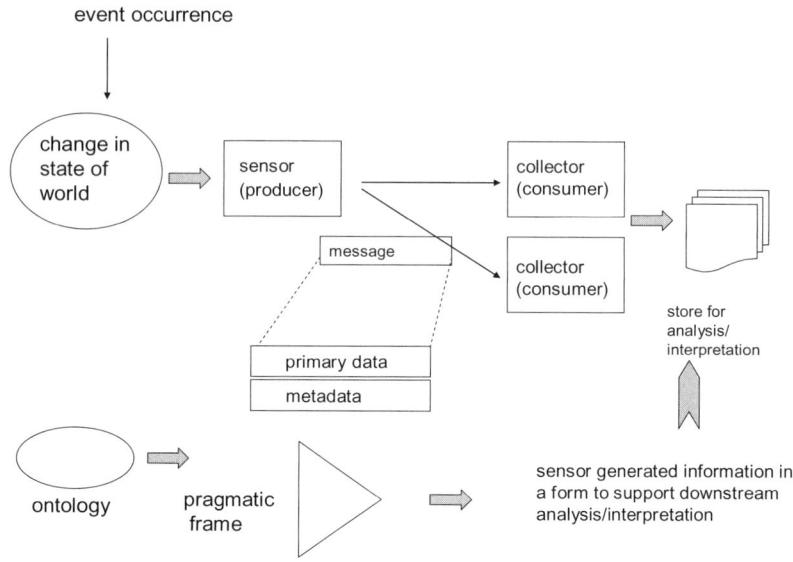

Figure 3.1

Applying the Pragmatic Framework to Sensors

analysis and interpretation. For example, a satellite camera transmits imagery data to a receiver on the earth's surface for storage and subsequent processing.

This sensor example fits neatly into our framework — the sensor plays the role of producer, and the collector is the consumer. Moreover, the pragmatic frame that governs the nature of the transmitted data concerns downstream interpretation and analysis. The figure also suggests that the same data may be sent to different collectors, and (not shown) a collector may receive data from several sensors. The possibility that the consumer may not be known at the time of sensing opens up the requirement for a publish/subscribe mechanism by which the interface required for receiving data is specified, allowing consumers that satisfy the interface requirements to receive the data. The possibility that a consumer can receive data from multiple sensors suggests that the pragmatic frame should take into account the requirement for a common approach to formatting data from multiple sources. Note that the message in this application consists of two components: primary data and metadata. By *primary data* we mean the raw data output of a sensor, while *metadata* are data that provide the basis for correct interpretation of the primary data.

Table 3.1. Contrasting Different Downstream Uses of Satellite-Transmitted Weather Data

Event causing world state change	Producer	Consumer	Pragmatics: subsequent use	Message contents
Rotating wind structure possibly indicating nascent hurricane	Weather satellite-borne wind-speed sensor	Local weather bureau	Determine strength and direction of storm	Wind velocity spatial pattern
Same	Same	Model development center	Development of more accurate models for prediction	Wind velocity spatial pattern and other sensor data such as temperature or pressure

An ontology that is selected in accordance with the pragmatic frame should generate the metadata. As a consequence, the primary data is also subject to the pragmatic frame because it fits the constraints of the ontology. Such data may differ in domain type (e.g., weather or terrain), resolution (how fine-grained the data are), and precision (how many significant figures are given in numerical fields). Domain type, resolution, precision, and other dimensions of interest must be selected in accordance with the pragmatic frame, i.e., with the intended use of the data by downstream processors.

Table 3.1 contrasts two different downstream uses of satellite-transmitted weather data. In the first row, the pragmatic frame concerns the use of such data for immediate weather prediction. The second row concerns use of data for the construction of models with increased powers of accurate prediction — in this case, the data are more extensive and higher in resolution than for the first frame.

Geospatial Imagery Sensors Background

According to Ref. 1, remote sensing sensors can be classified by their mounting platforms (such as spaceborne and airborne), by their physical properties (such as optical and electro-optical), or by the energy sensed (e.g., optical radiation, microwave energy, and sonar [acoustic] energy).[1] A sensor collects radiation and reports its intensity, usually as a function of position. Separate components generally perform the functions that are part of the measurement process. A *detector* provides a measurement of the incident radiation. An *optical system* collects the radiation incident on the instrument and directs it to the detector. A *pointing system* determines the direction from which radiation will come. Each type of sensor has its own components to perform these functions. Specific types are Linear scanner (line scanning sensor, conic scanning sensor, and whiskbroom scanning sensor), Pushbroom sensor, Frame camera, Radar, Hydrographic Sonar, and Lidar.

In System Entity Structure (SES) terms (Chapter 4), the above description can be captured as:

Sensor can be spaceborne or airborne in mountingPlatform
Sensor can be optical or electro-optical in physicalProperty
Sensor can be light, microwave, or acoustic in energy
From structure perspective, Sensor is made of collector, pointer, and detector

The resulting SES has the following form:

entity: Sensor
–specialization: Sensor-physicalPropertySpec
—entity: optical
—entity: electro-optical
–specialization: Sensor-energySpec
—entity: microwave
—entity: light
—entity: acoustic
–specialization: Sensor-mountingPlatformSpec
—entity: airborne
—entity: spaceborne
–aspect: Sensor-structureDec
—entity: detector
—entity: pointer
—entity: collector

Figure 3.2 shows the graphical representation. Later we will show how the SES helps to perform the data engineering tasks associated with sensors.

 Case Study: The National Imagery Transmission Format Migration to XML

The National Imagery Transmission Format (NITF) specification. Ref. 2 describes the file format for the transmission of both raw

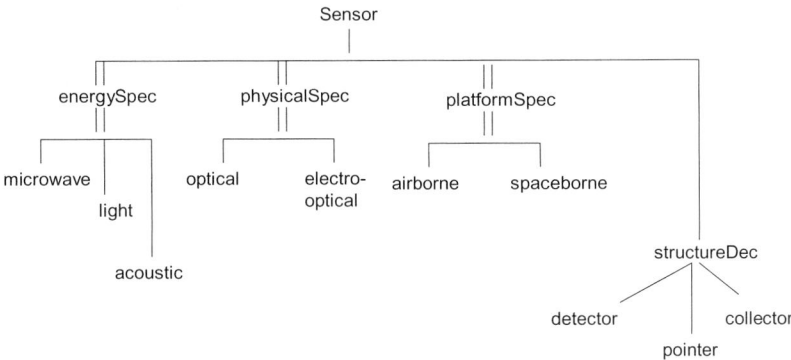

Figure 3.2

SES Diagram for Sensor Description

data images and exploited images. The raw data images are gathered using several different types of sensors. The producer operating the sensor places raw image data and metadata into an NITF file that is sent to an exploiter (consumer) who analyzes and/or uses the raw data and metadata to extract mission-specific information. NITF is a bit-level format specification, which is to say, it prescribes the actual layout of the 1's and 0's in a file as a function of the data to be encoded. This approach requires that exploiter programs be tailored to work with the exact location of data in a file. This data-location dependence makes it very costly to modify or expand the format because every such change is likely to require a major rewriting of the interpreting program. A solution is to provide an XML interface that can be employed by any exploiter program to access the data.

Figure 3.3 shows an approach to XML-based processing and objectives for the XML characterization of NITF. The process begins with NITF-encoded imagery sent by the producer to a ground station. The assumption here is that bandwidth constraints coupled with legacy encoding systems will work to retain the NITF-based transmission of data from the source. Upon arrival at the ground station, an NITF-to-XML translation process is executed, allowing XML-based transmission to exploiters, both initial and downstream. One objective is therefore to characterize the NITF information items and cast them into an XML framework. Design of NITF-to-XML translators can follow from such a characterization. A secondary objective is to examine such issues as 1) whether such translation can be done at the more-resource constrained source rather than at a ground station, 2) what are the bandwidth overhead requirements of the XML-

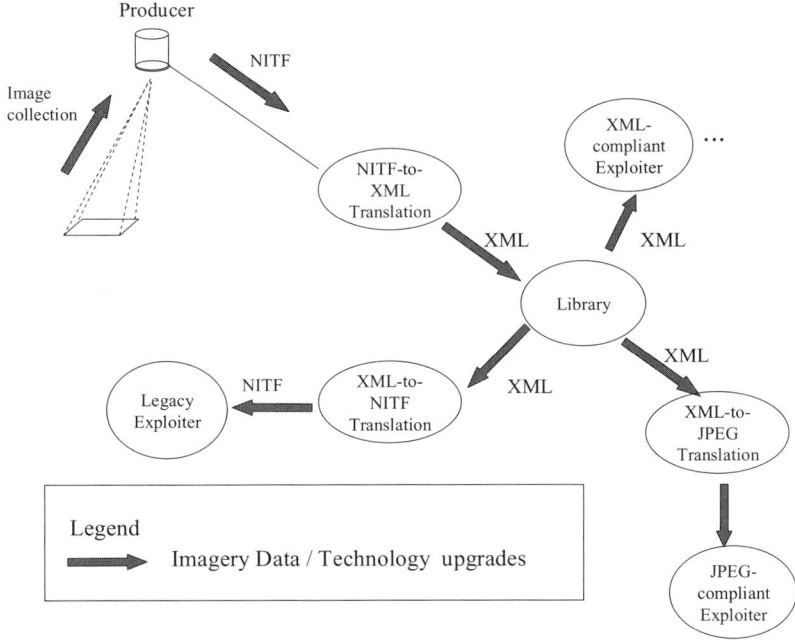

Figure 3.3

Approach to XML-based Characterization of NITF

based transmission, and 3) to what extent satellite-to-ground bandwidth limitations may prohibit such transmission. To allow for the inevitable inertia in retention of legacy software, we also see the need for reversing the XML-to-NITF translation capability. As shown in the figure, this would allow the coexistence of both XML-compliant exploiter software and NITF-based software in a transition period while older assets are phased out to be replaced by newer ones.

Standardization among producers/consumers of the XML-based metadata characterization will enable much faster and less costly upgrading of their software. Such upgrading is expected to be increasingly needed as new improvements in sensing technology continue to be introduced. Upgrades to an XML characterization of a sensor's metadata can be sent to all XML-compliant software systems. Vendors that desire to implement the change can do so quite easily given the modularity and machine-independence that XML enables. Vendors not desiring to upgrade can continue to provide service at the old level because additional XML elements can be ignored by the XML parser that serves to interface the software. Additionally, as each new version of the

XML file is deployed, upgrades to the exploiter can be pushed down as funding permits without making older exploiters unable to process any information. The addition of new tags and metadata definitions would not inhibit the exchange of image data between exploiters. The only drawback to this type of system would be the inability of older systems to use the new functionality included in the newer editions of the XML transmission. In contrast, a change in an NITF-based metadata specification most likely requires expensive reprogramming because of the intrinsic position-based interdependence of data elements.

The objective of characterizing NITF within an XML framework requires formulating such a characterization to be as compliant as possible with existing relevant standards. Such standards include *sensorML* for sensor descriptions, *transducerML* for transduced data, and a large set of International Standards Organization (ISO) geospatial metadata standards (19130/19115). A single XML standard adopted by all vendors would allow sending one and the same change item to each one, obviating the need for translating the new version into many variants. Furthermore, compliance with existing relevant standards will place the imagery community of interest in a better position to share data with other communities of interest (such as the cartography map makers) having related data activities.

Pragmatic Frame for Downstream Image Processing

With the NITF-to-XML transition case study as background, let's consider the nature of downstream processing that pragmatic frames have to account for. As illustrated in Figure 3.4, the following initial levels of downstream processing of raw imagery are described by NITF:[2]

1. Pre-Processed (also known as Radiometrically Corrected) — Basic level of image processing to correct for radiometric (radiation-related) and sensor conditions.
2. Geo-Referenced (also known as Geo-Rectified) — Adds image processing to provide geometric corrections and mapping to a cartographic projection.
3. Ortho-Rectified — Adds image processing to remove terrain relief displacement with ground control points and/or digital elevation models.

To support such image processing, the pragmatic frame must specify that the metadata include a variety of information con-

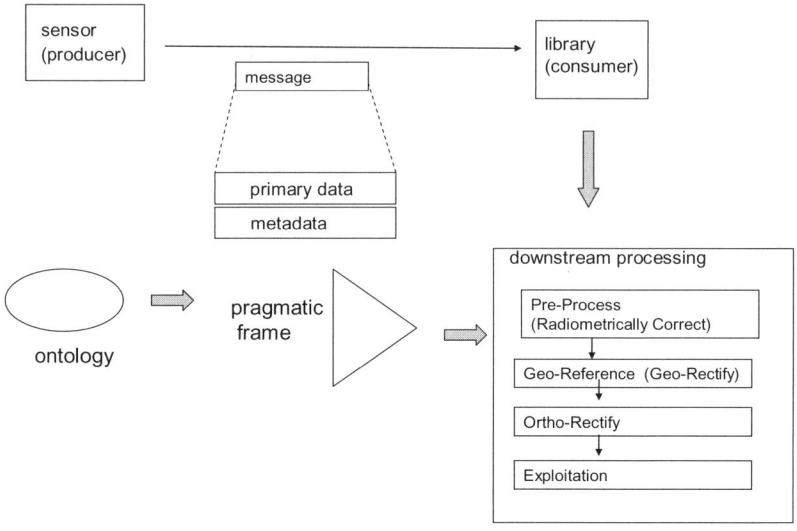

Figure 3.4

Downstream Processing of Imagery Data

cerning the sensor characteristics, its position and orientation while taking the images, control points for ortho-rectification, etc. To provide an overview of the relevant information, Table 3.2 summarizes the background discussion in Ref. 1.[1] The table is organized by the minimum information necessary for interoperable geometric correction of the sensor data: its rows concern such items as raw sensor data, data types and organization, the geolocation information, the geometric and radiometric calibration information, and the description of how to apply the geometric/radiometric information to the sensor data.

As indicated, the table was prepared from the preamble to a report of the ISO group in charge of developing standards for imagery and gridded data.[1] This report describes a sensor data model that provides semantic definitions of a set of data objects and of the relationships among them. The data model provides for:

- minimum content requirements for the raw imagery and gridded data
- methods for describing the relationship between components of the data content
- treatment of optional auxiliary information

Table 3.2. Summary of Considerations in Downstream Processing of Sensor Data

Information needed for interoperable correction of sensor data	Description and comments
Raw sensor data	Consist of the sensor readings and, optionally, the position of the sensor data in the sensor coordinate system. Stored in digital form, have the following properties: • data type: e.g., ASCII representation of numerical values, binary integers, floating point numbers • data units: e.g., Digital numbers (DN) read from the sensor unit • physical units, e.g., radiance in $W/sr/m^2$ • transformed values to reduce memory requirements
Sensor data types and organization	*Types:* • *matrix:* built from a grid values matrix of points, include a set of orientation parameters, which are valid for all pixels of the matrix • *swath:* produced when an instrument scans perpendicular to the trajectory of a moving point, consists of a series of instrument scans perpendicular to the ground track of a satellite • *point:* e.g., spot heights, the raw data of a laser scanning flight, or points on the seabed as the result of a hydrographical measurement *Organization:* • array • Riemann hyperspatial structure • table
Geolocation information	Geographical location of the sensor measurements used to describe a georeferenceable grid, e.g., orientation parameters, location ground control points, or functional fitting. Can be provided by: • sensor model parameters • ground control points • grid interpolation • polynomials • ratios of polynomials • universal real time information Data quality information, e.g., accuracy and precision of the georeferencing
Methods to georeference an imagery dataset	• *Image* reference method: e.g., uses a two-dimensional polynomial function based on any number of common points that relate the image to the surface of the Earth • *Sensor* reference method: uses a mathematical geometrical model of a sensor (requires many more initial parameters, e.g., calibrated focal length of the sensor and the position and attitude of the sensor while the image was taken; needed for ortho-rectification)

Continued

Table 3.2. Summary of Considerations in Downstream Processing of Sensor Data—cont'd

Information needed for interoperable correction of sensor data	Description and comments
Polynomial fitting methods	Required by image and sensor reference methods to fit polynomials to represent Earth coordinate positions to points on the image. Some aspects are: • functional fitting • mathematical location model • normalized model • grid interpolation • ground control points • quality information
Calibration parameters and how to apply them to the sensor data	• *Geometric:* For instruments used in precision photography, the relative position and height of each sensor element in the sensor focal plane must be provided. • *Radiometric:* The relation between the physical quantity measured and the instrument reading must be supplied with raw data

The sensor data model was defined at the conceptual level and does not specify either the encoding format for the data or the interface to access the data online. However, the committee's report[1] asserts that such a model provides a solid basis for developing interoperable data formats for encoding and common software interfaces for accessing the data.

Sensor Model

The sensor model takes the form of a Unified Modeling Language (UML) model as in Figure 3.5. A sensor model is made up of sensor parameters and possible platform parameters. Basic platform parameters are the heading and the yaw angle of the platform. Basic sensor parameters include the position, orientation, and operational mode at a given time and properties of the imagery. Associated are optical properties of the system, including distortions.

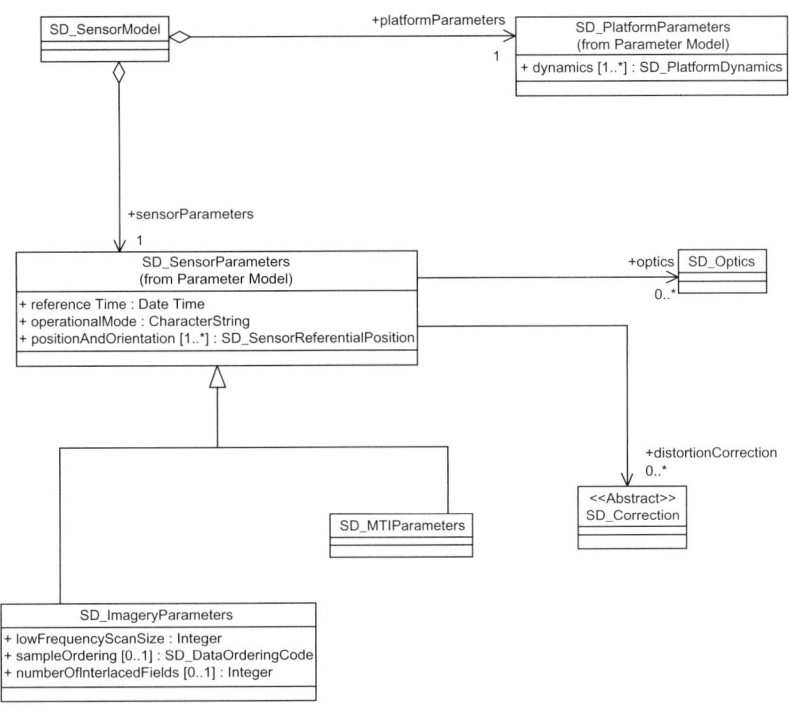

Figure 3.5

Excerpt of Sensor Model in UML (taken from Ref. 1)

Pragmatic Frame Principles: Maxims for Effective Conversations

We have seen that the ISO standard allows for different ways of providing geolocation information. It goes on to say, "The data provider should consult prospective data users when selecting which method or methods should be used to provide geolocation information. For example, for cloud studies where the areas of surfaces or volumes of observed elements must be known, enough information must be provided to allow the user to calculate this information." We see here a concern for providing primary data and metadata to support the intended downstream processing. In this section, we extend our pragmatic frame concepts toward a practitioner's methodology for data engineering.

In order for data engineering to be truly effective, the engineering team needs to examine the producing and consuming systems in depth to gain an understanding of the intent of the designers, the intent of the current consumer(s) and the possible intent of consumers of the future. *"Intent" here is equivalent to pragmatic frame, which delineates a data engineer's domain of interest.* Indeed, the pragmatic frame offers a kind of checklist of items to understand the intent of the producers and consumers. Consider the following questions that must be answered to formulate a pragmatic frame:

- Who are the relevant producers and consumers involved currently or in the future with the data under consideration?
- What are the processing stages or levels that are going to be applied to the data/metadata as they flow downstream?
- What are the actual data requirements for the current processing stages? What constraints are placed by the legacy systems in place? What should be the tradeoff between meeting current, legacy, and future uses of the data?

We'll address these questions in more depth later (Chapter 15). For now, let's assume that you have specified one or more pragmatic frames. The following provides a practical guide to formulating an ontology or data model that fits the frames you have identified.

A Guide to the Data Perplexed

To design the ontology/data model for a pragmatic frame it helps to use a metaphor in which the producing and consuming systems are seen as engaged in a dialogue or conversation. In other words, imagine that you are talking with a friend and want to succeed in effective communication with him/her. What are the principles that you use to do this? The following certain recognized maxims apply to exchange of information, whether between friends, or between producer and consumer systems generally:

- *Quantity* — match the amount of information to the needs of the intended consumer (pragmatic frame)
- *Quality* — provide high quality information if possible; if only flawed data is available, include necessary caveats
- *Relation (relevance)* — provide all, and only, information that is relevant to the pragmatic frame
- *Manner* — be concise, to-the-point, orderly; avoid ambiguity and obscurity

While the principles listed here may seem obvious, their consistent application across organizational boundaries and over time is challenging. We'll examine each of the aspects in turn.

Pragmatic Frame Principles: Maxims for Effective Conversations

Quantity:

1. Make the metadata/data quantitatively as informative as is required for the current purposes of the exchange.
   Example: If the consuming system requires a double precision real number, don't use an integer or a string.
2. Do not make the metadata/data more informative than is required.
   Example: If the consuming system expects an integer, don't complicate or burden the processing with a double that must be truncated or rounded to a value that may actually be misinterpreted or wrong.

Quality: Be aware of quality in your contribution. Be truthful.

1. Do not include metadata or data that you believe to be flawed, improper, or incorrect. If you must include flawed data, create a tag or a field that calls it into question by subsequent processors.
   Example: If a specification says that five digits of precision are required, and the producing system only supplies two digits of precision, do not create a structure that adds zeroes to the lesser precision. A subsequent system may read the extra zeroes as precise.
   Example: Do not use legacy fields for new information. Create new fields that clearly and truthfully describe how sources have been combined to form a new product. For example, for an imagery product that uses multiple images from multiple sources, provide metadata that clearly shows the sources and how they have been combined. Be sure that image idenitfications are unique and traceable.
2. Do not include metadata or data for which you lack adequate evidence that such data is correct.
   Example: If two fields have contradictory information in them, attempt to resolve the contradiction before passing the values on. Let's say that one field shows Time of Event = 1330 GMT, and later in the same document the same field has a different value — Time of Event = 1430 GMT. It may be a new event, or it may be a mistake.

Relation:

1. Be relevant, that is, include metadata/data that is likely to be consumed by other systems based on your understanding of the pragmatic frames. Ensure that scale and intent for a given frame are appropriate and applicable to a new frame.

Example: Don't rely on information intended for a generally used road map to generate precise targeting information. In this case neither the intent nor scales are likely to be transferable from one pragmatic frame to another.

Manner: Be perspicuous.

1. Avoid obscurity.
   Example: "Position" is an obscure term. For clarification, an engineer could decompose Position into latitude, longitude, elevation, time, and velocity.
2. Avoid ambiguity.
   Example: Degrees from north may be interpreted as either counterclockwise or clockwise unless specified. Don't assume that everyone follows the same convention or standard.
3. Be concise.
   Example: A field name of "Latitude_of_Observation_number_ 2_of_5" may be reduced more clearly and concisely to a group of fields in a hierarchical relationship that places the number of latitude observations as child elements under Observations.
4. Be orderly.
   Example: If there is a group of five successive Observations, list them in order, *not* as 3,4,2,1,5. This improves not only readability but automated processing capabilities.

Summary

We have illustrated the framework concepts of pragmatic frames and ontologies within the context of geospatial sensing and interpretation. An enormous variety of sensors (producers) and interpretation devices (consumers) have emerged in this domain. Most such systems are highly inflexible in what they do and how they do it. This leads to single purpose "stove-piped" chains in which very little of the available information from all the systems can be fused together. The need to formulate the underlying pragmatic frames and develop appropriately flexible ontologies for this domain motivates the data engineering approach we have presented. This situation is hardly unique to geospatial sensing and interpretation. It pervades most areas of life where systems have developed with little coordination among their parts. After illustrating the concepts, we closed with a set of principles to guide pragmatic frame and ontology/data model development.

The basic idea here is: In data engineering, keep in mind as a guide what you would do to effectively communicate in ordinary conversations.

References

1. ISO/TC 211/WG 6/CD 19130 Editing committee, "CD 19130.2 Geographic information — Sensor data model for imagery and gridded data," 1998
2. National Geospatial-Intelligence Agency (NGA), "National Imagery Transmission Format (NITF) Version 2.1 Commercial Dataset Requirements Document," 21 October 2004

Part II

System Entity Structure Concepts and Operations

Introduction to the System Entity Structure

This chapter provides a brief introduction to the *System Entity Structure* (SES). We will employ this ontological framework as the basis for modeling and simulation based data engineering in subsequent chapters. In the final chapter, having explored this basis, we return to consider the relationship between the SES and other ontological frameworks. Figure 4.1 provides a quick overview of the elements and relationships involved in a System Entity Structure. Entities represent things that exist in the real world or sometimes in an imagined world. Aspects represent ways of decomposing things into more fine-grained ones. Multi-aspects are aspects for which the components are all of one kind. Specializations represent categories or families of specific forms that a thing can assume. Pictorially, entities will be shown as boxes, while aspects, specializations, and multi-aspects will be shown as graphical objects with one, two, or three ovals. An alternative display, which we will use more often, will be a tree structure as illustrated in Figures 4.3–4.5. In this depiction, the items (entities, aspects, specializations, and multi-aspects) are displayed as nodes and the relationships of aspects, specializations, and multi-aspects to entities are shown as vertical lines, arrows, and a triple of parallel lines, respectively. The salient

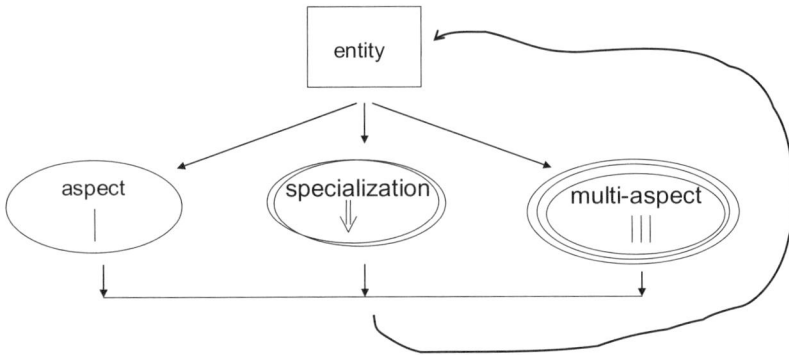

Figure 4.1

Overview of SES Items and Relationships

feature of an SES, made explicit in these graphical representations, is that entities alternate with the other items. For example, a thing is made up of parts; therefore, its entity representation has a corresponding aspect which, in turn, has entities representing the parts.

Variables

A *variable* is a slot attached to an entity that can be assigned a value from a given range set; when the value is a string, the variable is often called an *attribute*. Since variables are not the most salient of distinguishing features of the SES framework, we'll concentrate on those that are (aspects, specializations, multi-aspects) and return later to consider the role of variables.

Aspects

The concept of aspect is illustrated in Figure 4.2. An aspect expresses a way of decomposing an object into parts. More correctly, an *aspect denotes the relationship between the object and the parts into which it has been broken*. If we consider the way a book is constructed from physical pieces, it would consist of a front cover, a back cover, and the internal core (not shown in the figure; we will return later to describe the composition of the core in terms of its pages). Often there are many ways in which an object can be decomposed. Consequently, we use different aspects

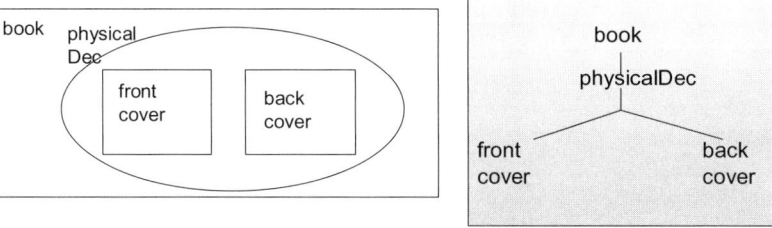

Figure 4.2

Illustrating the Aspect for Physical Decomposition of Book

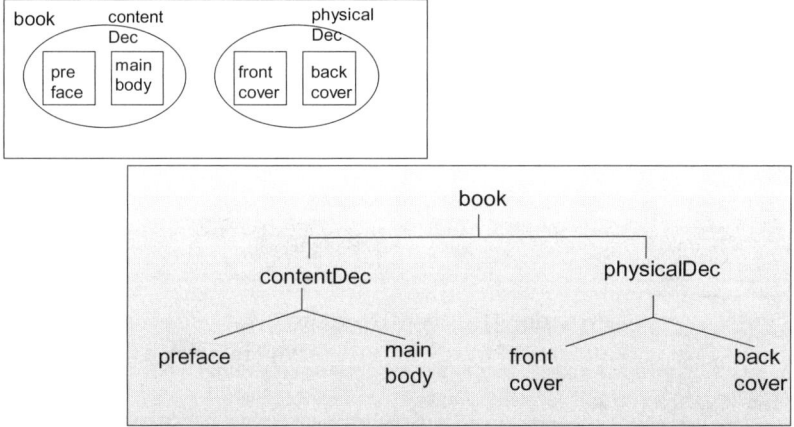

Figure 4.3

Illustrating the Physical and Content Aspects of Book

to keep track of the different decompositions that have been thought of. Thus, we can label the aspect representing the physical constitution of a book, physicalDec.

Another way of breaking a book down is through its content, for example, into its preface and main body. This is illustrated in Figure 4.3 with the aspect, contentDec. The figure shows the two items, physicalDec and contentDec, as aspects of the entity book.

For practice in this way of thinking we provide a software tool accessible from the web site (www.devsworld.org) that generates SES from natural language input. The natural language is actually restricted to a few suggestive forms. For declaring the aspect relationship we have, for example:

From the physical perspective, a book is made of a front cover and a back cover

which generates the physicalDec aspect in the SES for book. Similarly, we write

From the content perspective, a book is made of a preface and a main body

The two sentences taken together generate the SES shown in Figure 4.3.

Of course, a decomposition can involve more than two components. Likewise, an aspect can have several entities. For example, we can write

From the functional perspective, a house is made of a living room, bedroom, and kitchen.

Exercise

Provide the sentence to the natural language tool (www.devsworld.org) and observe the output.

Specializations

Real-world things can often be described using categories such as flavors, colors, and sizes. In object-oriented design, this concept is partially realized in the concepts of classes and subclasses. In the terminology of C++ and Java, we have base and derived classes, where a derived class object represents a special case of its base class. In the SES, specialization implements the concept of base and derived classes. As with aspects, specializations are labeled to represent the different categories into which derived classes can fall. Thus, a *specialization denotes the relationship between a general object and its variants belonging to a given category*. For example, in Figure 4.4, the colorSpec specialization denotes the colors that a book cover can take on. In the restricted natural language we can write:

A cover can be red or blue in color

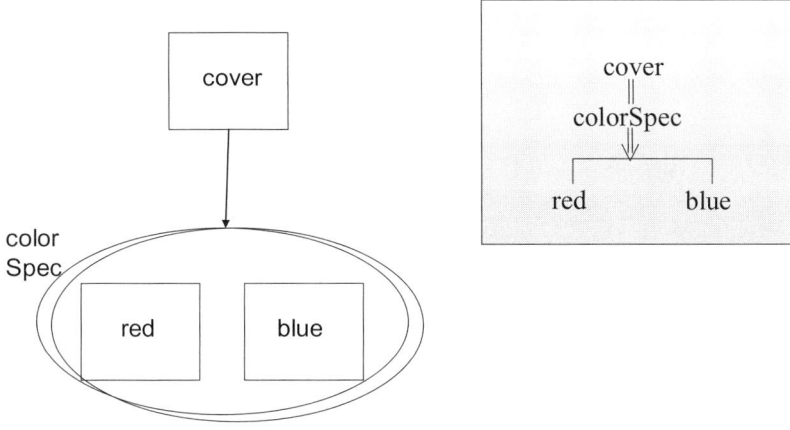

Figure 4.4

Illustrating SES Specialization

to generate the colorSpec relationship between the generic entity cover and the variant entities, red and blue.

Similarly, we can write:

A cover can be cardboard or paper in material

to generate the materialSpec specialization shown in Figure 4.5 which represents two different types of book cover. Again, there can be more than two entities in a specialization. For example, a printer might offer several colors for a book cover, as might be expressed in:

A cover can be red, green, blue, or black in color

Exercise

Provide the sentence to the natural language tool (www.devsworld.org) and observe the output.

Interaction between Aspects and Specializations

Consider an SES for a book that is generated by the following:

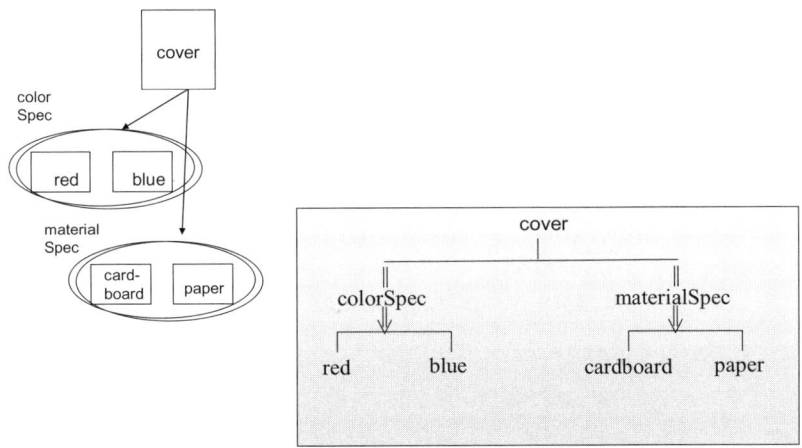

Figure 4.5

Illustrating Multiple Specializations for the Same Entity

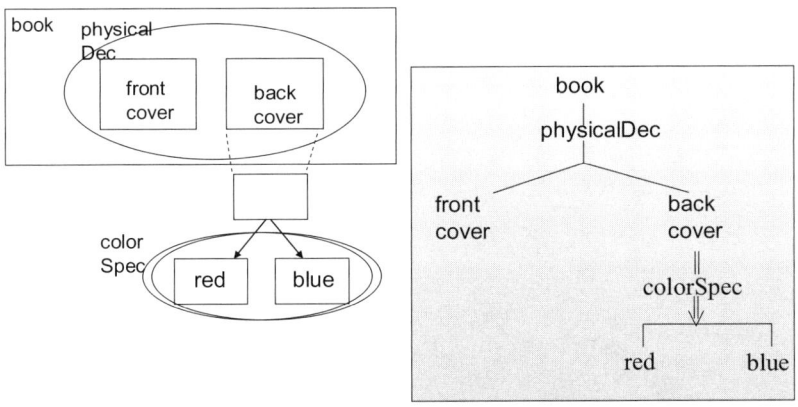

Figure 4.6

SES for Book that Has an Aspect and a Specialization

From the physical perspective, a book is made of a front cover and a back cover
The back cover can be red or blue in color

As illustrated in Figure 4.6, in the resulting SES, the entity book has an aspect labeled physicalDec, which has entities, frontCover and backCover, where the latter has a specialization labeled colorSpec having entities red and blue. The figure suggests that we can obtain two different kinds of book — one having a front cover coupled with a red back cover, and the other having a front cover coupled with a blue back cover. In other words, the members of a specialization represent different alternatives that can be chosen to fill the slot represented by the parent entity.

Interaction among Specializations

Let's include the material that a book cover might have as well as its color. We can write:

From the physical perspective, a book is made of a front cover
    and a back cover
The back cover can be red or blue in color
The back cover can be cardboard or paper in material

to generate an SES for a book. Since there are two color possibilities and two material possibilities, we must have $2*2 = 4$ possible alternative back covers. In general, as illustrated in Figure 4.7, the specializations under an entity combine through a cross product of their members to give a single compound specialization that we could have defined directly. As before, each combination represents an alternative that can be substituted for the entity in an aspect containing it.

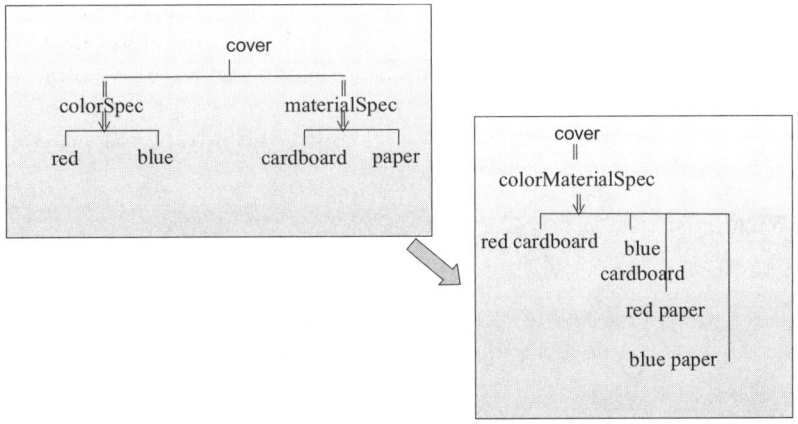

Figure 4.7

Cross Product of Specialization

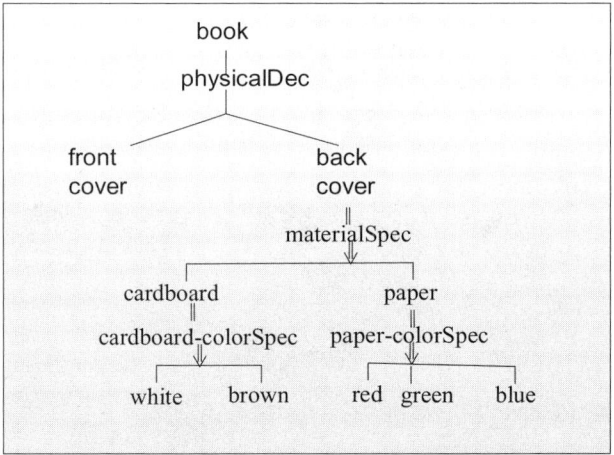

Figure 4.8

Illustrating Different Specializations for Color

The SES generated by the following:

From the physical perspective, a book is made of a front cover and a back cover
The back cover can be cardboard or paper in material
paper can be red, green or blue in color
cardboard can be white or brown in color

is illustrated in Figure 4.8. Note that the color specializations of paper and cardboard are differentially labeled representing the fact that their material composition allows quite different sets of dyes. Thus paper covers can come in three colors while cardboard covers can come in only two colors for a total of five possibilities. Ref. 5 shows how to count the number of possibilities there are for a given SES.

Exercise

Provide the sentence to the natural language tool (www.devsworld.org) and observe the output.

multiAspects

Let's return to physical decomposition of the book and deal with the core material between the front and back covers. As

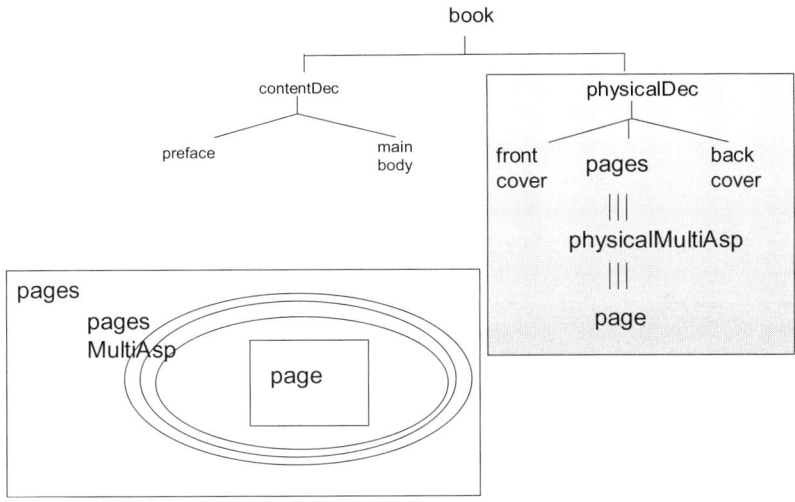

Figure 4.9

SES for Book with a Core Composed of Pages

illustrated in Figure 4.9, we'll consider the core to consist of a collection of pages. In natural language, this can be expressed as:

From the physical perspective, a book is made of a front cover, pages, and a back cover
From the physical perspective, pages are made up of more than one page

This generates the part of the SES shown in the shaded box. The multiAspect, pagesMultiAsp, captures the relationship between the entity, namely, pages, and its constituents, which are all instances of the same entity, namely, page. To show that a multi-aspect represents the relationship of a whole to its parts — which happen to be uniform in composition — we can just as well write the following:

From the physical perspective, a book is made of a front cover, a core, and a back cover
From the physical perspective, the core is made of more than one page

Exercise

Provide these sentences to the natural language tool (www.devsworld.org) and observe the output.

Indeed, a *multiAspect can be regarded as a special case of an aspect in which the entities of the aspect are homogeneous in nature*. This implies, as in Figure 4.10, that aspects and multi-Aspects can offer alternative decompositions of an entity. For example, we can consider the core as a part of the physical decomposition of a book. Then we attach two different aspects: the multiAspect decomposition into pages and the decomposition by way of content into preface and main body.

Exercise

Write sentences for the natural language tool (www.devsworld .org) that generate the SES shown in Figure 4.10.

Sometimes it is useful to refer to the repeatable entity in a multiAspect as the *generating* entity. For example, page is the generating entity of physicalMultiAsp. The reason is that we can generate instances that are substantially different even though they come from the same template. Later we will see that this occurs when the generating entity has a substructure that contains variables or specializations. For example, if we want to differentiate the pages of a book, we can provide a substructure under page that provides for page numbers, and that might distinguish first, middle, and last pages of chapters.

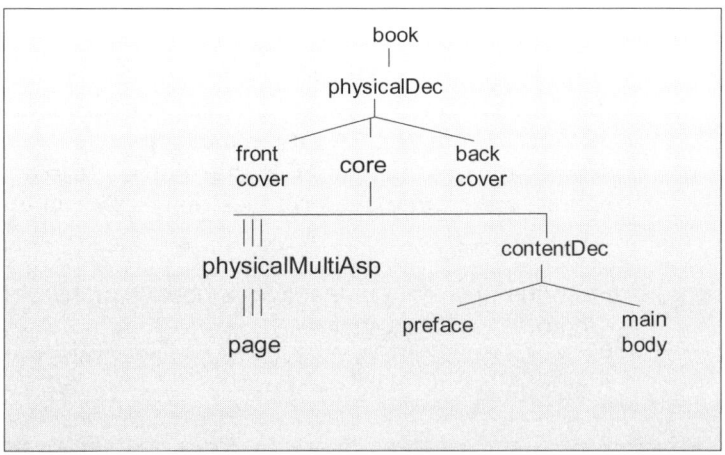

Figure 4.10

Illustrating that an Entity Can Have Both multiAspects and Aspects

Exercise

Write sentences for the natural language tool (www.devsworld.org) that extend the SES in Figure 4.10 to include substructure of a page to distinguish different instances.

Variables and Range Specifications

Aspects, multiAspects, and specializations support the hierarchical structuring capability of the SES. This is important for organizing data elements into groups (aspects, specializations) and recurring structures (multiAspects). However, much of the content of an SES is carried by its attached variables. Pragmatics enters here. In applications, it often is critically important to specify the values that variables can take on since data sent from one system to another may have different uses in each. For example, different targeting systems may employ different precisions for the same variable, e.g., location. Let's follow the steps to specify variables and their range specifications for a bicycle.

First, we describe the structure of the bicycle in Figure 4.11:

From the structure perspective, a bicycle is made of a front wheel, a back wheel, and a handlebar
From the equipment perspective, the handlebar is made of a bell, a brake, and speedometer

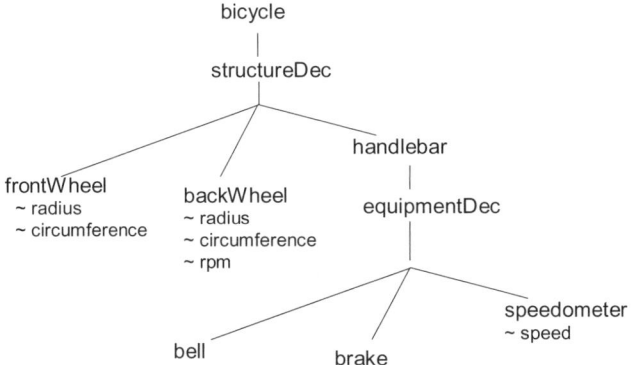

Figure 4.11

SES for a Bicycle with Attached Variables

Next, we assign variables to the entities:

The front wheel has a radius, and a circumference
The back wheel has a radius, a circumference, and an rpm
The speedometer has a speed

Now we can refer to variables of specific entities. For example,

the front wheel's radius

refers to the radius of the front wheel as opposed to

the back wheel's radius

which refers to the radius of the back wheel.
  We also use the dot notation to identify a variable of an entity, e.g.,

backWheel.radius

Later in pruning, we will be able to assign values to variables, e.g.,

the back wheel's radius is 36 or backWheel.radius = 36.

Range Restrictions

As just mentioned, range specifications are important in pragmatics of information exchange. Moreover, semantically, values assigned usually make sense only when we know more about the possible range and units of the variable's values. At the most basic level, we can describe the data type of a variable. For example, in the natural language specification (www.devsworld.org) we can say

the crosswalk has a controlLight
the range of the crosswalk's controlLight is string with values
   Stop and Go

which states that there is a variable called controlLight whose data type is string and whose values are limited to the strings, "Wait," "Stop," and "Go."
  For the bicycle in Figure 4.11, we can specify:

The range of backwheel's radius is double

This states that the values for the variable, backWheel.radius must be real numbers with 64 bits floating point precision. Often, the range of values that can make sense for a real-world object is quite restrictive. For example, a radius of a circle can only be a positive real number, and for a real bicycle, the minimum and maximum radius might realistically be related to rider's dimensions such as leg length. This kind of information can be conveyed in the form:

The range of backwheel's radius is double with values [10,30]

Note the use of the mathematical concept of interval. As in math notation, open and closed interval ends can be expressed using parentheses — "[" (open) and "]" (closed).

EXAMPLE: To illustrate the use of range restrictions, consider extending the bicycle in Figure 4.11 as follows:

A bicycle can be childBike or adultBike in rider
A childBike has a height
The range of a childBike's height is double with values [5,10]
An adultBike has a height
The range of an adultBike's height is double with values [20,30]

This results in the SES shown in Figure 4.12.

Note that the specialization into children and adult bikes allows us to associate appropriate ranges for height of the bar.

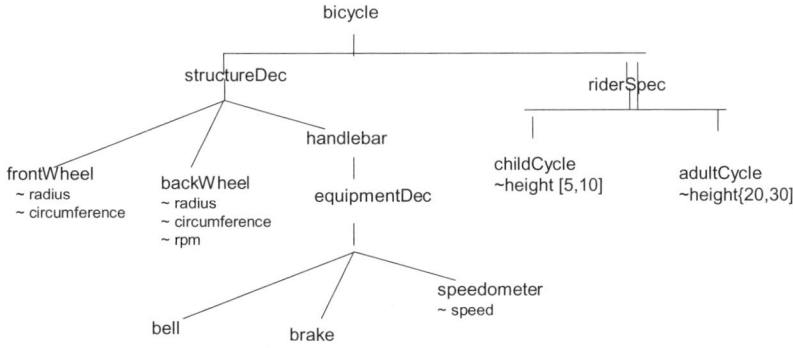

Figure 4.12

SES for Bicycle with Rider Specialization

As we discuss later (Chapter 7), this is to be preferred over supplying the same information within a single range specification for height. As expressed in XML, the SES generated by the bike-rider specialization appears as follows:

```
<entity name = "bicycle">
<specialization name = "bicycle-riderSpec">
    <entity name = "childBike">
        <var name = "height" rangeSpec =
        "double510">
        </var>
    </entity>
    <entity name = "adultBike">
        <var name = "height" rangeSpec =
        "double2030">
        </var>
    </entity>
</specialization>
</entity>
```

Computed Values and Formulas

Often, the value of one variable is computed from the values assigned to other variables. For example, the speed registering on the speedometer depends on the circumference of the back wheel and the rpm (rotations per minute). This computation may be expressed by a complex algorithm or a simple formula. For the speedometer we have:

speedometer.speed = backWheel.circumference × backWheel.rpm

To express such formulas, we assume that a computation can be represented by a black box that computes a single output from one or more inputs. For example, the *times* (×) just used can be represented as a multiplier in Figure 4.13(a):

In general, we can incorporate a model base in which functions can be defined, such as:

$$\text{output} = f(\text{input1}, \text{input2}, \ldots, \text{inputN})$$

and then refer to the inputs and outputs of the function using possessive or dot notation.

Computed Values and Formulas

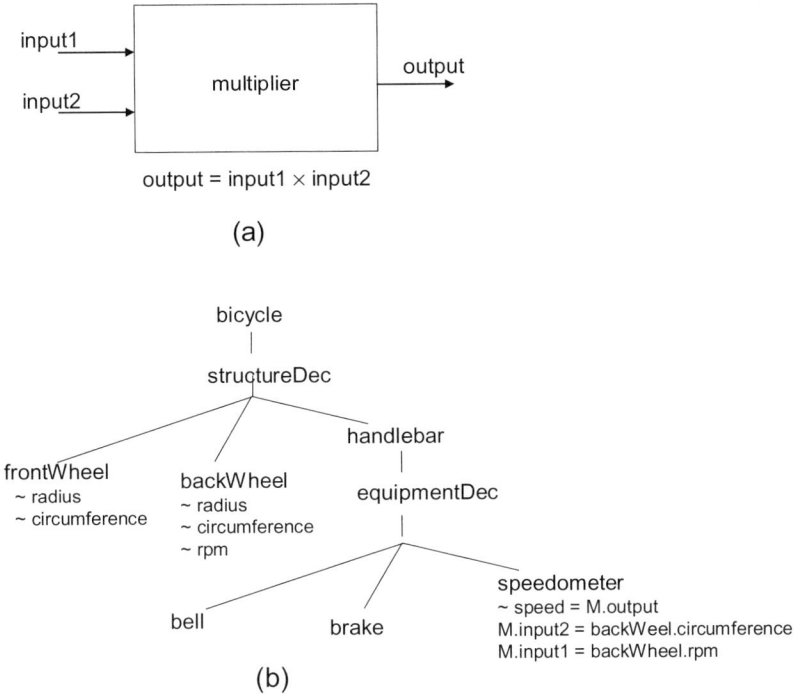

Figure 4.13

Representation of Multiplication as a Black Box

For example, we refer to the multiplier's inputs and outputs as: multiplier's input1, multiplier's input2, and multiplier's output.

To establish a computed relationship between a variable and some others, we attach the appropriate function to the entity having that variable and assign the appropriate variables to the inputs and output of the function.

For example, in the natural language tool, we can say that:

The speedometer's speed comes from the multiplier's output where
The multiplier's input1 comes from the back wheel's circumference where
The multiplier's input2 comes from the back wheel's rpm

This sets up the coupling specification (discussed in Chapter 11):

speedometer.speed = multiplier.output
multiplier.input2 = backWeel.circumference
multiplier.input1 = backWheel.rpm

The extended SES is shown in Figure 4.11(b).

Summary

Table 4.1 summarizes the basic concepts underlying the SES that were introduced in this chapter.

As may have become apparent, often there is more than one way to develop an SES for the same domain. For example, you may have to decide whether to represent the variants of an entity by a specialization or simply as the values of an attached variable. Although there are guidelines and approaches that tend to limit the number of possible alternative SESs that can be developed,

Table 4.1. Summary of Basic Concepts of SES

Item	Denotes	When to use
Entity	A thing in the real or modeled world	Use to represent a thing that stands alone or is a part or variant of another thing.
Aspect	The relationship between a thing and its components when decomposed from a certain perspective	Use when you want to represent an "and" connective among sub-things of a thing — where the "and" denotes the necessity that all of the sub-things must appear together to comprise the thing.
Multi-aspect	A special kind of aspect in which all the components are homogeneous in nature	Use for the same objective as an aspect except that the components are all from the same class.
Specialization	The relationship between a thing and its variants from a given family	Use when you want to represent an "or" connective among sub-things of a thing — where the "or" denotes the fact that a choice of one of the variants can replace the original.
Variable	A property, quality, or attribute of an entity to which it is attached	Use to distinguish and characterize different instances, in space, time, or other dimensions. Can provide range specifications for data types and value constraints.

there is still the undeniable fact that multiple, equally viable, and quite distinct alternatives might exist. As the saying goes, "there is more than one way to skin a cat." We'll return to discuss this issue later (in Chapter 7). For now, we note that the lack of unique representation is characteristic of ontology development in general. This intrinsically equivocal nature has led to a variety of different ways of representing what amounts to the same domain of discourse. The problem then arises that the lack of common understandings hinders effective exchange of information among participants who are using different ontologies. Trying to rectify this situation has become known as the harmonization problem. We'll consider it also in the same discussion of guidelines for good SES development — the use of such guidelines in the future might reduce the need for harmonization to rectify the "stovepipes" of the past. For a review of the SES and an extended example see Refs. 1 and 2.

The system entity structure concepts were first presented in Ref. 11. Ref. 10 subsequently extended and implemented these concepts in a knowledge-based design environment. Application to model base management originated with Refs. 6 and 12. Subsequent formalizations and implementations were developed in Refs. 7, 8, and 13. Automatic SES pruning is discussed in Ref. 5. Applications to various domains are given in Refs. 3, 4, 9, and 12.

References

1. Zeigler, B. P., and J. W. Rozenblit, "Design and Modeling Concepts," *Encyclopedia of Robotics*, John Wiley & Company, pp. 308–322, 1988
2. Sevinc, S., and B. P. Zeigler, "Entity Structure Based Design Methodology: Application to LAN Protocol Design," *IEEE Trans. on System Engineering*, Vol. 14, No. 3, pp. 375–383, 1988
3. Chi, S. D., J. Lee, and Y. Kim, Using the SES/MB Framework to Analyze Traffic Flow. Trans. of SCS, 14(4), pp. 211–221, 1997
4. Cho, T. H., B. P. Zeigler, and J. W. Rozenbit, "A Knowledge-Based Simulation Environment for Hierarchical Flexible Manufacturing," *IEEE Trans. Sys. Man & Cyber*. Part A: Systems and Humans, Vol. 26, No. 1, pp. 81–91, January 1996
5. Couretas, J., "System Entity Structure Alternatives Enumeration Environment (SEAS)," Ph.D. Thesis, Dept. of ECE, University of Arizona, 1998
6. Kim, T. G., C. Lee, B. P. Zeigler, and E. R. Christensen, "System Entity Structuring and Model Base Management," *IEEE Trans. Sys. Man & Cyber.*, Vol. 20, No. 5, pp. 1013–1024, 1990

7. Luh, C., and B. P. Zeigler, "Model Base Management for Multifaceted Systems." *ACM Trans. on Modeling and Comp. Sim.* 1(3): pp. 195–218, 1991
8. Park, H. C., W. B. Lee, and T. G. Kim, "RASES: A Database Supported Framework for Structured Model Base Management," *Simulation Practice and Theory*, Vol. 5, pp. 289–313, 1997
9. Hyu C. Park, and Tag G. Kim, "A Relational Algebraic Framework for VHDL Models Management," *Trans. of SCS*, Vol. 15, No. 2, pp. 43–55, June 1998
10. Rozenblit, J. W., J. Hu, B. P. Zeigler, and T. G. Kim, "Knowledge-Based Design and Simulation Environment (KBDSE): Foundational Concepts and Implementation," *J. Operations Research Society*, Vol. 41, No. 6, pp. 475–489, 1990
11. Zeigler, B. P., Multifacetted Modelling and Discrete Event Simulation, Academic Press, London, 1984
12. Zeigler, B. P., Object-Oriented Simulation with Hierarchical, Modular Models: Intelligent Agents and Endomorphic Systems, San Diego, CA: Academic Press, 1990
13. Zeigler, B. P., and G. Zhang, "The System Entity Structure: Knowledge Representation for Simulation Modeling and Design," in: *Artificial Intelligence, Simulation and Modeling* (Eds. L. A. Widman, K. A. Loparo, and N. Nielsen), John Wiley, pp. 47–73, 1989

System Entity Structure Axioms: Interpretations and Applications

Now that we have some familiarity with the elements and semantics of the System Entity Structure (SES), we turn to a more formal characterization that will give its formation rules and help to apply its concepts. The SES was originally characterized by its axioms by Ref. 1 and later by Ref. 2. In that axiomatization, the SES was defined as a labeled tree with attached variable types which satisfies the following axioms:

- *uniformity*: Any two nodes that have the same labels have identical attached variable types and isomorphic subtrees.
- *strict hierarchy*: No label appears more than once down any path of the tree.
- *alternating mode*: Each node has a mode that is either entity, aspect, or specialization; if the mode of a node is entity, then the modes of its successors are aspect or specialization; if the mode of a node is aspect or specialization, then the modes of its children are entity. The mode of the root is entity.
- *valid brothers*: No two brothers have the same label.

- *attached variables*: No two variable types attached to the same item have the same name.
- *inheritance*: The parent and any child of a specialization combine their individual variables, aspects and specializations when pruning is activated (see discussion later).

In this book, we will provide an alternative formulation of the SES as a relational structure in the manner of Park, see Ref. 3. This formulation will offer an insight into the roles of the elements and how the axioms are satisfied, particularly the axioms of uniformity and alternation. Moreover, it is the basis for our later discussion of equality tests and other comparison operations that allow us to evaluate how much SESs have in common.

Relational Specification of the SES

We start with a simplified version in which multiAspects, variables, and other enhancements are omitted. An SES is specified by a structure:

$$SES = <$$
$$Entities,$$
$$Aspects,$$
$$Specializations,$$
$$rootEntity,$$
$$entityHasAspect,$$
$$entityHasSpecialization,$$
$$aspectHasEntity,$$
$$specializationHasEntity,$$
$$entityHasVariable,$$
$$variableHasRange$$
$$>$$

where

- *Entities, Aspects, Specializations, Variables* are finite mutually disjoint sets
- *rootEntity* ∈ *Entities*

and the following are relations:

$$entityHasAspect \subseteq Entities \times Aspects$$
$$entityHasSpecialization \subseteq Entities \times Specializations$$

Relational Specification of the SES

$$aspectHasEntity \subseteq Aspects \times Entities$$
$$specializationHasEntity \subseteq Specializations \times Entities$$
$$entityHasVariable \subseteq Entities \times Variables$$
$$variableHasRange \subseteq Variables \times RangeSpec$$

The SES generated by such a specification is a directed graph whose nodes are the entities, aspects, and specializations (more correctly, the items that are accessible from the rootEntity) and whose edges are determined by the given relationships. Variables are attached to entities and have range specifications. Let's see how such a graph emerges from the specification.

We need to put some restrictions on these elements to avoid pathological cases later on when we manipulate the structure.

We use the following notation:

$R \subseteq A \times B$ indicates a relation over A and B; recall that a relation is simply a set of ordered pairs.

$R_{range}(a) = \{b \mid (a,b) \in R\}$ denotes the elements in B that are associated with a

$R_{domain}(b) = \{a \mid (a,b) \in R\}$ denotes the elements in A that are associated with b

$range(R) = \cup_{a \in A} R_{range}(a)$ denotes the subset of B that appears anywhere in a pair

$domain(R) = \cup_{b \in B} R_{domain}(b)$ denotes the subset of A that appears anywhere in a pair

We'll require that specializations and aspects are not empty, which means that they represent actual choices and decompositions respectively. Furthermore, there shouldn't be any aspects or specializations that are free-standing — every such item should have at least one parent entity. In relational terms, this comes out as:

$$range(entityHasAspect) = domain(aspectHasEntity)$$
$$range(entityHasSpecialization) =$$
$$domain(specializationHasEntity)$$

The first equation succinctly states that every aspect mentioned in *entityHasAspect* is also mentioned in *aspectHasEntity* (so every aspect that has a parent entity also has a child entity) and conversely (so every aspect that has a child entity also has a parent entity). The second equation places the same requirements on specializations. Note that this does not require all entities to have aspects or specializations. Indeed, the *leaf* entities are entities that have no children.

An SES is said to be *well-formed* if it satisfies the above requirements. From now on, we'll assume all such structures are well-formed unless stated otherwise.

In Algorithm 5.1, as we unfold new items from the relationships, we add lines to the graph as well. We borrow the notation from the C family of programming languages in the form: X+=Y, which means update the set X by including the set Y in it.

ALGORITHM 5.1

Generate the directed graph from an SES specification

//initialize
$accessibleEntities = \{rootEntity\}$
$accessibleAspects = \phi$
$accessibleSpecializations = \phi$
$newEntities = \{rootEntity\}$
$newAspects = \phi$
$newSpecializations = \phi$

do {
$accessibleEntities = +newEntities$
$accessibleAspects = +newAspects$
$accessibleSpecializations = +newSpecializations$

//clear the new aspects and specializations
$newAspects = \phi$; $newSpecializations = \phi$
//remove already seen entities from the new entities
$newEntities- = entities$
//generate new aspects and specializations
For each $entity \in newEntities$:
$newAspects+ = entityHasAspect(entity)$
[draw a line from each entity to its related aspects]
$newSpecializations+ = entityHasSpecialization(entity)$
[draw a line from each entity to its related specializations]

//generate new entities
For each $aspect \in newAspects$
$newEntities+ = aspectHasEntity(aspect)$
[draw a line from each aspect to its related entities]
For each $specialization \in newSpecializations$
$newEntities+ = specializationHasEntity(specialization)$
[draw a line from each specialization to its related entities]

> //repeat until there are no new aspects or specializations
> if ($newAspects = \phi \wedge newSpecializations = \phi$)
> break
> }

EXAMPLE: A natural language description (www.devsworld.org) has the form:

From the physical perspective, a book is made of a front cover and a back cover
The back cover can be red or blue in color
The front cover is like the back cover in color

It describes the following SES:

SesForBook = <
Entities = {*book,red,blue*}
Aspects = {*physicalDec*}
Specializations = {*colorSpec*}
rootEntity = *book*
entityHasAspect = {(*book,physicalDec*)}
entityHasSpecialization = {(*frontCover,colorSpec*),(*backCover, colorSpec*)}
aspectHasEntity = {(*physicalDec,frontCover*),(*physicalDec, backCover*)}
specializationHasEntity = {(*colorSpec,red*),(*colorSec,blue*)}
>

Executing Algorithm 5.1, we go through the following steps:

1. Starting with book, there is only one aspect, *physicalDec* — add it to *accessibleAspects* and draw a line from book to *physicalDec*.
2. *physicalDec* has two entities, *frontCover* and *backCover* — add them in to *accessibleEntities* and draw a line from *physicalDec* to each one.
3. *frontCover* has one specialization, *colorSpec* — add it to *accessibleSpecializations* and draw a line from *frontCover* to *colorSpec*.
4. *colorSpec* has two entities, *blue* and *red* — add them to *accessibleEntities* and draw a line from *colorSpec* to each one.

5. *backCover* has one specialization, *colorSpec* — it is already in *accessibleSpecializations*; draw a line from *frontCover* to *colorSpec*.

Notice that both *frontCover* and *backCover* have the same specialization, *colorSpec*. Thus, in the directed graph of Figure 5.1(a) there are converging lines from the entities to the specialization. This is an example of the *uniformity* property — both covers have the same selection of colors in this case. Figure 5.1(b) shows the way such uniformity is depicted in the tree-like graphs we usually employ. Going from left to right, once an item is displayed with its entire substructure, any further occurrence of the item does not have to show the substructure because it is already known.

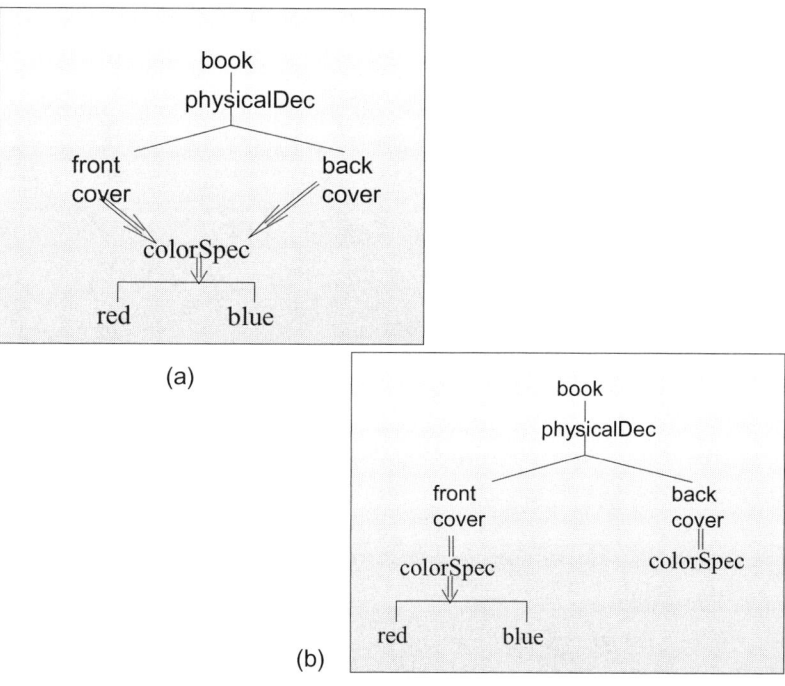

Figure 5.1

Uniformity in Graph and Tree Depictions

Application of Uniformity

The SES in Figure 5.2 is described by the natural language form:

Application of Uniformity

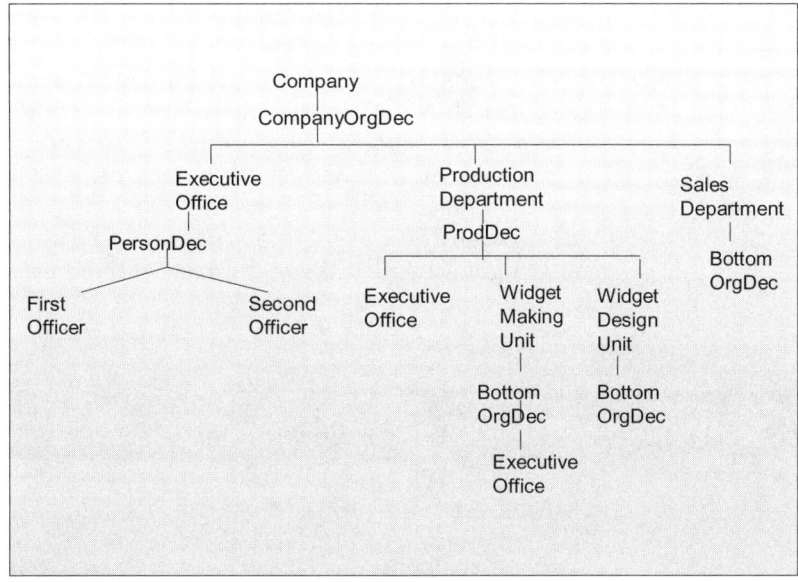

Figure 5.2
Illustrating the Uniformity Axiom

From the organization perspective, a Company is made of an Executive Office, a Production Department, and a Sales Department

From the personnel perspective, the Executive Office is made of a First Officer and a Second Officer

From the production perspective, the Production Department is made of an Executive Office, a Widget Making Unit, and a Widget Design Unit

From the bottom organization perspective, the Widget Making Unit is made of an Executive Office

From the bottom organization perspective, the Widget Design Unit is like the Widget Making Unit

From the bottom organization perspective, the Sales Department is like the Widget Making Unit

There are several examples of uniformity in the SES in Figure 5.2:

- The entity Executive Office (as well as its substructure) is used for describing the management group at each organizational level. Of course, the titles may be different at each level, such as Chief Executive Officer at the Company level and Lead

Designer at the Widget Design Unit level. We can accomplish this with the use of variables attached to entities.
- The BottomOrgDec aspect is defined to enable the use of the Executive Office structure for bottom-level units that have only executives but no other workers. This aspect is introduced in the Widget Making Unit and is then reused in the Widget Design Unit and the Sales Department. For example, in the Sales Department, the First and Second officers might be called Sales Representative and Assistant Sales Representative.

Satisfaction of the Axioms

Let's consider the status of the axioms stated earlier in the relational formulation of an SES. We have seen that the uniformity axiom is satisfied in our interpretation of the relational formulation. We will show that the alternating mode and valid brothers axioms are satisfied as well. However, the strict hierarchy axiom is not satisfied automatically, and we will show how to test whether it is satisfied as a requirement by an arbitrary SES.

Valid Brothers Axiom

This follows from the set theoretic properties of the relational specification. In a relation

$$R \subseteq A \times B$$
$$R(a) = \{b \mid (a,b) \in R\}$$

denotes the set of children of a which are brothers, or siblings, of each other. Since $R(a)$ is a set, there are no repetitions of elements in it, hence no items with the same labels.

For example, in the SES specification for book, if we tried to add another child to the *physicalDec* aspect with the same name as *backCover*, we would get:

SesForBook = <
. . .
aspectHasEntity = {(*physicalDec,frontCover*), (*physicalDec,backCover*),(*physicalDec,backCover*)}
. . .
>

But then the axioms of set theory would cause any interpretation of this to ignore the repeated pair.

Alternating Mode Axiom

This follows from the structure of the relations in the specification. We note that to get from one entity to another, we need to expand the *entityHasAspect* or *entityHasSpecialization* relation to get an aspect or specialization, respectively, and then expand the *aspectHasEntity* or *specializationHasEntity* from the result.

Strict Hierarchy Requirement

Recall that the strict hierarchy axiom stated that no label appears more than once down any path of the tree.

In the relational formulation, the structure generated by Algorithm 5.2 is a directed graph. For each node, consider all the paths from the root to it and replicate the node for each path, calling each replication, an *occurrence* of the label of the node. We call this graph, the *expanded* graph. For example, in Figure 5.1(b) we have replicated colorSpec for each of the paths to it from the root. Strict hierarchy requires that the expanded graph of an SES be a finite tree. Basically, if there is a cycle from a node back to itself, then there would be an infinite number of paths from the root to that node, thus rendering the expanded graph an infinite tree.

Algorithm 5.2 checks for strict hierarchy by keeping track of all the paths as they are generated by unfolding the SES relations. Starting from the root entity, with the initial path consisting of just it, we expand an entity's aspects and specializations, checking if each has appeared earlier in the path leading to it. If no repetition has occurred, we continue checking the entities of the new aspects and specializations. If at any point we spot a repetition of an item earlier in the path leading to it, we stop and declare the strict hierarchy requirement to be violated.

ALGORITHM 5.2

Check for strict hierarchy in an SES specification

//initialize
checkHierEntity("*rootEntity*")
//
checkHierEntity(path){
entity = *lastItemOnPath(path)*

Chapter 5 System Entity Structure Axioms

> $newAspects = entityHasAspects(entity)$
> For each $aspect \in newAspects$
> if $isAnywhereOn(aspect, path)$, break "repeat aspect
>   found";
> else $checkHierAspect(path + \text{"}aspect\text{"})$
> }
> $checkHierAspect(path)\{$
> $aspect = lastItemOnPath(path)$
> $newEntities = aspectHasEntity(aspect)$
> For each $entity \in newEntities$,
> if $isAnywhereOn(entity, path)$, break "repeat entity
>   found";
> else
> $checkHierEntity(path + \text{"}entity\text{"})$
> }
> //similarly for specializations

EXAMPLE: Consider the SES for book, in Figure 5.3, where we add *book* as one of the entities of *physicalDec*. This, in effect, states that *book* is made up of parts, one of which is itself. Indeed, we would declare,

From the physical perspective, a book is made of a front cover, a back cover, and a book.

The SES specification would now have a modified relation:

$SesForBook = <$
. . .
$aspectHasEntity = \{(physicalDec, frontCover),$

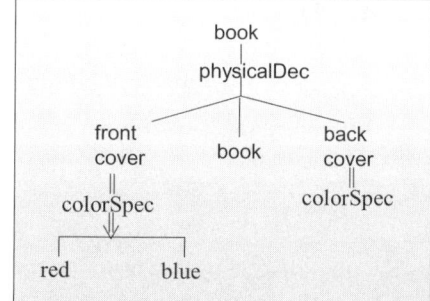

Figure 5.3

An Example Where Strict Hierarchy Is Violated

(*physicalDec,backCover*),(*physicalDec,book*)}
...
>

In applying Algorithm 5.2, upon coming to the second occurrence of *book*, we would see that it is found earlier on the path *"book"* + *"physicalDec"* and declare that to be a violation of strict hierarchy.

Unique Path Labeling

Strict hierarchy applied to the SES gives the following property:

Unique path labeling property: there is a uniquely labeled path from the root to every node in the expanded graph.

The proof proceeds as follows: Modify Algorithm 5.2 so that, for each item, it produces the paths ending in that item. String hierarchy assures that there is a finite number of such paths. Use the valid brothers axiom to show that no two paths have the same sequence of labels.

This property allows us to distinguish the different occurrences of the same label in the expanded graph.

The SES examples in Figures 5.1 and 5.2 have unique path labeling property.

In Figure 5.1,

"book, frontCover" is the path from root to the first occurrence of colorSpec
"book, backCover" is the path from root to the second occurrence of colorSpec

In Figure 5.2,

"Company, CompanyOrgDec, ExecutiveOffice, PersonDec" is the path from root to the first occurrence of FirstOfficer
"Company, CompanyOrgDec, ProductionDepartment, ProdDec, ExecutiveOffice, PersonDec" is the path from root to the second occurrence of FirstOfficer

Something less than the full path from the root to a label may be sufficient to disambiguate it. We'll return to consider how to characterize the minimal information needed to distinguish all occurrences of a label later (Chapter 10).

Exercise

In Figure 5.2,

a. Identify the occurrences of ExecutiveOffice.
b. For each such occurrence, write the path from the root to it.

Pruning Process: Brief Introduction

The process of pruning the SES is that of assigning values to variables and reducing the choices represented by the *entityHasAspect* and *specializationHasEntity* relations. More specifically, an entity may have several aspects to select from representing several ways of decomposing it. Likewise, a specialization may have several entities to select from representing different ways of specializing an entity. Ultimately, in a completely pruned structure there are no choices left. In other words, in such a structure, every entity (that had a choice) now has exactly one aspect and every specialization has exactly one entity. The process of pruning is that of making selections in some computer-assisted manner. Using the XML representation (to be discussed in Chapter 6) takes the form of creating instances of DTDs or Schemas that represent the SES. We'll return to an in-depth discussion of pruning in Chapter 8.

System Entity Structure with multiAspects

We now extend the original definition of the SES to include multi-Aspects. Recall that a multi-aspect can be regarded as a special case of an aspect in which the entities of the aspect are homogeneous in nature. Thus, the extended SES includes sets and relations involving entities and multiAspects that parallel those involving entities and aspects. The main difference is that the *multiAspectHasEntity* relation is functional, i.e., there is only one entity associated with a *multiAspect* — the pruning process must take care to generate copies of the same entity as needed. In addition, we attach a variable to a multiAspect to allow representing the number of replications of the generic entity as well as bounds on this number.

An SES is specified by a structure:

$$SES = <$$
$$Entities,$$
$$Aspects,$$

System Entity Structure with multiAspects

Specializations,
rootEntity,
entityHasAspect,
entityHasMultiAspect,
entityHasSpecialization,
aspectHasEntity,
multiAspectHasEntity,
multiAspectHasVariable,
specializationHasEntity,

entityHasVariable,
variableHasRange
>

where

- *Entities, Aspects, multiAspects, Specializations, Variables* are finite mutually disjoint sets
- *rootEntity* ∈ *Entities*

and the following are relations:

entityHasAspect ⊆ *Entities* × *Aspects*
entityHasMultiAspect ⊆ *Entities* × *multiAspects*
entityHasSpecialization ⊆ *Entities* × *Specializations*
aspectHasEntity ⊆ *Aspects* × *Entities*
multiAspectHasEntity ⊆ *multiAspects* × *Entities* (is a functional relation)
multiAspectHasVariable ⊆ *multiAspects* × *Variables*
specializationHasEntity ⊆ *Specializations* × *Entities*
entityHasVariable ⊆ *Entities* × *Variables*
variableHasRange ⊆ *Variables* × *RangeSpec*

EXAMPLE: The SES specification for Figure 4.9 is:

SesForBook = <
Entities = {*book,red,blue,core,page*}
Aspects = {*physicalDec*}
multiAspects = {*physicalMultiAsp*}
Specializations = {*colorSpec*}
rootEntity = *book*
entityHasAspect = {(*book,physicalDec*)}
entityHasMultiAspect = {(*core,physicalMultiAsp*)}
entityHasSpecialization = {(*frontCover,colorSpec*),(*backCover, colorSpec*)}
aspectHasEntity = {(*physicalDec,core*),(*physicalDec, frontCover*), (*physicalDec,backCover*)}

multiAspectHasEntity = {(*physicalMultiAsp,page*)}
multiAspectHasVariable = {(*physicalMultiAsp,numpages*)}
specializationHasEntity = {(*colorSpec,red*),(*colorSpec,blue*)}
>

Exercise

Show that if an SES has the strict hierarchy property, then each of the SESs associated with its entities also has the strict hierarchy property. In particular, each of the entities under a multiAspect has this property. However, show that the converse is not necessarily true.

Inheritance

Recall that the inheritance axiom asserts that the parent and any child of a specialization combine their individual variables, aspects, and specializations. How all these substructure elements are combined is not manifested in the SES itself but becomes explicit when pruning is activated. We introduce the concept of pruning here and will pursue it in greater depth later.

Multiple Inheritance of Specializations and Variables

The first thing to notice here is that inheritance actually accounts for the interaction of specializations under the same entity discussed before. Consider Figure 5.3, replicated and decorated in Box 1 of Figure 5.4.

Starting with the SES for cover, we select *red* from *colorSpec* (pruning is the selection of choices presented in the SES). The result is illustrated in Box 2 of Figure 5.4. We now have a cover that is red, and can be given the name *red_cover*. Further, it inherits *materialSpec*, the parent cover's other specialization. Now selecting *cardboard* from *materialSpec* results in an entity called *red_cardboard_cover* that is red and cardboard. Now consider variables attached to the entities. They are accumulated as the pruning process and inheritance proceeds. For example, suppose that *red* has a variable, *frequency*, and *cardboard* has a variable, *thickness*. Then, as shown in Box 3 of Figure 5.4, *red_cardboard_cover* inherits the variables, *frequency* and *thickness*.

Inheritance 85

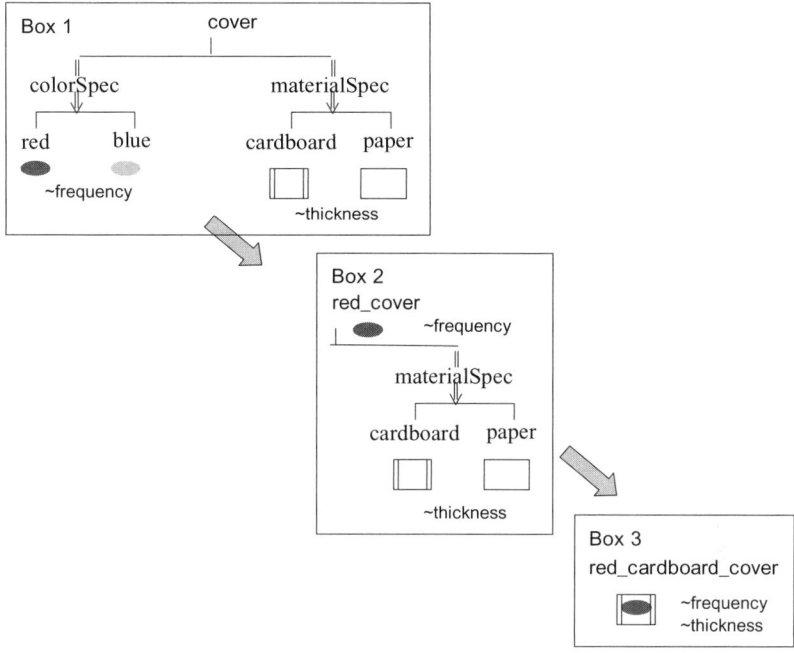

Figure 5.4

Illustrating Multiple Inheritance

Inheritance of Aspects

Let's consider an SES in which an entity has both a specialization and an aspect. For example, Figure 5.5 illustrates an SES specified by

From the physical perspective, a book is made of a frontCover, a core, and a backCover
A book can be biography or fiction in genre

During pruning, inheritance proceeds as illustrated at the bottom of Figure 5.5, so that a biographical book inherits the physical decomposition of a generic book. Inheritance of aspects accumulates aspects just as inheritance of variables accumulates variables. For example, the physical decomposition of a generic book is added to aspects of a biography which already includes an aspect that describes its content.

Note that fictional works have different content structures than do biographies. This means that the result of pruning fiction from genreSpec in Figure 5.5 would produce a fictional work

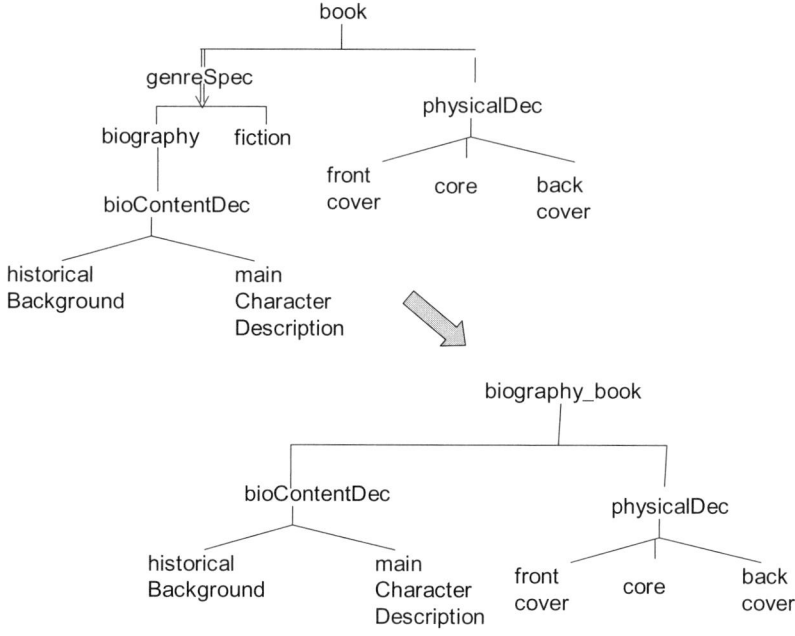

Figure 5.5

Inheritance of Aspects

having the physical decomposition of any book as well as the fiction-specific structuring of its content.

Exercise

Using natural language input, specify an SES that is similar to Figure 5.5 with the content decomposition of novels (fictional books) also specified.

Inheritance of multiAspects

The same inheritance rules apply for multiAspects as for aspects. For example, consider toppings in the pizza SES in Figure 5.6. Here topping not only has a specialization but also a multiple decomposition into ingredients as well. Notice that both multi-Aspects have the associated variables to represent the number of replications of their generating entity in a pruned entity structure.

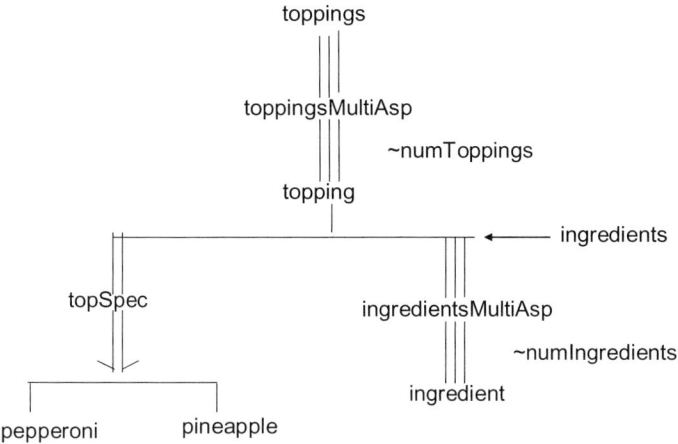

Figure 5.6

Inheritance of multiAspects

How does inheritance act in this case? Consider pruning a pizza topping. Suppose the pepperoni choice is selected and we have *pepperoni_topping*, an entity that inherits *ingredientsMultiAsp*. This means that *pepperoni_topping* is decomposed into one or more ingredients. Likewise, a choice of *pineapple* would result in a similar inheritance of ingredients. Of course, *pepperoni* and *pineapple* each have different ingredients (although some may be common). We'll return to the inclusion of constraints on selection from specializations later (see Chapter 9).

Inheritance Clashes

Inheritance carries with it the possibility of clashes in which structure items with the same name are inherited through different paths. This problem is well-known in object-oriented programming, and various solutions have been proposed and implemented. One approach, in the application area of geographic information, is covered later in Chapter 18. At this point, we assume that the user creates system entity structures in which clashes either do not occur or are managed in a manner that is specific to the domain or application.

Summary

We formulated the SES as a relational structure and showed how this formulation offered insight into the roles of the elements.

Also, we showed how the axioms, that were laid down in early formulations, are satisfied in the relational structure, particularly, the axioms of uniformity and alternation. The relational structure and axioms provide the foundation for developing the concepts and operations that will be discussed throughout the book.

Problems

1. Use the SES to describe the class of all sandwiches that you like eating. *Hint*: From a physical perspective, consider what makes a sandwich different from other kinds of food preparations. Then, starting with some sandwiches that come to mind, ask yourself what ingredients can be replaced by others (e.g., peanut butter by margarine?) or co-exist together. Are there constraints on combinations of ingredients? If so refer to Chapter 9 to specify them.
2. A song or other music composition has two types of decompositions, one based on the arrangements of notes within measures, and the other on melodic, harmonic, and rhythmic themes. Develop an SES for the class of simple nursery rhymes. Can you restrict its prunings to just those that are likely to be acceptable to children?
3. Describe the players on the field of a baseball game. Rather than treating each player individually, use aspects, multiAspects, and specializations to capture natural categories such as infielders. Develop similar representations for football, and basketball games. Write an SES for "game" that has baseball, football, and basketball explicitly within a specialization. Can you come up with a generic SES that can be pruned to describe any one of the games without using the explicit approach? In other words, are there commonalities that can be factored out of the individual SESs to provide a deeper representation from which each can be pruned?

References

1. Zeigler, B. P. "Multifaceted Modelling and Discrete Event Simulation," London, UK: Academic Press, 1984
2. Zeigler, B. P., and Zhang, G. "The System Entity Structure: Knowledge Representation for Simulation Modeling and Design," in: *Artificial Intelligence, Simulation and Modeling* (Eds. L. A. Widman,

K. A. Loparo, and N. Nielsen), Hoboken, NJ: John Wiley, pp. 47–73, 1989

3. Park, H. C., Lee, W. B., and Kim, T. G. "RASES: A Database Supported Framework for Structured Model Base Management," *Simulation Practice and Theory*, vol. 5, pp. 289–313, 1997

System Entity Structure: Computational Representations

This chapter discusses various ways to construct and represent System Entity Structures (SES) in computational form. Figure 6.1 depicts some of these SES specification protocols and their interrelation. The *sesRelation* at the upper right is a Java class that closely parallels the formal definition given in Chapter 5. Another object-oriented representation is a subset of Document Object Models (*DOM*), the core class for representing eXtensible Markup Language (XML) documents in computer memory.[2] The implementation language we use is Java and we employ Sun's implementation of the DOM specification.[3] We'll discuss the sesRelation and the DOM classes and a mapping from the first to the second. Once it is in DOM form, using available methods in the implementation, SES/JAVA, an SES can be written out as an XML Document Type Definition (DTD) or an XML Schema. For simplicity, we'll stay with the DTD alternative in the next sections, but everything that holds for DTDs also holds for schemas. From a DTD, instances of the SES (viz., pruned entity structures) can be derived using a standard XML approach. Starting

Creating an Instance of Class sesRelation

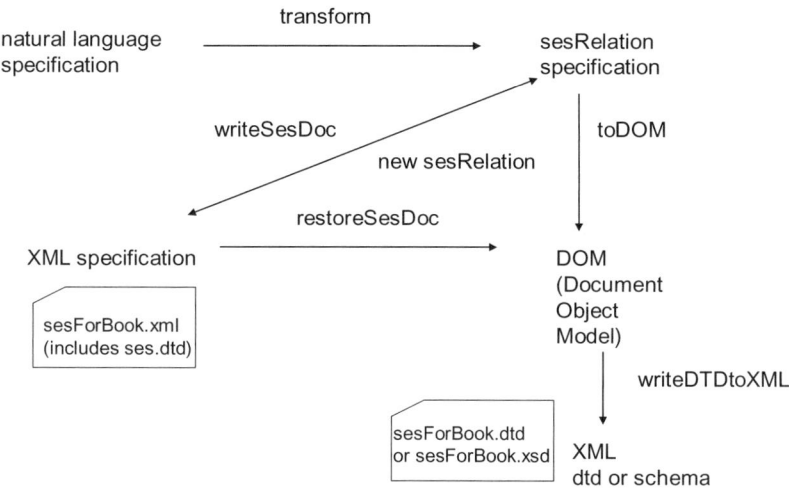

Figure 6.1

Some Ways of Specifying an SES and Their Interrelation

from the lower left of Figure 6.1, we specify an SES as an XML file, such as sesForBook.xml. This file can then be read into memory and a DOM representation made of the SES using restoreSesDoc. Finally, at the upper left, a way of expressing an SES in natural language is provided (and has been used in earlier chapters). This representation can be transformed into an instance of the class sesRelation. Then by drawing upon the just discussed writeDTD to XML, we can generate the XML DTD (or Schema) that operationalizes it.

Creating an Instance of Class sesRelation

The most direct way of representing an SES in computational form parallels that of the formal definition in Chapter 5. As illustrated in Figure 6.2, we have a class sesRelation whose instances represent SESs.

The class sesRelation has an instance variable for the root entity of an SES as well as instance variables to reference each of the relations that make up an SES according to the definition such as entityHasAspect, entityHasSpec, aspectHasEntity, etc. We can create an instance of sesRelation and set the root entity's name using the method, create, as in:

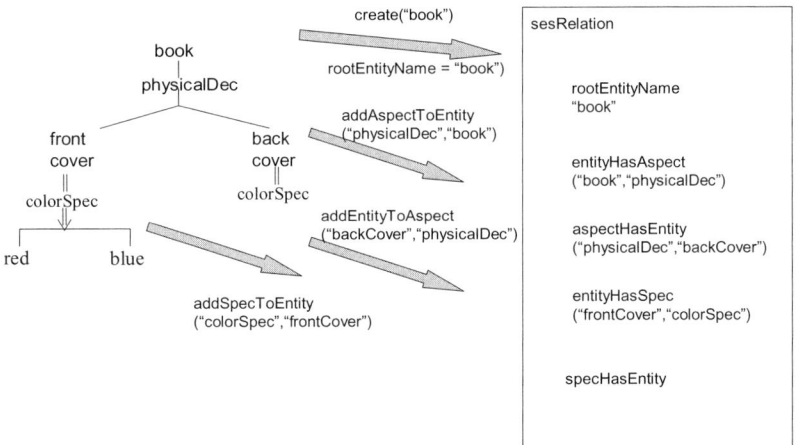

Figure 6.2

Constructing an SES by Creating and Modifying an Instance of sesRelation

sesRelation ses = sesRelation.create("book");

Now, for each of the relations there is a method for adding a pair to that relation. For example, to create the SES shown in Figure 6.2, we can invoke the following sequence of method calls:

ses.addAspectToEntity("physicalDec", "book");
ses.addEntityToAspect("frontCover", "physicalDec");
ses.addEntityToAspect("backCover", "physicalDec");
ses.addSpecToEntity("colorSpec", "frontCover");
ses.addEntityToSpec("red", "colorSpec");
ses.addEntityToSpec("blue", "colorSpec");
ses.addSpecToEntity("colorSpec", "backCover");

Notice that there is no need to sequence the method calls in any particular order. Each call basically adds a pair of strings to a relation in a stand-alone manner. Of course, there are natural orders such as the one just given that reflect top-down constructions. However, because all information is in the form of relations, no objects are constructed and there is no requirement to construct a parent object before its child.

Where order does show up, it is in the generation of the reachable items using the algorithm for generating the directed graph in Chapter 5. This algorithm starts with the root and successively retrieves new items as dictated by the relations. Therefore, it will generate the graph in a top-down manner. The generated items are displayed with the method:

ses.generateItemsGraph();

which for the above will display:

```
==================================================
Items for book
==================================================
entities [red, backCover, frontCover, book, blue]
aspects [physicalDec]
specializations [colorSpec]
```

The SES spanned from the root is displayed with:

ses.printTree();

which prints:

entity: book
–aspect: physicalDec
—entity: backCover
——specialization: colorSpec
———entity: red
———entity: blue
—entity: frontCover
——specialization: colorSpec
———entity: red
———entity: blue

Exercise

Place the first method call in the above sequence of method invocations last. What is the resulting printed tree? Now remove the method call entirely. What is the resulting printed tree? Explain.

Since printTree() can get into a never-ending loop if the SES does not obey strict hierarchy, it checks for strict hierarchy before attempting to print. The strict hierarchy property is checked using Algorithm 5.2 and is implemented in the method:

ses.checkStrictHier();

Exercise

Is strict hierarchy violated in the example just given?

Add each of the following invocations to the construction sequence:

ses.addEntityToSpec("book", "colorSpec"); and
ses.addSpecToEntity("colorSpec", "blue");

Is strict hierarchy violated in these examples?

Notice that in the alternating mode, uniformity, and valid brothers' axioms are intrinsically satisfied in the sesRelation representation. This is because the formulation is in one-one correspondence with the set theoretic structure that intrinsically respects these axioms as discussed in Chapter 5. This will be in marked contrast to the specification as an XML file as discussed below.

Representing the SES as a DOM

Let us now consider the mechanics of representing the SES within the XML environment. The first step is to represent an SES as a data structure that conforms to the XML specification[1] and can be manipulated by Java XML tools.[3] We discuss how this representation process can work in Java using its Sun's interfaces and classes for XML representation and manipulation.

In Sun's Java packages, we have the following points:[1]

- An instance of class Document is a representation of an XML document as a tree structure.
- The nodes of the tree are instances of class Element that represent tags in the XML document. Method createElement("tag") of Document creates an Element instance to represent "tag."
- The first Element instance created by createElement becomes the root of the tree.
- Method appendChild (tagElement) of Element appends tagElement as a child of the calling Element instance (it does so by adding it to its NodeList).

---

[1] We simplify the discussion by treating concepts as classes rather than interfaces.

- Method setAttribute(attribute, string) of Element creates an attribute slot (if not already created) for the calling Element instance and sets it to string.

As illustrated in Figure 6.1, the method toDOM maps the sesRelation instance into a DOM. A fragment of the algorithm used in this method is shown in the following:

ALGORITHM

convert SES to DOM representation

//initialize
createSesDoc();//creates a *Document* object, *sesDoc* with root, *sesRoot*
sesRoot.setAttribute("name", entity);
convertEntity(sesRoot, rootEntity)
//
convertEntity(entityElement, entity){
aspects = entityHasAspect(entity)
For each *aspect* ∈ *aspects* :
    aspectElement = sesDoc.createElement("aspect");
    entityElement.appendChild(aspectElement);
    aspectElement.setAttribute("name", aspect);
    convertAspect(aspectElement, aspect)
}
convertAspect(aspectElement, aspect){
newEntities = aspectHasEntity(aspect)
For each *entity* ∈ *newEntities*,
    entityElement = sesDoc.createElement("entity");
    aspectElement..appendChild(entityElement);
    entityElement.setAttribute("name", entNm);
    convertEntity(entityElement., entity);
}

The process is illustrated in Figure 6.3. The initialization part of the algorithm creates a Document, sesDoc. Calling sesDoc's createElement, we create the root Element, sesRoot, which represents an entity tag. Calling sesRoot's setAttribute method, we set the attribute name of sesRoot to "book." The process continues recursively. For each aspect of an entity, we create an Element with aspect tag and set its attribute name to the aspect's name.

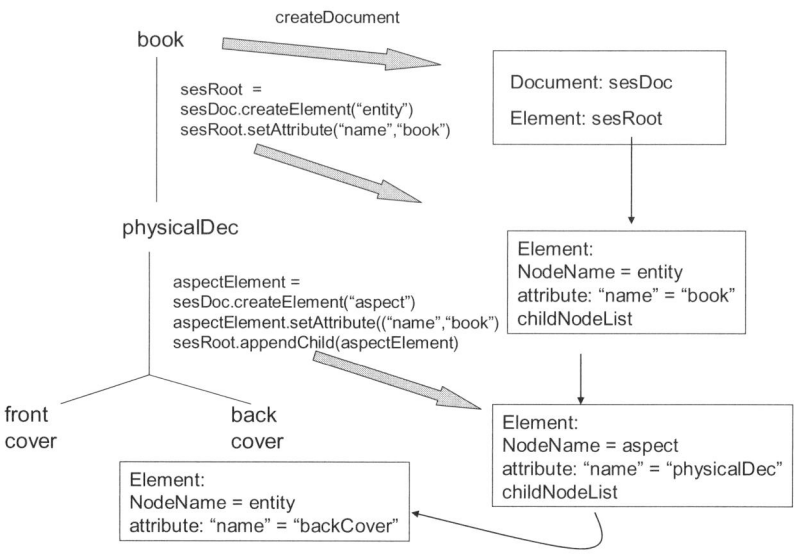

Figure 6.3

Converting an SES into a Document Instance

This Element is appended to the childNode list of the entity. For each entity of an aspect, we go through a similar mapping process. MultiAspects, specializations, variables, and other attachments of the SES are treated similarly.

Note that as we proceed, we pass down a pair consisting of an SES item and its DOM Element representative. This pattern will apply to many transformations of an SES structure into other forms — define a transformation for each of the SES items and, at the end of each one's work, have it call the transformation methods of the items that are attached to it or are in its substructure.

Specifying the SES as an XML File

As illustrated in Figure 6.4, as an alternative to creating a direct representation of an SES (using sesRelation), we can create a file such as sesForBook.xml that specifies an SES as given below:

Specifying the SES as an XML File

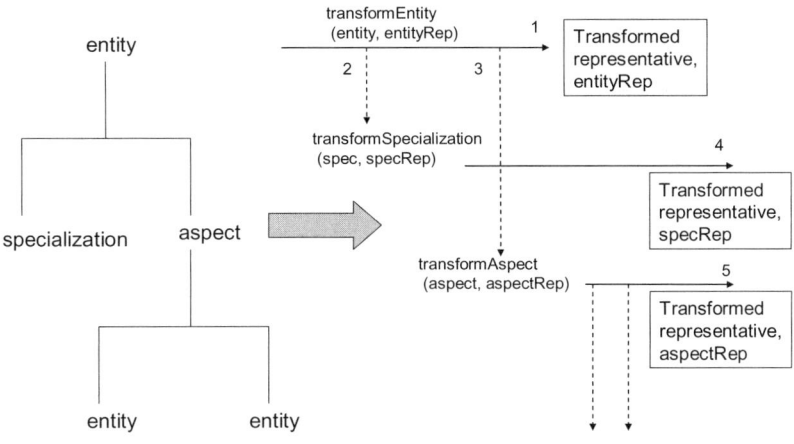

Figure 6.4

Illustrating the Recursive Pattern in Which Each Item Has Its Own Transformation and Calls Those of Its Children

```
<?xml version='1.0' encoding='UTF-8'?>
<!DOCTYPE entity SYSTEM "ses.dtd" []>
<entity name = "book">
  <aspect name = "physicalDec">
    <entity name = "backCover">
      <specialization name = "colorSpec">
        <entity name = "red"/>
        <entity name = "blue"/>
      </specialization>
    </entity>
    <entity name = "frontCover">
      <specialization name = "colorSpec"/>
    </entity>
  </aspect>
</entity>
```

This file must obey the structure imposed by the ses.dtd, a document type definition that captures the alternating mode axiom of the SES specification given in Chapter 5. For example, as illustrated in Table 6.1, you cannot place an entity directly within the scope of another one — only aspects, multiAspects, specializations, and variables are allowed under an entity. Likewise, only entities can be placed within the scopes of aspects, multiAspects, and specializations.

However, in a moment, we will see that the other SES axioms are not checked through the standard XML validation tools.

Table 6.1. Representing an SES as a DTD

SES item	Maps to XML specification
Entity	`<!ELEMENT entity ((aspect | specialization | multiAspect)*,var*)>` `<!ATTLIST entity name CDATA #REQUIRED>`
Aspect	`<!ELEMENT aspect (entity*)>` `<!ATTLIST aspect name CDATA #REQUIRED>`
multiAspect	`<!ELEMENT multiAspect (numberOfComponents,entity)>` `<!ATTLIST multiAspect name CDATA #REQUIRED>`
Specialization	`<!ELEMENT specialization (entity*)>` `<!ATTLIST specialization name CDATA #REQUIRED>`
Variable	`<!ELEMENT var (#PCDATA)>` `<!ATTLIST var name CDATA #REQUIRED>`

Creating a DOM from an SES XML File

An XML file that specifies an SES such as "sesForBook.xml" can be placed into computer memory as a Java DOM using:

SESOps.restoreSesDoc(path+ "sesForBook.xml");

where the path string provides the appropriate path to each file. This method calls upon a standard XML parser which analyzes the incoming file and constructs a Document tree hierarchy of Elements such as described earlier.

Writing from a DOM for an SES into a DTD

To operationalize an SES within the XML environment, we must map it into a DTD or schema. Once an SES has been mapped into DOM form (either originating as an sesRelation or as an XML file), it can be written as a corresponding DTD, e.g., "BookDTD.dtd" using:

SESOps.writeDTDToXML(path+ "BookDTD");

In the case of originating from "sesForBook.xml," the output, "BookDTD.dtd," would look like the following:

```
<?xml version='1.0' encoding='us-ascii'?>
<!-- DTD for a book -->

<!ELEMENT book (aspectsOfbook)>
<!ELEMENT aspectsOfbook (physicalDec )>
<!ELEMENT physicalDec ( frontCover , backCover )>
<!ELEMENT backCover (colorSpec)>
<!ELEMENT frontCover (colorSpec)>
<!ELEMENT colorSpec (red | blue )>
<!ELEMENT red ( #PCDATA)>
<!ELEMENT blue ( #PCDATA)>
```

To do the mapping into a DTD, we use a specialization of the general recursive pattern for transforming an SES into another structure (Figure 6.3). The name attributes of SES elements become elements of the DTD. For example,

```
<entity name = "book">
```

becomes

```
<!ELEMENT book...>
```

The specializations of an entity are listed in the order they are discovered and separated by commas as obligatory occurrences. At the end of the list, the element aspectsOfentity appears if there are indeed any aspects for the entity. The aspects and multiAspects of the entity are collected together and separated by "|" to require that one and only one must be selected. For example, if book had two aspects, physicalDec and contentDec, and two specializations, colorSpec and genreSpec, it would be transformed to:

```
<!ELEMENT book (colorSpec, genreSpec,
  aspectsOfbook)>
<!ELEMENT aspectsOfbook (physicalDec |
  contentDec )>
```

Processing then migrates to handle each of the specializations, aspects, and multiAspects with corresponding transformations. Note that this XML representation imposes an arbitrary sequential order on elements that are unordered in the SES representation. For example, a statement that a book is made of a front cover and a back cover does not necessarily imply that the first is always listed before the second. However, the translation into XML:

```
<!ELEMENT physicalDec ( frontCover , backCover )>
```

does impose this requirement. The reason is that XML is a syntax which is much easier to process when all elements are fixed in specific sequences. On the other hand, the SES is a semantical concept that, being stated in set theory, deals most naturally with unordered collections.

Exercise

Consider a house made of a living room, bedroom, and kitchen. Would it be a different house if it were made of a bedroom, kitchen, and living room? Both descriptions describe the same house for the SES but different houses for the XML representation.

Transforming an SES into a Schema representation is discussed after first considering the relation between SES and XML validation.

Testing an SES for Validity

We have seen that an XML specification such as "sesForBook.xml" must obey the structure imposed by ses.dtd, a document type definition that captures the alternating mode axiom of the SES definition as given in Chapter 5. However, other important axioms such as uniformity, valid brothers, and strict hierarchy cannot be enforced by the DTD or schema constructs of XML. Thus, the situation is quite different from that of the sesRelation for which the strict hierarchy axiom is the only axiom that is not intrinsically satisfied.

Exercise

Use a validation tool such as XMLSpy to check that the above sesForBook.xml is valid. Modify the file to violate the uniformity, valid brothers, and strict hierarchy axioms, and check whether the tool still reports that the file is valid.

To repeat, an SES specification as an XML file would not necessarily result in a valid SES even if it passes the well-formedness and validation checks of standard XML tools. The axioms that specifically characterize an SES — as opposed to any other structure that might be represented within XML — must be checked

Testing an SES for Validity

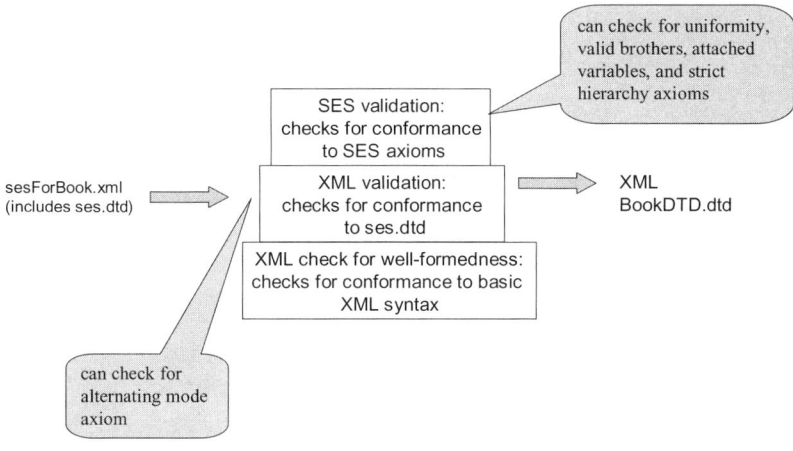

Figure 6.5

Illustrating the Layers of Validation that an SES Specification May Go Through

by a more demanding validation tool. Such a method is provided and operates on a DOM that represents the SES in memory. So, for example,

validateSes.validate(path+"sesForBook.xml")

checks the SES axioms of the file "sesForBook.xml." Since to be parsed into a DOM, the XML specification first has to be validated, we can say that SES-specific validation is at a layer higher than generic XML validation, as shown in Figure 6.5.

Exercise

Using validateSes. (www.devsworld.org), check that the above sesForBook.xml is valid. Modify the file to violate the uniformity, valid brothers, and strict hierarchy axioms, and check whether the tool still reports that the file is valid.

The method validateSes.validate first checks for strict hierarchy using the approach discussed earlier. It then examines uniformity by checking that at most one of the collection of items with the same name has a nonempty substructure. It also checks for duplicates that would violate the valid brothers and attached variables axioms; it checks that specializations and aspects are not empty as we required in Chapter 5. (The hierarchical DOM structure already guarantees that there are no free-standing aspects or specializations.) Finally, the validation process examines a number of other sources of error such as undefined names and names that are used for more than one item or variable.

Schema Representation of SES

Table 6.2 provides the basic approach to mapping from an SES that satisfies the axioms to a Schema representation for it.

In the schema context we have seen that uniformity is not automatically enforced, so this mapping needs to be augmented with a mechanism to satisfy the uniformity axiom. We can do so by using a reference to a global element. For example, a global element Book may be defined and then all occurrences of Book can be referred to this global version using the *ref* attribute of element. We have implemented a transformation that does this automatically by maintaining a set of element names that have been encountered so far. When processing a new item, its name is checked for membership in the set; if so, then the string:

```
<xs:element ref ="[item.name]" />
```

is generated; otherwise, a string starting with

```
<xs:element name ="[item.name]" >
```

is generated according to the recursion in the table.

SimpleTypes

In Chapter 4 we discussed the specification of variables and their ranges. The construct, simpleType is the corresponding XML concept. We show how the natural language tool (www.devsworld.org) allows variable range specifications to be specified in the SES and checked in the resulting Schema. Recall the specification,

the crosswalk has a controlLight
the range of the crosswalk's controlLight is string with values
  Stop and Go

Table 6.2. Representation of SES as an XML Schema

SES item	Maps to XML specification	Comment

Entity	`<xs:element name="[entity.name]">`
	`<xs:complexType>`
	`<xs:sequence>`
	`...`
	
	`...`
	`</xs:sequence>`
	`</xs:complexType>`
	`</xs:element>`
		The substructure of the entity is sandwiched between opening and closing `<xs:complexType>` `<xs:sequence>` tags

Aspect	`<xs:element name="[aspect.name]">`
	`<xs:complexType>`
	`<xs:sequence>`
	`<xs:element name="[child.entity.name]"/>`
	`...`
	`</xs:sequence>`
	`</xs:complexType>`
	`</xs:element>`
		Aspects and multi-aspects are collected together into a choice similar to a specialization; each aspect is represented as an `<xs:sequence>`

multiAspect	`<xs:element name=="[multiAspect.name]">`
	`<xs:complexType>`
	`<xs:sequence>`
	`<xs:element name="[entity.name]"`
	`minOccurs="[numberComponentsVar.min]"`
	`maxOccurs="[numberComponentsVar.max]">`
	`<xs:element>`
	`<xs:sequence>`
	`</xs:complexType>`
	`</xs:element>`
		Within the collected aspects, each multiAspect is represented as an `<xs:sequence>` with multiply occurring element

Specialization	`<xs:element name="[spec.name]">`
	`<xs:complexType>`
	`<xs:choice>`
	`<xs:element name="[child.entity.name]"/>`
	`...`
	`</xs:choice>`
	`</xs:complexType>`
	`</xs:element>`
		A specialization is represented as `<xs:choice>`

Variable	`<xs:complexType>`
	`<xs:attribute name="[var.name]"`
	`type="xs:+[var.rangeSpec]"/>`
	`</xs:complexType>`
		A variable is represented as an attribute, where its rangeSpec value determines its data type and restrictions (see below)

which states that there is a variable called controlLight whose data type is string and whose values are limited to the strings, "Wait," "Stop," and "Go." This is translated into the following fragment in the XML representation.

```
<entity name="crosswalk">
<var name="controlLight" rangeSpec=
  "enumWaitStopGo">
<simpleReference name=" enumWaitStopGo "
  restrictionValuePairs="enumeration, Wait,
  enumeration, Stop, enumeration, Go"
  restrictionBase="string"/>
</var>
</entity>
```

This fragment takes form allowed by an expanded ses.dtd which includes specification of the simpleReference element.

The fragment is translated into a corresponding Schema as:

```
<xs:simpleType name=" enumWaitStopGo">
      <xs:restriction base="xs:string">
      <xs:enumeration value="Wait"/>
      <xs:enumeration value="Stop"/>
      <xs:enumeration value="Go"/>
   </xs:restriction>
</xs:simpleType>

<xs:element name=" crosswalk">
      <xs:complexType>
        <xs:attribute name="light "
        type=" enumWaitStopGo"/>
      </xs:complexType>
</xs:element>
```

Another example illustrates specification of intervals. The natural language input:

The range of backwheel's radius is double with values [10,30]

is translated to:

```
<entity name="backwheel">
<var name="radius" rangeSpec="double1030">
<simpleReference name="double1030" restriction
  ValuePairs="minInclusive, 10, maxInclusive, 30"
restrictionBase="double"/>
</var>
</entity>
```

which then appears in a Schema as:

```
<xs:simpleType name=" double1030">
    <xs:restriction base="xs:double">
        <xs:minInclusive value="10"/>
        <xs:maxInclusive value="30"/>
    </xs:restriction>
</xs:simpleType>
```

ComplexTypes

Global elements that are created in the Schema can be easily converted to complexTypes. For example, starting with:

```
<xs:element name="[entity.name]">
<xs:complexType>
<xs:choice>
<xs:element name="[child.entity.name]"/>
</xs:choice>
</xs:complexType>
</xs:element>
```

we create new complexType by providing a name to complexType, and then remove it from enclosing element, letting the later point to it using the type attribute.

```
<xs:element name="[ entity.name]" type =
  [entity.name]Type"/>
<xs:complexType name = [entity.name]Type>
<xs:choice>
<xs:element name="[child.entity.name]"/>
</xs:choice>
</xs:complexType>
```

Exercise

Consider representing locations of objects on the Earth. The coordinates for location are heterogeneous, i.e., longitude and latitude are in degrees while elevation is in meters. A shape of a region is given by a polygon whose vertices are homogeneous, i.e., each vertex has the same pair of coordinates, longitude and latitude. Locations and vertices have identifications such as names, station numbers, etc.

Consider general alternatives for embedding data in XML schema:

- Use attributes to carry both data values and identifications.
- Use the text content of an element to carry the important value of the entity it encodes. Attributes are to be used for identification only.

Discuss various approaches to mapping an SES for locations and regions to an XML schema.

Schema Validation of SES

The schema created for an SES will detect invalid brothers, but it will not check for uniformity. The schema in Figure 6.6 violates uniformity for a as well as strict hierarchy; thus it cannot represent an SES.

Nevertheless, the instance below shows that this structure can have instances and therefore is meaningful within XML. Thus XML allows many non-SES compliant structures.

```
<?xml version="1.0" encoding="UTF-8"?>
<a xmlns:xsi="http://www.w3.org/2001/XML
  Schema-instance" xsi:noNamespaceSchema
  Location=" nonSES.xsd">
<a>
<c></c>
</a>
<b></b>
</a>
```

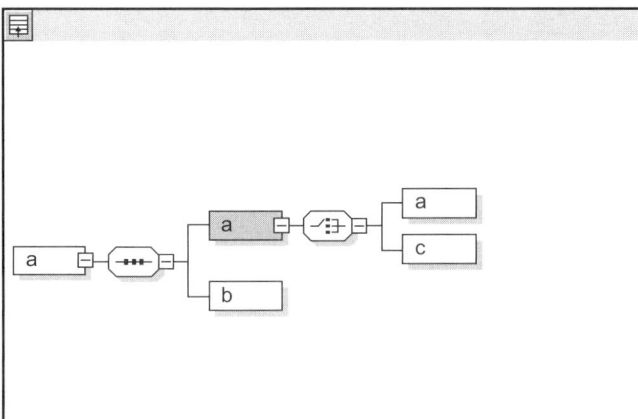

Figure 6.6

XML Schema (illustrated using XMLSpy®)[4]

Natural Language Specification of an SES

In this section, we briefly discuss the natural language specification supported by the web site www.devsworld.org. Such inputs can be transformed using a transform method in a form illustrated by:

sesRelation ses = natLangToSes.transform(path + "sesForBook.txt")

where the input text file satisfies the syntax given in Appendix A. Examples of such texts have been sprinkled throughout earlier chapters.

The resulting sesRelation object can be written out as an XML file specification, e.g., as:

ses.writeSesDoc(path + "bookSes.xml");

Later this file can be restored to memory using restoreSesDoc and then written as a DTD using writeDTDToXML as mentioned earlier. Alternatively, because the method toDOM maps the sesRelation object into a DOM, we have included it within the sesRelation's version of the DTD writer — amounting to the composition of the two methods at the right side of Figure 6.1. In this way, for example,

ses.writeDTDToXML(path + "BookDTD");

generates BookDTD.dtd from the ses directly.

It is also possible to create a new sesRelation instance corresponding to one specified in XML, using the constructor sesRelation with the XML files as argument.

Summary

Various ways to construct and represent System Entity Structures (SES) in computational form are depicted in Figure 6.1. In particular, an SES may be represented as an instance of a particular DTD and then transformed into an XML DTD or schema. The instances of the latter then represent the pruned entity structures of the SES. We showed that, distinct from its XML implementation, the SES, as an axiomatically structured object, gives rise to its own validation concepts. On this basis, a restricted form of natural language input was developed to support SES definition.

References

1. Extensible Markup Language (XML) 1.0, http://www.w3.org/TR/2004/REC-xml-20040204 (accessed Nov. 2006)
2. Document Object Model Core, http://www.w3.org/TR/2004/REC-DOM-Level-3-Core-20040407/core.html#ID-1590626202 (accessed Nov. 2006)
3. Working with XML: The Java/XML Tutorial, http://java.sun.com/xml/tutorial_intro.html (accessed Nov. 2006)
4. Altova XMLSpy® Enterprise Edition, http://altova.com/ (accessed Nov. 2006)

Appendix A: Syntax for Natural Language Specification of an SES

General

1. Copula:
"is" and "are" are treated the same

2. Compounds:
Follow English style:
x and y;
x, y, and z; NOTE: the commas are mandatory for 3 or more constituents
x, y, z, and w;
x or y;
x, y, or z;
x, y, z, or w;

3. Determiners:
"a," "the," ... are removed from the input before processing

4. End of Sentence:
use "!" instead of "."

5. Sentence order:
- The entity mentioned first in the first sentence becomes the root of the SES.
- A variable must be attached to an entity before giving it a range specification (see below).
- Otherwise sentences can be in any order.

Forms

In the following, CAPITALs indicate variables, and lowercase indicates mandatory key words in the order shown.

Specialization:
THING can be VARIANT1, VARIANT2, or VARIANT3 in CLASSFAMILY!

results in:

entity: THING
—specialization: THING-CLASSFAMILYSpec
——entity: VARIANT1
——entity: VARIANT2
——entity: VARIANT3

Violations of the "or" clause, e.g.,

THING can be VARIANT1, VARIANT2, or VARIANT3 in
 CLASSFAMILY!
THING can be VARIANT1 in CLASSFAMILY!
THING can be VARIANT1, in CLASSFAMILY!

result in:

entity: unknown

Aspect:
From VIEW perspective, THING is made of COMPONENT1, COMPONENT2, and COMPONENT3!

results in:

entity: THING
—aspect: THING-VIEWDec
——entity: COMPONENT2

――entity: COMPONENT3
――entity: COMPONENT1

Violations in "and" clause result in "entity: unknown"

multiAspect:
From multiple perspective, THINGS are made of more than one THING!

results in:

entity: THINGS
—multiAspect: THINGS-multipleMultiAsp
――numComponentsVar: numTHINGS, min: 0, max: 10,
――entity: THING

From multiple perspective, THINGCOLLECTION is made of more than one THING!

results in:

entity: THINGCOLLECTION
—multiAspect: THINGCOLLECTION-multipleMultiAsp
――numComponentsVar: numTHINGCOLLECTION, min: 0, max: 10,
――entity: THING

Attached Variables:
THING has VAR1, VAR2, and VAR3!

results in:

entity: THING
—var: VAR1,rangeSpec: null
—var: VAR3,rangeSpec: null
—var: VAR2,rangeSpec: null

Restriction: THING must be an entity in some other sentence.

THING has VAR1! NOTE: single variable treatment is allowed.

entity: THING
—var: VAR1,rangeSpec: null

Range specifications of variables:
The range of THING's VAR1 is RANGE!

results in:

entity: THING
—var: VAR1,rangeSpec: RANGE

Restriction: VAR1 must have been attached to THING earlier.

Advanced Forms

Uniformity:

1. Giving an entity another's specialization:
THING is like ANOTHER in CLASSFAMILY!

results in:

entity: THING
—specialization: ANOTHER-CLASSFAMILYSpec

Restriction: ANOTHER must be given the CLASSFAMILY specialization in some other sentence (also ANOTHER and THING must be reachable from the root).

2. Giving an entity another's decomposition:
From VIEW perspective, THING is like ANOTHER!

results in:

entity: THING
—aspect: ANOTHER-VIEWDec

Restriction: ANOTHER must be given the VIEWDec aspect in some other sentence (also ANOTHER and THING must be reachable from the root).

Formula:
Follow the example:
   The speedometer's speed comes from the multiplier's output where the multiplier's input1 comes from the backwheel's circumference where the multiplier's input2 comes from the backwheel's rpm!

entity: bicycle
—aspect: bicycle-structureDec
——entity: backwheel
———var: radius,rangeSpec: real
———var: rpm,rangeSpec: null
———var: circumference,rangeSpec: null
——entity: handlebar
———aspect: handlebar-equipmentDec
————entity: speedometer
—————var: speed,rangeSpec: double
—————var: function: speedometer.speed = multiplier.output ;multiplier.input1 = backwheel.circumference ;multiplier.input2 = backwheel.rpm ;,rangeSpec: null
————entity: bell
————entity: brake
——entity: frontwheel
———var: radius,rangeSpec: real
———var: circumference,rangeSpec: null

7

Mappings: Transformations and Restructurings

As depicted in Figure 7.1, mappings, in the form of transformations and restructurings, play a major role in model-based data engineering. *Transformations* are mappings from one representation to another.[1] For example, Chapter 6 discussed transformations from a restricted natural language format to an SES expressed within XML and from the latter to the XML Schema that it specifies. *Restructurings* are mappings whose domain and range are the same. As the name suggests, a restructuring changes the structure of an object without changing the form in which it is expressed. In this chapter, we'll discuss some restructurings that will set the basis for applications, such as harmonization, to be discussed later. If the representation doesn't change, only its details, why restructure? Consider harmonization — different SESs may appear to be representing the same information in different ways. A method behind this madness may be revealed

---

[1] Often transformations refer to mappings in general. We make the distinction between transformations and restructurings as defined here.

113

114 Chapter 7 **Mappings: Transformations and Restructurings**

Figure 7.1

Transformations and Restructurings

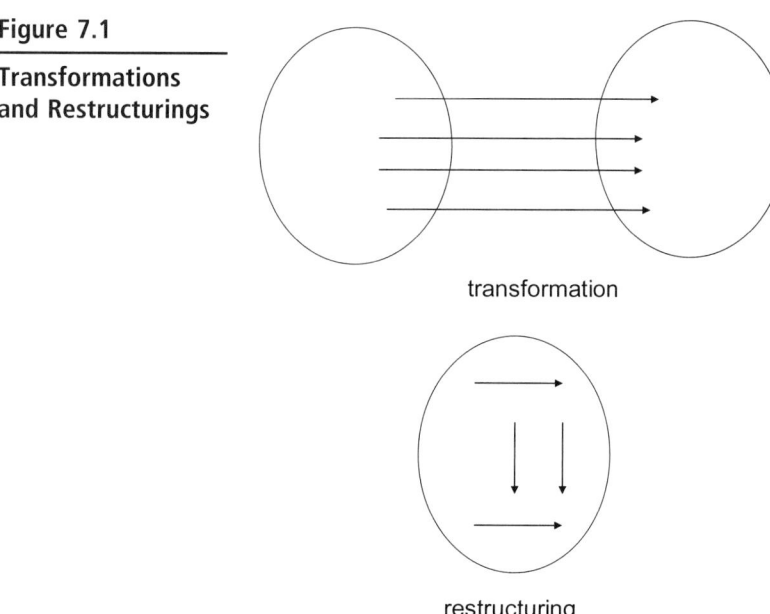

transformation

restructuring

if we discover that there is a standard form of restructuring that actually maps one into the other (Chapter 12). We'll start with basic tree operations such as removing a node and connecting up its neighbors. Such restructurings apply to reducing the size of a tree, such as a Document Object Model (DOM) (Chapter 6). This points to a common reason for restructuring, viz., optimization — finding, from a given perspective, the best representation of some given information within a representation domain. A concept of equivalence must support such restructurings, i.e., the before and after structures must be equivalent with respect to some aspect of interest to the modeler. Actually, there is a more fundamental reason for learning about the restructurings to be discussed below. It will improve your methodology for building SESs. Once you understand restructurings, you will naturally consider the alternative representations they bring out when you encounter situations to which they apply. Indeed, having a range of options to consider for any given situation in a subject area is the sine qua non of an expert in the area.

Tree Operations: Eliminating Specialization and Aspect Labels

The general process of removing a node in a tree is illustrated in Figure 7.2. We first find the node to be eliminated, marked p for

Tree Operations: Eliminating Specialization and Aspect Labels 115

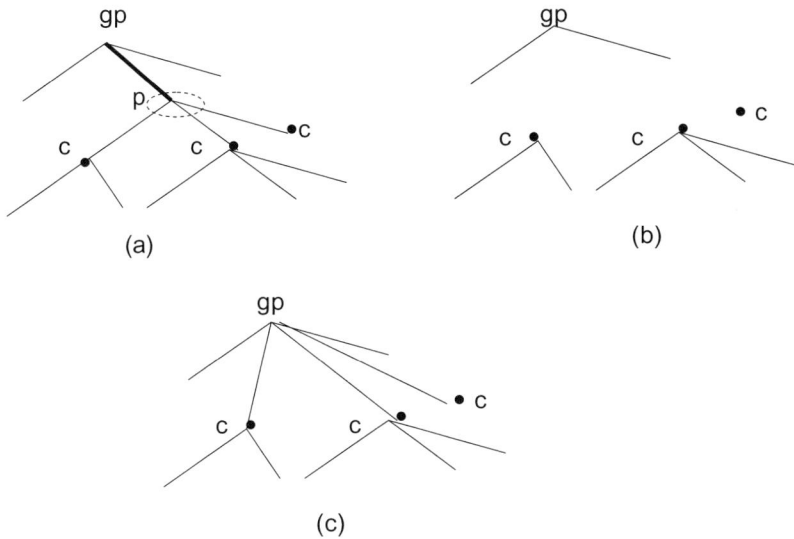

Figure 7.2

Removing Parent and Attaching Children to Grandparent

parent in Figure 7.2(a). We then find its parent (or parents), marked *gp* for grandparent, and its children marked *c*. To eliminate *p* we attach each child *c*, to the grandparent *gp*, and remove *p* from *gp*. The same kind of manipulation can be done with higher order ancestors and/or descendents. For example, we can eliminate a node and its parent by attaching children to their great-grandparent. We may also have to distribute some children to different ancestors depending on where they fit.

For the tree structure representation of a DOM, there are some additional complications to account for:

- An existing node must be cloned before appending it to a new parent.
- The replace method has to be used to replace an existing child by a new node (appending and removing will not work).
- Pointers to nodes that have been cloned (including their subordinates, if cloning is deep) are outdated so their targets have to be discovered afresh if needed.

We apply this general restructuring process to the eliminated labels, including those of aspect, multiAspect, and specialization. Eliminating such labels in a Schema for an SES reduces the amount of overhead in carrying payload information. The resulting SES is equivalent to the original in the sense that the same family of pruned entity structures is defined. At the same time, this mapping is not reversible — a problem, since such restructuring removes information that may be needed in downstream processing of the transmitted data.

As an example, here are the steps needed to remove a specialization.

1. Given the specialization element, get its parent sequence element and the list of its choice grandchildren.
2. Create clones of the grandchildren and attach them to the parent sequence.
3. Remove the specialization element from the sequence.

These steps are illustrated in Figure 7.3.

We may further optimize by removing redundant elements. For example, a sequence of one choice is a choice — therefore we can eliminate all sequences of this nature. Similarly, since a choice of one sequence is a sequence, this allows us to remove the sequence. The following steps illustate this optimization:

1. Given the sequence element, get its parent complexType, and the list of its choice children.
2. If there is only one choice in this list, create a clone of this choice and attach it to the complex parent.
3. Remove the sequence from the complex parent.

This is illustrated in Figure 7.4.

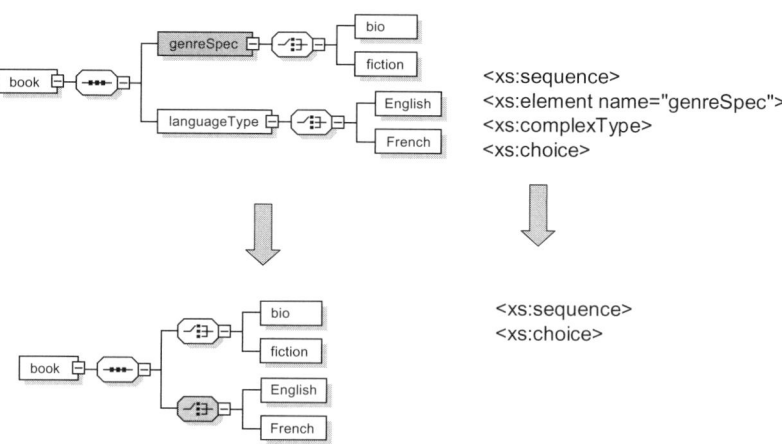

Figure 7.3

Removing Specialization Labels

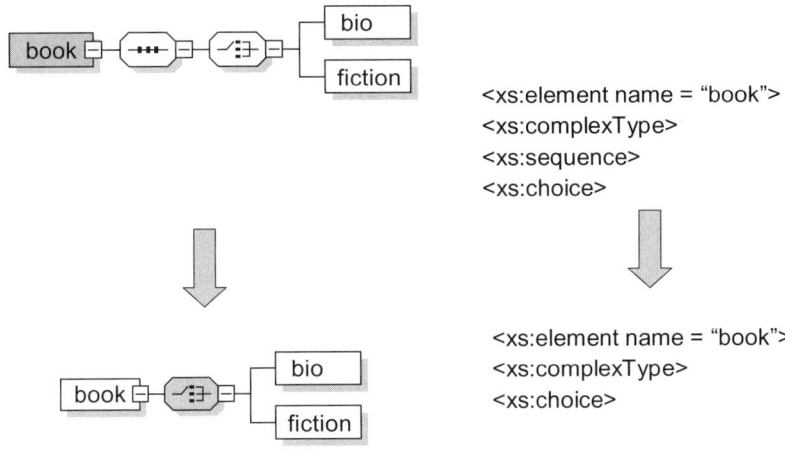

Figure 7.4
Removing a Sequence Having Only a Single Choice

In SES/JAVA, a class writeSchemaWithOptions provides the option of removing all labels for aspects, specializations, and multiAspects from SES Schema before writing the Schema to an xsd file.

Restructuring of multiAspects

The rest of this chapter presents various restructurings of the SES that will be important in applications such as improving representation utility and harmonizing different SESs for the same domain. The first restructuring under consideration is to expand multiAspects that have a specialization under the generating entity.

Consider the natural language specification

From physical perspective, toppings are made of more than one topping
A topping can be Pepperoni or Pineapple in flavor

which results in the SES in Figure 7.5(a) and the following relational representation:

SesForToppings = <
Entities = {*toppings,topping,Pineapple,Pepperoni*}
Aspects = {}

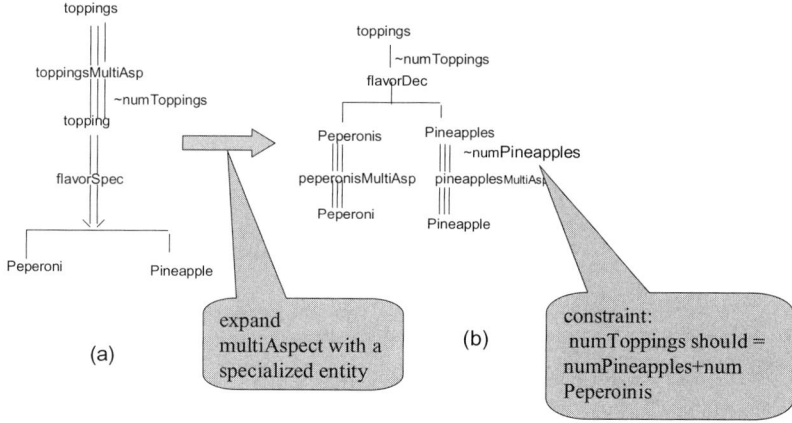

Figure 7.5

Expanding a multiAspect

$multiAspects = \{physicalMultiAsp\}$
$Specializations = \{flavorSpec\}$
$rootEntity = toppings$
$entityHasAspect = \{\}$
$entityHasMultiAspect = \{(toppings, physicalMultiAsp)\}$
$entityHasSpecialization = \{(topping, flavorSpec)\}$
$aspectHasEntity = \{\}$
$multiAspectHasEntity = \{(physicalMultiAsp, topping)\}$
$multiAspectHasVariable = \{(physicalMultiAsp, numToppings)\}$
$specializationHasEntity = \{(flavorSpec, Pineapple),$
   $(flavorSpec, Pepperoni)\}$
$>$

Note that we have a multiAspect, toppingsMultiAsp, with generating entity, topping, that has a specialization, flavorSpec. The result of expansion is shown in Figure 7.5(b). Notice that each entity of the specialization is given its own multiAspect and these form a new decomposition of the root entity.

The SES with the expanded multiAspect is expressed in the following:

$SesForToppingsExpand = <$
$Entities = \{toppings, Pineapples, Pepperonis,$
   $Pineapple, Pepperoni\}$
$Aspects = \{flavorDec\}$
$multiAspects = \{PineapplesMultAsp, PepperoniMultAsp\}$
$Specializations = \{\}$

Restructuring of multiAspects

$rootEntity = toppings$
$entityHasAspect = \{toppings, flavorDec\}$
$entityHasMultiAspect = \{(Pineapples, PineapplesMultAsp),$
    $(Pepperonis, PepperonisMultAsp)\}$
$entityHasSpecialization = \{\}$
$aspectHasEntity = \{(flavorDec, Pineapples),$
    $(flavorDec, Pepperonis)\}$
$multiAspectHasEntity = \{(PineapplesMultAsp, Pineapple),$
    $(PepperonisMultAsp, Pepperoni)\}$
$multiAspectHasVarible = \{(PineapplesMultAsp, numPineapples),$
    $(PepperonisMultAsp, numPepperonis)\}$
$specializationHasEntity = \{\}$
$>$

We add the constraint:

$numToppings = numPineapples + numPepperonis$

i.e., that total number of toppings is the sum of the special types.

Exercise

Describe the transformation in general terms. *Hint*: Suppose that you are given an SES:

$SesForXs = <$
$Entities = \{Xs, X, x, y, z, \ldots\}$
$Aspects = \{\}$
$multiAspects = \{XsMultiAsp\}$
$Specializations = \{XSpec\}$
$rootEntity = Xs$
$entityHasAspect = \{\}$
$entityHasMultiAspect = \{(Xs, XsMultiAsp)\}$
$entityHasSpecialization = \{(X, XSpec)\}$
$aspectHasEntity = \{\}$
$multiAspectHasEntity = \{(XsMultiAsp, X)\}$
$multiAspectHasVariable = \{(XsMultiAsp, numXs)\}$
$specializationHasEntity = \{(XSpec, x), (XSpec, y), (XSpec, z) \ldots\}$
$>$

Write the formal version of the SES for the expanded version, *SESForXs*.

See also Figure 7.6, below.

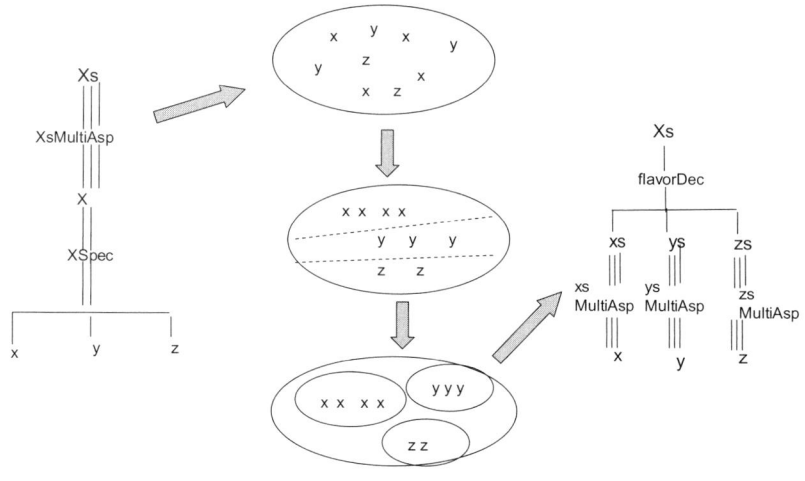

Figure 7.6

Illustrating the Expansion of a multiAspect

 ### Establishing That the multiAspect Restructuring Preserves Semantic Interpretation

As will be discussed in Chapter 11, the domain of discourse of the SES ontology is ultimately simulation models. A restructuring preserves the semantic interpretation if the transformed SES specifies the same family of pruned entity structures and hence the same family of simulation models. Figure 7.6 suggests a method of proof for this assertion.

After restoring an SES into memory, the method,

SESOps.expandMultiAsp(multaspNm)

will restructure the multiAspect with name multaspNm to the expanded SES described in Figure 7.6.

Figure 7.5 makes clear that the original unexpanded form of a multiAspect representation with specialization (Figure 7.5(a)) is more succinct than is its restructured counterpart (Figure 7.5(b)). However, the latter is more explicit in displaying the kinds of prunings that are possible. Therefore the restructured version is good for conveying the semantics of a multiAspect. Since it can be automatically performed, the restructuring is convenient to invoke at the time of pruning to provide the user with explicit choices. We return later (Chapter 9) to discuss how rules and relations can support such pruning.

Restructuring Variables and Specializations

Compare the following declarations of variables and ranges:

A person has a height
The range of a person's height is double with values [4.0,7.5]

A building has a use
The range of a building's use is string with values residential, commercial, and recreational

The first defines a variable, height, with a continuous range of values in some interval representing human height measurements in units of feet. We'll return in a moment to discuss such continuous range specifications. The second defines a variable that takes on discrete values describing the uses of a building. Now, consider the declaration:

A building can be residential, commercial, or recreational in use

which defines a specialization, useSpec. Notice that the entities of this specialization correspond to the discrete values of the variable of the first declaration. Figure 7.7(a) illustrates the restructuring that replaces a variable with a specialization for the same entity, where the discrete values of the variable reappear as entities under the specialization. Using specializations instead of discrete-valued variables has significant benefits in all but the simplest applications in which simple labels are sufficient for descriptive purposes. For instance, Figure 7.7(b) illustrates how each of the specialized entities of the use specialization can be developed further to have a substructure of its own. For example, residential, commercial, and recreational buildings can have different physical decompositions to represent the way they are built — houses with one or two floors, office structures with multiple levels each with work cubicles, and clubs with halls and sports facilities.

Exercise

Flesh out Figure 7.7(b) to show physical decompositions along the lines just discussed.

The extra expressive power facilitated by the specialization approach becomes even more apparent when the entity in

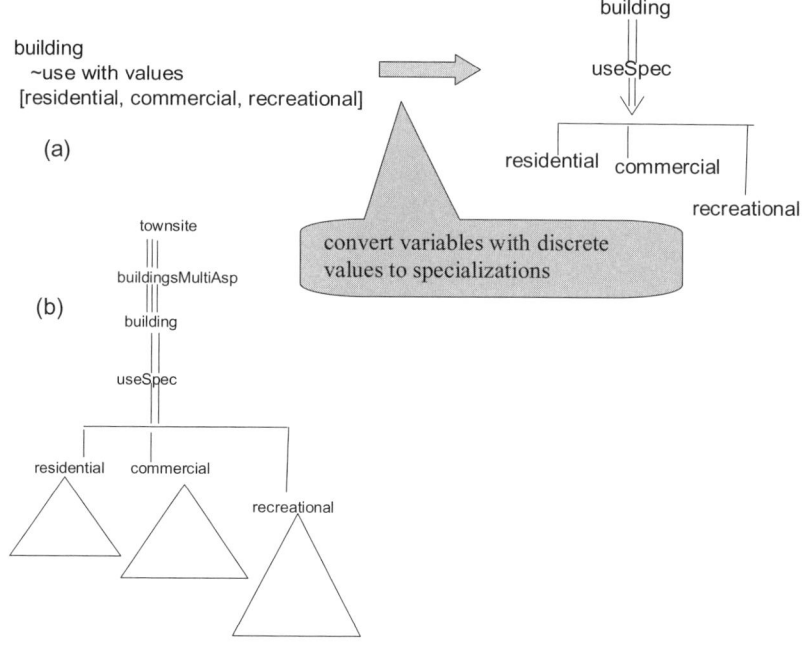

Figure 7.7

Restructuring a Variable into a Specialization

question becomes included in a larger structure. For example, planning for a new town site development might be represented by the SES in Figure 7.7(b). Here applying the multiAspect transformation discussed earlier allows us to generate desired numbers of the different types of buildings that will populate the new development. The resulting SES could then be mapped to a simulation model to study alternatives for numbers and placement of buildings. Later when the development is open for sale, the SES can be used for keeping track of the buildings sold and for other inventory purposes.

Exercise

Apply the multiAsp expansion to restructure Figure 7.7(b) into a town site development that can be populated with multiple types of buildings.

Converting between Variable Values and Specializations

Restructuring methods can be defined that operate on an SES in relational form to convert between variable range sets and specialization representations as just discussed. Methods that work on an sesRelation in SES/JAVA are:

changeSpecToAttr(specialization)
changeAttrToSpec(entity, variable)

Exercise

Formalize the restructuring that converts variables with discrete values to specializations with corresponding entities. Likewise formalize the reverse restructuring that undoes the first. Apply these methods to an sesRelation for the SES in Figure 7.7(a).

Partitioning Continuous Ranges Using Specializations

The use of variables to capture continuous ranges of values is often more appropriate to the semantics and pragmatics than the use of variables for discrete values. However, specializations can also be introduced to partition the range set of a continuous variable. For example, in Chapter 4 we saw how bike sizes can be partitioned into intervals that represent children and adult ranges. Figure 7.8 illustrates an example in which the interval [−1,1] is partitioned into positive and negative intervals and a singleton set containing zero. The specialization having the corresponding entities is shown as well.

Exercise

Use the natural language tool (www.devsworld.org) to specify the SES for the specialization in Figure 7.8.

Converting between Variables and Aspects

Just as a discrete variable can be restructured into a specialization, so can a group of variables be replaced by an aspect. Consider how to describe a geographical location. There are various

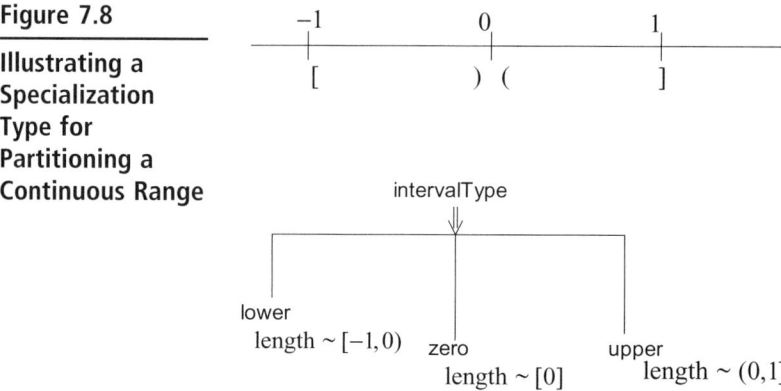

Figure 7.8

Illustrating a Specialization Type for Partitioning a Continuous Range

coordinate systems to structure the information that needs to be provided. One way to represent a coordinate system in an SES is to attach a variable for each coordinate to an entity such as geographicalLocation. For example, to convey latitude and longitude, we would have two corresponding variables with range sets of angle units. Alternatively, we can create an aspect, coordinateDec, that has two entities, Longitude and Latitude, as illustrated in Figure 7.9. Each such entity could then have an aspect, angleDec to break down angle units into the commonly used degrees, minutes, and seconds. Two advantages of this approach are 1) using decomposition, we can explicitly show the breakdown of units into their more fundamental components and 2) using uniformity (Chapter 5), the aspects, coordinateDec and angleDec, can be employed under various entities to express the needed information in a highly reusable manner.

Exercise

Use the natural language tool (www.devsworld.org) to specify the SES in Figure 7.9.

Exercise

Restructure the SES of Figure 7.9 so that it employs variables for coordinates, longitude, and latitude.

Increasing Specialization Specificity

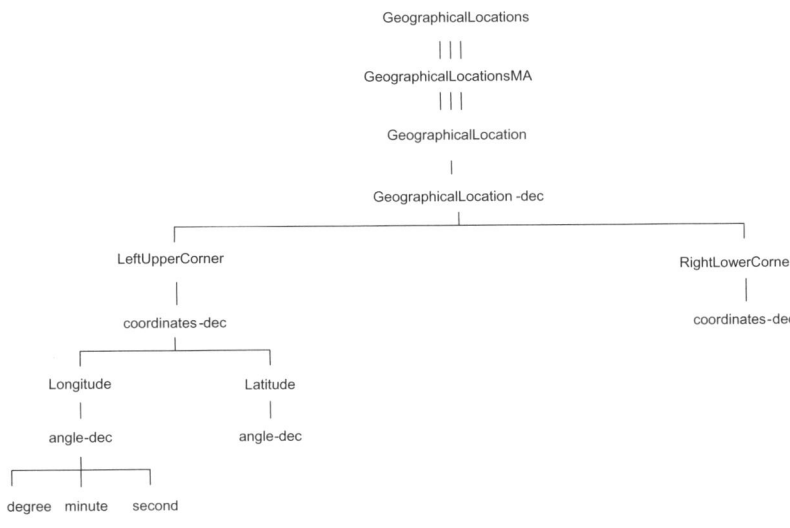

Figure 7.9

Illustrating the Use of Aspects to Replace Variables

Exercise

Use the natural language tool to specify the SES that you restructured in the previous exercise.

Increasing Specialization Specificity

Sometimes in a specialization may differ in only a few respects. If so it is possible to increase the specificity of the specialization to apply only to the respects in which they differ. For example, a full name decomposes into a first and last name with a middle segment. However, English and Spanish differ in the number of middle names that can be in the middle segment. We might start by stating that there are two different decompositions of a full name, as in Figure 7.10(a).

However, we would notice that EnglishNameDec and SpanishNameDec differ only in the middle segment, the first having EnglishMiddle, the second having SpanishMiddle. Accordingly, as in Figure 7.10(b), we can move the specialization down from Full Name to Middle. This increases its specificity while providing the same basis for pruning.

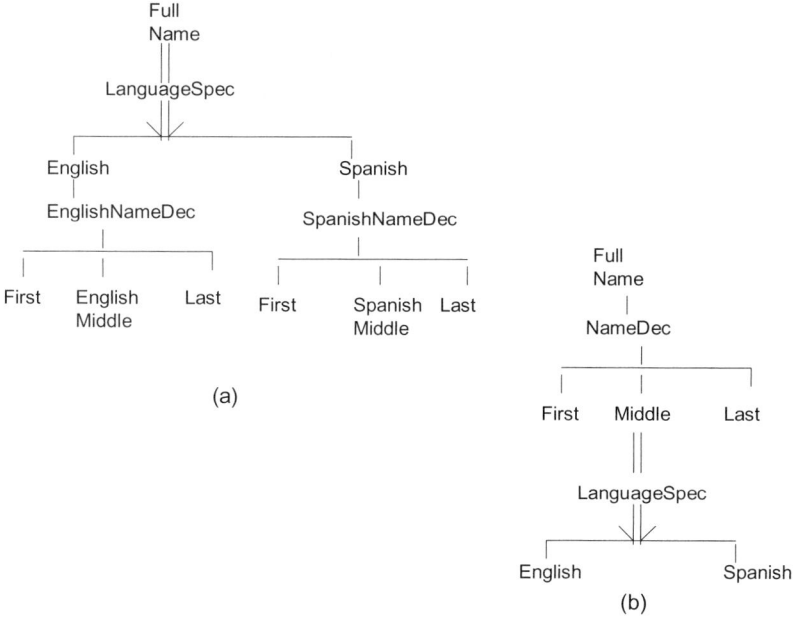

Figure 7.10

Lowering Specializations to the Lowest Possible Level

Summary

Restructurings are changes to the structure of an SES that do not change the family of the pruned entity structure it generates. They are useful in harmonization and in optimization. Methodologically, they provide the modeler with alternative patterns to develop SESs. This kind of domain knowledge is characteristic of advanced modelers.

The restructurings that we discussed included:

- expanding multiaspects via their entity specializations,
- replacing a variable by a specialization whose entities represent values in the range set,
- replacing a variable by a specialization whose entities have variables representing subintervals of the range set,
- replacing a group of variables by an aspect whose entities correspond to the variables,

- replacing specializations by their product specialization, and
- making a specialization more specific and placing it lower down in the structure.

8

Pruned Entity Structures and XML Schema Instances

So far we have introduced the System Entity Structure (SES) as the basis for our simulation-based ontology engineering methodology. Examining Figure 6.1 we can see that two levels are evident there, ontology and implementation. As depicted in Figure 8.1, at the ontology level, the modeler creates an SES to satisfy a pragmatic frame of interest in a given application domain. The SES is expressed as an XML schema/document definition (XSD or DTD) at the implementation level (Chapter 6). In this chapter, we introduce the family of pruned entity structures (PES) of an SES that represents a logically possible set of world states. Each such state is like a snapshot of the world taken from the point of view of a particular pragmatic frame. In this approach, a family of XML document instances of a schema is the concrete encoding of the family of pruned structures. An XML instance encodes data that directly represent a *static* world state. Later we will extend this concept to the *dynamic* case (Chapter 11).

Figure 8.2 recasts the information exchange framework (Chapter 1) in the context of the SES ontology framework. As

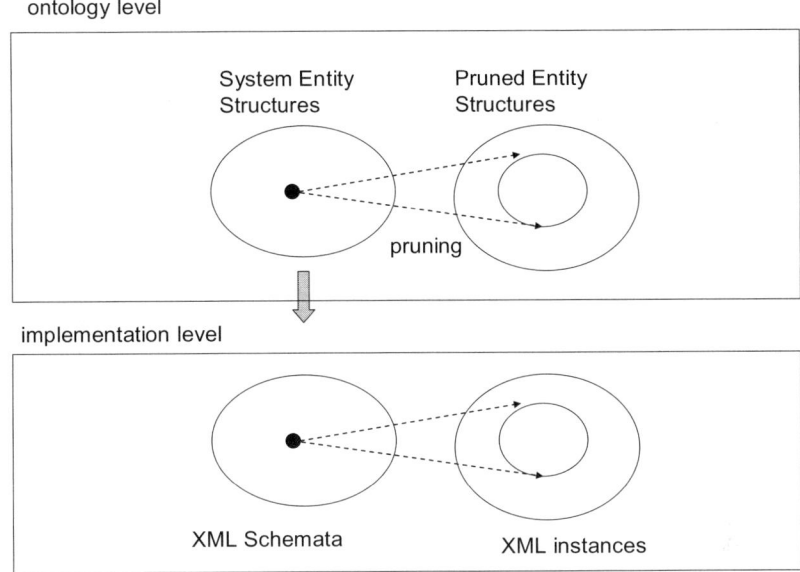

Figure 8.1

Relating SES and XML Through Ontology and Implementation Levels

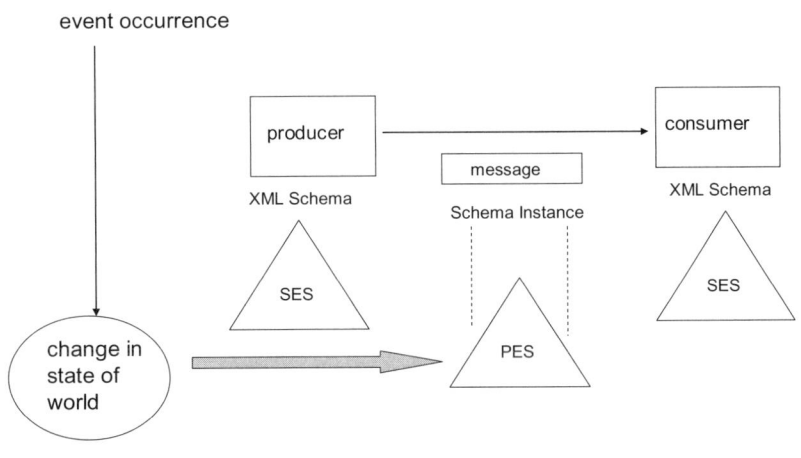

Figure 8.2

Restating the Information Exchange Framework in the SES Context

just discussed, the SES is implemented as an XML schema that is the basis for document instances. The producer sends these instances in messages such as those transported by SOAP in a web-service environment (Ref. 1). The consumer receives and interprets these messages using the same schema in which they were sent. The following discussion elucidates the SES pruning process, its products, and its implementation in XML.

Pruned Entity Structures

Recall that a simplified SES is specified by a structure:

$$SES = <$$
$$Entities,$$
$$Aspects,$$
$$Specializations,$$
$$rootEntity$$
$$entityHasAspect,$$
$$entityHasSpecialization$$
$$aspectHasEntity$$
$$specializationHasEntity$$
$$>$$

$entityHasAspect \subseteq Entities \times Aspects$
$entityHasSpecialization \subseteq Entities \times Specializations$
$aspectHasEntity \subseteq Aspects \times Entities$
$specializationHasEntity \subseteq Specializations \times Entities$

We also required that such a structure is well-formed so that there are no free-standing or empty aspects or specializations.

The process of pruning the SES is that of reducing the choices represented by the *entityHasAspect* and *specializationHasEntity* relations. Note that there are no choices to be made in the *aspectHasEntity* relation, since each aspect represents a decomposition in which all the components must appear. Ultimately, in a completely pruned structure there are no choices left. In other words, in such a structure, every entity (that had a choice) now has exactly one aspect and every specialization has exactly one entity.

Thus in a completely pruned entity structure (PES) we have that *entityHasAspect* and *specializationHasEntity* relations are respectively equal to the functions:

entityHasAspectFn: *Entities* → *Aspects*
specializationHasEntityFn: *Specializations* → *Entities*

Recall that a functional form of a relation associates at most one range element with a domain element. For example,

entityHasAspect = {(e,a) | *entityHasAspectFn*(e) = a}

We note that *entityHasAspectFn* is a partial function since it does not mention the entities that had no aspects in the original SES. On the other hand, *specializationHasAspectFn* is a total function since we required that all specializations be accounted for in the original *specializationHasEntity* relation.

In the case where there are multiple occurrences of items, we have to use the unique identifiers discussed in Chapter 5 to allow different selections for each occurrence.

Let's look at some examples:

Example 1

The SES for Figure 8.3 is described by:

SesForCover = <
Entities = {book,red,blue,paper,cardboard}
Aspects = {}
Specializations = {colorSpec,materialSpec}
rootEntity = cover
entityHasAspect = {}
entityHasSpecialization = {(cover,colorSpec),(cover,materialSpec)}
aspectHasEntity = {}
specializationHasEntity = {(colorSpec,red),(colorSpec,blue),
  (materialSpec,paper),(materialSpec,cardboard)}
>

There are four complete prunings of which two are shown in Figure 8.3(b) and (c). They are given by the following functions, respectively:

specializationHasEntity = {(colorSpec,red),
  (materialSpec,cardboard)}

and

specializationHasEntity = {(colorSpec,blue),
  (materialSpec,paper)}

Chapter 8 Pruned Entity Structures and XML Schema Instances

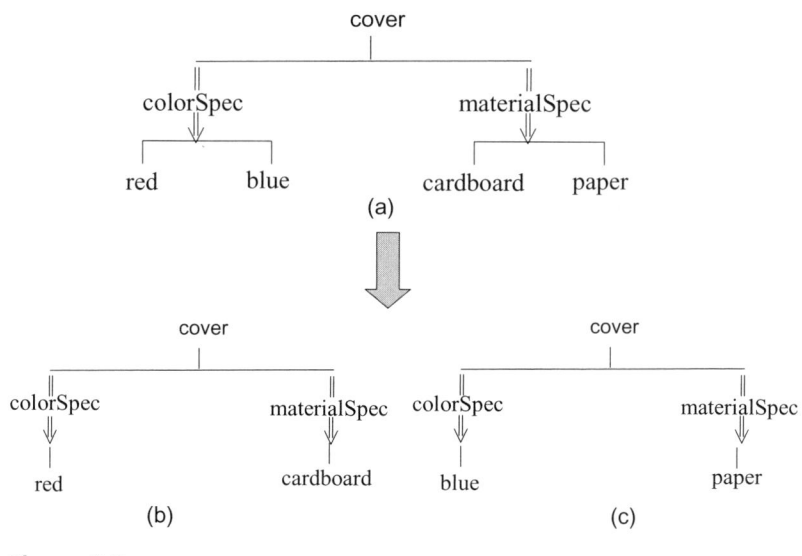

Figure 8.3

SES in (a) is pruned to PESs in (b) and (c)

Example 2

Recall the example in Figure 5.1 in which both *frontCover* and *backCover* have the same specialization, *colorSpec*. Replicated in part in Figure 8.4, we have to use the unique identifiers for the occurrences: *frontCover.colorSpec* and *backCover.colorSpec*.

By replacing *colorSpec* with each of its versions, the relation in the SES takes the form:

specializationHasEntity = {(*frontCover.colorSpec,red*),
 (*frontCover.colorSpec,blue*) (*backCover.colorSpec,red*),
 (*backCover.colorSpec,blue*)}

Now there are four completely pruned SESs. The two PESs shown in Figure 8.4(b) and (c) are represented, respectively, by:

specializationHasEntity = {(*frontCover.colorSpec,red*),
 (*backCover.colorSpec,blue*)}

and

Pruned Entity Structures

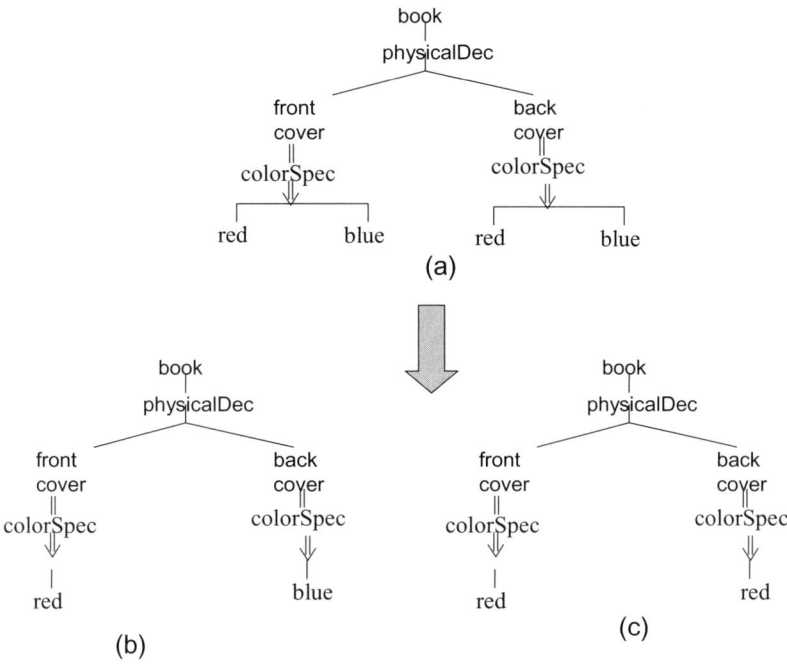

Figure 8.4

SES in (a) Is Pruned to PESs in (b) and (c)

specializationHasEntity = {(*frontCover.colorSpec,red*),
  (*backCover.colorSpec,red*)}

A special case of the general concept is where the uniformity of an item carries over from an SES to a PES. This is equivalent to declaring that the item is to be treated as *context free*; in other words, all occurrences are treated the same. Applied to *colorSpec* in Figure 8.4, this means that the item selected under *frontCover* must always agree with that under *backCover*. Thus, prunings are always of the form:

specializationHasEntity = {(*frontCover.colorSpec,x*),
  (*backCover.colorSpec,x*)}

Example 3

Figure 8.5(a) shows an SES for a book that provides two aspects, *contentDec* and *physicalDec*, corresponding to decompositions from the perspectives of content and physical constitution, respectively.

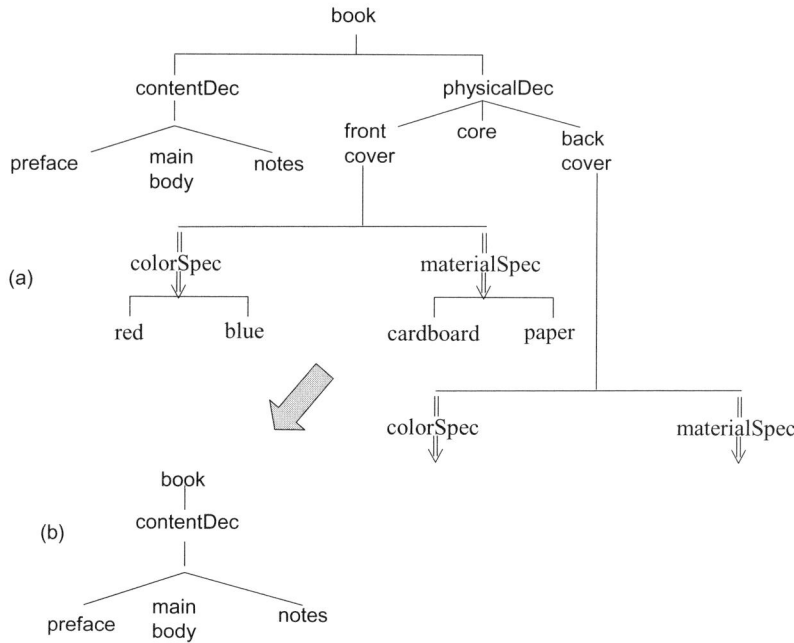

Figure 8.5.

SES with Multiple Aspects under Book in (a) Is Pruned to PESs in (b)

The SES is described by:

SesForBook = <
Entities =
{*book,red,blue,cardboard,paper,core,preface,mainBody,notes*}
Aspects = {*physicalDec,contentDec*}
Specializations = {*colorSpec,materialSpec*}
rootEntity = *book*
entityHasAspect = {(*book,physicalDec*),(*book,contentDec*)}
entityHasSpecialization = {(*frontCover,colorSpec*),
 (*frontCover,materialSpec*),(*backCover,colorSpec*),
 (*backCover,materialSpec*)}
aspectHasEntity = {(*physicalDec,frontCover*),
 (*physicalDec,backCover*),(*physicalDec,core*)(*contentDec,preface*),
 (*contentDec,mainBody*),(*contentDec,notes*)}
specializationHasEntity = {(*colorSpec,red*),(*colorSpec,blue*),
 (*materialSpec,cardboard*),(*materialSpec,paper*)}
>

Prunable Entity Structures

In this case, there are two aspects under book so that pruning is required to reduce the *entityHasAspects* relation to a function. If we select the *contentDec* aspect then we obtain

entityHasAspect = {(*book,contentDec*)}

and the pruned entity structure takes the form of Figure 8.5(b).

Pruning an SES with multiAspects

The process of pruning the SES with multiAspects has two different tendencies:

1. reducing the choices represented by the *entityHasAspect* and *specializationHasEntity* relations, and
2. increasing the basis for choices by expanding a *multiAspect* for pruning.

For example, in the SES for Figure 8.6, *core* is decomposed by a multiAspect into page entities. In pruning, we can decompose the book core into a desired number of copies *page1,...pageN*, all of which have the same substructure as the entity page.

Prunable Entity Structures

To recognize the structures that are somewhere between the original SES and a completely pruned entity structure, we'll define the concept of *prunable* entity structure.

Formally a Prunable Entity Structure (PES) is a structure:

PES = <
originalSES
Entities,
Aspects,
MultiAspects
Specializations
rootEntity
entityHasAspect
entityHasMultiAspect
entityHasSpecialization
aspectHasEntity

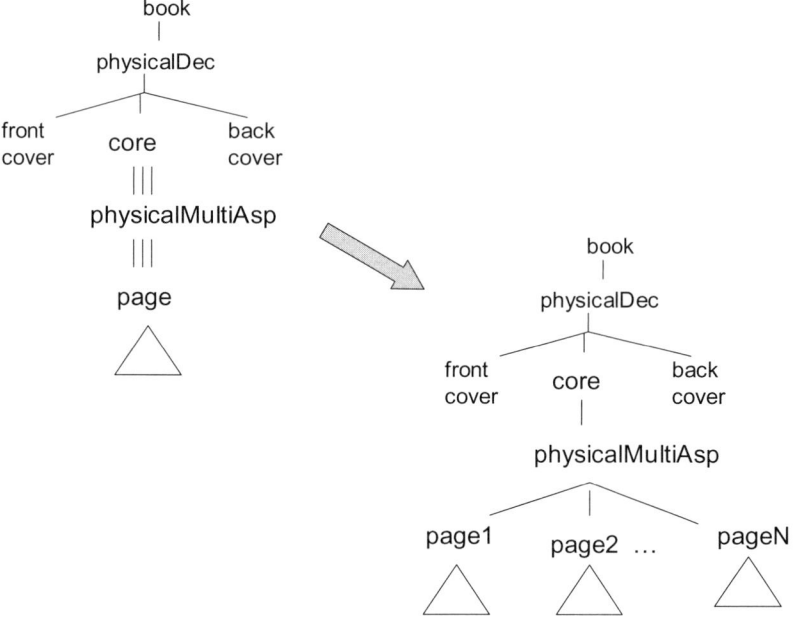

Figure 8.6

Opening Up a multiAspect for Pruning of Copies of Its Entity

multiAspectHasEntity
specializationHasEntity
>

with the same constraints as in the SES except that

$originalSes \subseteq SES^+$
$multiAspectHasEntity \subseteq multiAspects \times PES^+$

where

- SES^+ = set of all system entity structures and
- PES^+ = set of all prunable entity structures

Notice that there is a slot for retaining a link to the originating SES. To create a new PES we copy the elements and relations from the originating SES with the exception that the *multiAspectHasEntity* relation starts out empty. That is, we say

PES is derived from SES if

- $originalSes_{PES} = SES$
- all items and relations of *PES* are copies of corresponding elements in *SES*.
- $multiAspectHasEntity_{PES} = \varphi$

To describe how the *multiAspectHasEntity* relation gets filled in, consider the following:

Let *multiAspect* have a generating entity, e. Then, let

- $SES(e)$ denote the substructure of the *SES* under the entity, e
- $PES(e)$ is derived from $SES(e)$

We add $PES(e)$ under the multiAspect so that now we have:

$$(multiAspect, PES(e)) \in multiAspectHasEntity$$

Example 4

Corresponding to Figure 8.6, we have

SesForCoreBook = <
Entities = {*book,frontCover,backCover,core,page*}
Aspects = {*physicalDec,contentDec*}
Specializations = {*pageSpec*}
rootEntity = *book*
entityHasAspect = {(*book,physicalDec*)}
entityHasMultiAspect = {(*core,coreMultiAsp*)}
entityHasSpecialization = {(*page,pageSpec*)}
aspectHasEntity = {(*physicalDec,frontCover*),
  (*physicalDec,backCover*),(*physicalDec,core*)}
multiAspectHasEntity = {(*coreMultiAsp,page*)}
specializationHasEntity = {(*pageSpec,thick*),(*pageSpec,thin*)}
>

The substructure below entity, *page* is:

SesForCoreBook(page) = <
Entities = {*page,thick,thin*}
Aspects = {}
Specializations = {*pageSpec*}
rootEntity = *page*
entityHasAspect = {}
entityHasSpecialization = {(*page,pageSpec*)}

$aspectHasEntity = \{\}$
$multiAspectHasEntity = \{(coreMultiAsp,page)\}$
$specializationHasEntity = \{(pageSpec,thick),(pageSpec,thin)\}$
>

Creating a PES by deriving it from the SES, we have

PesForPage = <
originalSes = *SesFoCoreBook*(*page*)
Entities = {*page,thick,thin*}
Aspects = { }
Specializations = {*pageSpec*}
rootEntity = *page*
$entityHasAspect = \{\}$
$entityHasMultiAspect = \{\}$
$entityHasSpecialization = \{(page,pageSpec)\}$
$aspectHasEntity = \{\}$
$multiAspectHasEntity = \{\}$
$specializationHasEntity = \{(pageSpec,thick),(pageSpec,thin)\}$
>

This PES can now be pruned by selecting either of the thick or thin alternatives under the page specialization. We can continue to add new pages to the book (i.e., new prunable entity structures for page) and then prune each as desired.

The conformsTo Relation

A PES is a dynamic structure in the sense that we increase its pruning capacity whenever we expand one of its multiAspects. It is useful to have a definition against which expansions can be tested to see if the overall relationship of the PES to its originating SES is preserved. In fact, we can ask for conditions under which a PES conforms to an arbitrary SES (which should be satisfied by the originating SES as well).

As illustrated in Figure 8.7, the definition of *conformsTo* is recursive. To start, we have to assure that the top level elements of the PES are all consistent with those of the SES. If so, we then proceed to the included PESs for the entities hanging under the multiAspects. These in turn may have further multiAspects and prunable entity substructures, and so on. This process cannot go on forever, however, since we require the strict hierarchy axiom to be true of the top level SES.

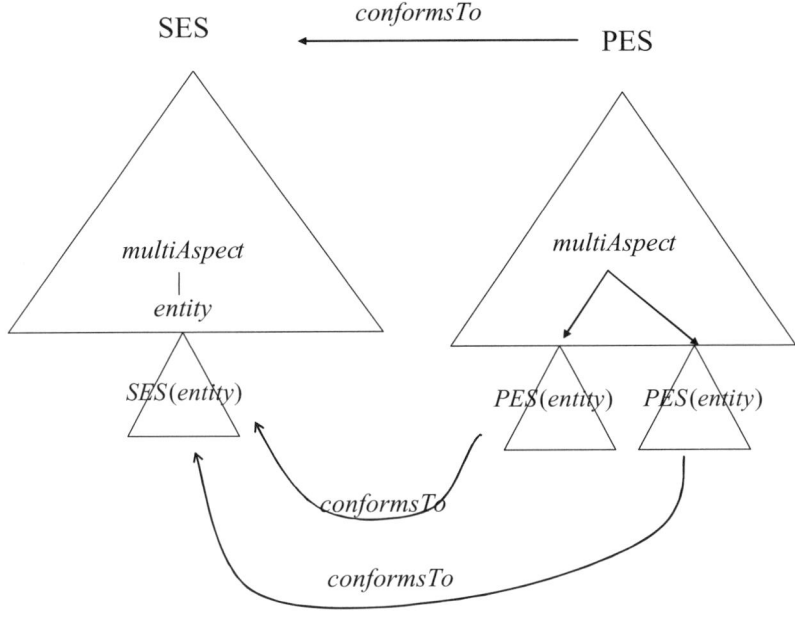

Figure 8.7

The Relation "conformsTo" between a PES and an SES

Formally, the definition of "conformsTo" is given by:

PES conformsTo SES if:

The elements of PES are included in the corresponding elements of SES, e.g.,

$$specializationHasEntity_{PES} \subseteq specializationHasEntity_{SES}$$

and

$$\forall (multiAspect, PES(e)) \in multiAspectHasEntity_{PES}$$

we have

$$(multiAspect, e) \in multiAspectHasEntity_{SES}$$

and

PES(entity) conformsTo SES(entity)

When Is Pruning Complete?

If the SES contains no multiAspects, pruning is complete when all selections have been made, i.e., when the *entityHasAspect* and *specializationHasEntity* relations have been reduced to functions. The same will be true if we limit pruning of any PES to selection of aspects from entities and entities from specializations. We can take various approaches to bring a PES into a state where no further multiAspects can be expanded. For example, we can limit the total number of entities, the total number of items, or other measures of the size of a PES. An approach that corresponds to that employed in XML is to place individual limits on the number of copies that any individual multiAspect can allow. To do this we associate a variable with a multiAspect that represents the maximum number of components (or copies of its entity) that are allowed. Actually, we can also place a lower bound on this number as well. For example, in the SES XML representation,

```
<multiAspect name="coreMultiAsp">
<numberComponentsVar name="numPages" min="100"
  max="200"/>
<entity name="page">
```

asserts that the name of the variable to count the number of components for *coreMultiAsp* is *numPages* and its lower and upper bounds are 100 and 200, respectively.

Exercise

Extend the definition of *conformsTo* to require that the upper and lower bounds on each multiAspect are respected.

Exercise

Provide a definition of "pruned entity structure" which asserts that a prunable entity structure has no additional selections or expansions to be made.

The *state of completion* of a PES can be defined as follows:

Let N_{Spec} be the number of specializations in the PES and all its added PESs

Let $N_{Spec,Pruned}$ be the number of specializations in the PES and all its added PESs which are functional (i.e., pruned)

Then the *specialization completion ratio* is $\dfrac{N_{Spec,Pruned}}{N_{Spec}}$

Similarly the *aspect completion ratio* can be defined.

Exercise

Define the aspect completion ratio to reflect the number of entities that have a single aspect.

The *multiAspect completion ratio* can be defined as follows:

Let $N_{multiAspect}$ be the sum of the multiAspect upper bounds over all the multiAspects in the *SES*.

Let $N_{multiAspect,copies}$ be the sum of the actual number of copies of the entity under each multiAspect in the *PES*.

Then the *multiaspect completion ratio* is $\dfrac{N_{multiAspect,copies}}{N_{multiAspect}}$.

Pruning Using SES/JAVA Tools

Corresponding to the various ways of creating System Entity Structures, there are several ways to construct Prunable Entity Structures. Figure 8.8 depicts a process that starts with an XML specification of an SES and finishes with a PES that conforms to it.

A pruning that starts from an XML specification of an SES and produces a PES is represented as an XML instance of the SES.

To start we have the SES for *CoreBook* given in XML as follows:

```
<?xml version='1.0' encoding='UTF-8'?>
<!DOCTYPE entity SYSTEM "ses.dtd" []>
<entity   name = "CoreBook">
<aspect   name = "contentDec">
      <entity   name = "backCover"/>
      <entity   name = "frontCover"/>
      <entity   name = "core">
      <multiAspect name="coreMultiAsp">
      <numberComponentsVar name="numPages"
        min="0" max="200"/>
            <entity name="page" >
                <specialization name="pageSpec">
                <entity name="thick">
                </entity>
                <entity name="thin">
```

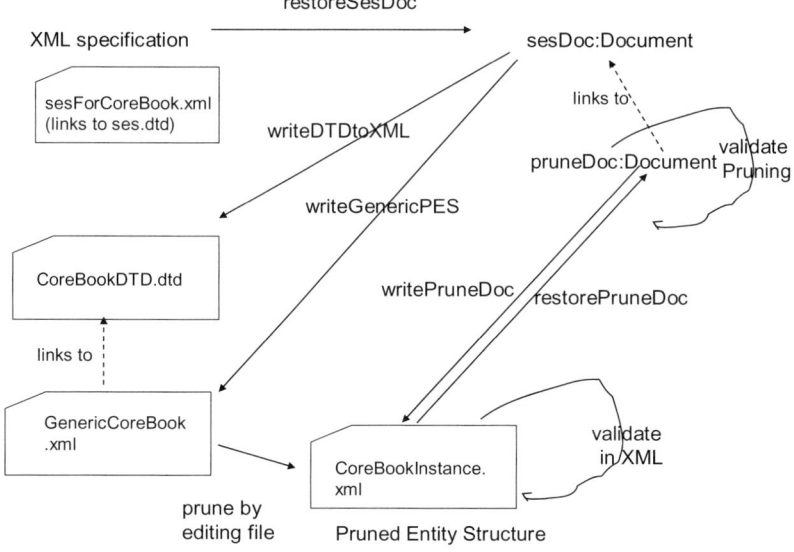

Figure 8.8

Operations Relating to Prunable Entity Structure

```
                </entity>
              </specialization>
            </entity>
        </multiAspect>
      </entity>
  </aspect>
</entity>
```

We restore this file to memory using:

restoreSesDoc(path + "sesForCoreBook.xml")

as a result of which it becomes a Document, sesDoc. As before we can write the file containing the corresponding DTD, using:

writeDTDToXML(folder + "CoreBookDTD")

which gives us:

```
<?xml version='1.0' encoding='us-ascii'?>
<!--   DTD for a CoreBook   -->
```

```
<!ELEMENT CoreBook (aspectsOfCoreBook)>
<!ELEMENT aspectsOfCoreBook (contentDec )>
<!ELEMENT contentDec ( frontCover , backCover ,
 core )>
<!ELEMENT frontCover ( #PCDATA)>
<!ELEMENT backCover ( #PCDATA)>
<!ELEMENT core (aspectsOfcore)>
<!ELEMENT aspectsOfcore (coreMultiAsp )>
<!ELEMENT coreMultiAsp (page*)>
<!ELEMENT page (pageSpec)>
<!ELEMENT pageSpec (thick | thin )>
<!ELEMENT thick ( #PCDATA)>
<!ELEMENT thin ( #PCDATA)>
<!ATTLIST coreMultiAsp
numPages CDATA #IMPLIED
>
```

We can now write an XML file for an instance of *CoreBook* by including a link to the file "*CoreBookDTD.dtd*" and proceed to make selections and expansions as dictated by the SES.

```
<?xml version='1.0' encoding='UTF-8'?>
<!DOCTYPE CoreBook SYSTEM "CoreBookDTD.dtd" []>
<CoreBook>
    <aspectsOfCoreBook>
        <contentDec>
            <frontCover/>
            <backCover/>
            <core>
                <aspectsOfcore>
                <coreMultiAsp  numPages = "1"/>
                    <page>
                        pageSpec>
                            <thick/>
                        </pageSpec>
                    </page>
                </aspectsOfcore>
            </core>
        </contentDec>
    </aspectsOfCoreBook>
</CoreBook>
```

Submitting this file to an XML validation tool will test its conformance to the DTD. We will discuss the relation of this test to conformance to the SES soon.

To write a file that can be edited to create a valid instance, we can use, for example:

writeGenericPES(path+ "PrunedCoreBook.xml", "CoreBookDTD.dtd")

which will write out a file PrunedCoreBook.xml with the link to CoreBookDTD.dtd and a *generic* structure in which all specializations and aspects are listed without any choices made. This lets the user prune by deleting entities from specializations and aspects from entities. Also each multiAspect is listed with a single copy of its substructure so that the user can copy and delete as desired.

Validation and Completion State of PES

Figure 8.9 shows the layers of verification and validation applicable to prunable entity structures. These layers contain both standard XML checks as well as checks of the conformsTo relation and computation of the completeness state of pruning.

Methods in the class, *validatePruning* carry out these tests. We note that the generic pruned entity structure will typically not pass the XML validation test for instances since choices will not necessarily have a single element within their scope. On the other hand, such structures pass the *conformsTo* test as discussed above.

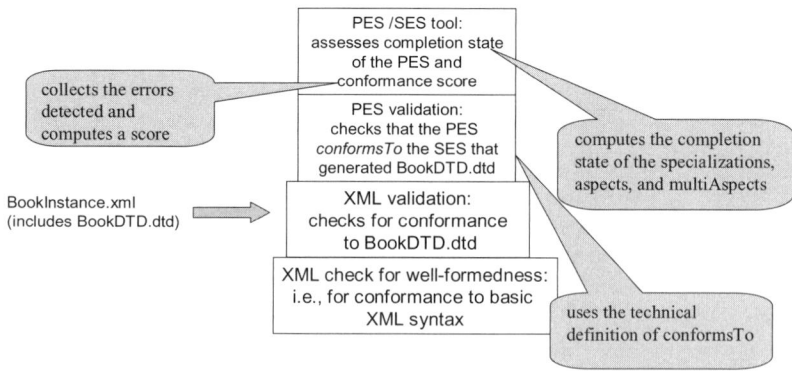

Figure 8.9

Layers of Verification and Validation

Summary

At the ontology level, the family of pruned entity structures (PES) of an SES represents the set of world state descriptions of the ontology. At the implementation level, the SES is represented as an XML schema and the set of all XML document instances of a schema is the concrete encoding of the family of PESs. An XML instance encodes data that directly represent a static world state. In Chapter 11, we will extend this concept to the dynamic case in which simulation models are the elements described by the PESs. We showed how the SES, as an axiomatically structured object, engenders its own concepts of validation distinct from those of the underlying XML implementation. In this vein, we developed the conformance relation between an SES and any of its partially, or fully, pruned entity structures, as well as associated metrics for the completion state of pruning.

Reference

1. SOAP, A Protocol for Accessing a Web Service, http://www.w3schools.com/soap/soap_intro.asp (accessed Nov. 2006)

9
Constrained Pruning

We have seen that pruning of an SES is the concept at the ontology level that corresponds to filling in the slots of an instance document at the implementation level (Chapter 8). Also, according to Chapter 2, ontology frameworks have varying capabilities to constrain the simultaneous values assigned to their slots. XML is strong in enabling the modeler to constrain the values that are acceptable to simple data types. However, it has no support for specifying multiple slot constraints. Such constraints are possible to state and validate in the SES framework. Constraint-based pruning of SESs originated in applications to system design and configurations where synthesis and composition rules were applied to constrain pruning to plausible alternatives.[1-4] In this chapter, we discuss how constraints can be imposed via rules and relations, as well as where constrained pruning is useful in the context of the information exchange framework.

Constraints on Specialization Selections

Often entities cannot be independently selected from a specialization. Where this is the case, it is because there is a constraint

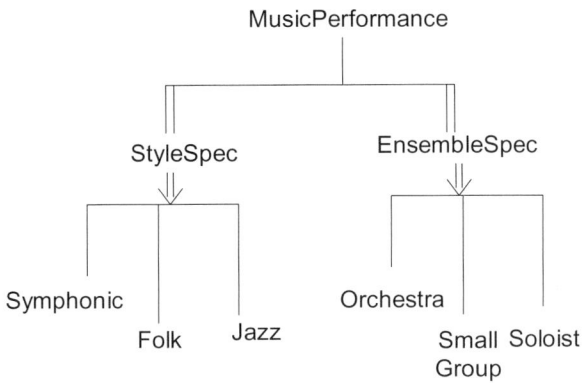

Figure 9.1

SES for Musical Performance

on the set of all simultaneous selections from an interacting collection of specializations. For example, consider the specification:

A musicPerformance can be Symphonic, Folk, or Jazz in Style
A musicPerformance can be Orchestra, SmallGroup, or Soloist in Ensemble

This specifies an SES for musical performance similar to that in Figure 9.1.

Not all combinations are likely. Indeed, some might be so unlikely as to be considered impossible. For example, Symphonic music can only be played by an Orchestra; Folk music is usually played by a small group and sometimes by a soloist; and Jazz is usually played by a Small Group and sometimes by an Orchestra.

Rule-based Approach

Let's look at how Table 9.1 might be enforced by rules of the form:

If select a from A then must select b1,b2, ... , b3 from B

Table 9.1. Possible Music Performances

StyleSpec/ EnsembleSpec	Orchestra	Small group	Soloist
Symphonic	x		
Folk		x	x
Jazz	x	x	

As an example, having selected Symphonic for style constrains the choice of ensemble:

Symphonic selection rule: If select Symphonic from StyleSpec then must select Orchestra from EnsembleSpec

Here

a = Symphonic and A = StyleSpec = {Symphonic, Folk, Jazz}

while

b1 = Orchestra and B = EnsembleSpec = {Orchestra, Small Group, Soloist}.

Similarly:

Folk selection rule: If select Folk from StyleSpec then must select SmallGroup or Soloist from EnsembleSpec

Exercise

Write the selection rule for Jazz to implement Table 9.1.

EXAMPLE: *XML Treatment of Use, Fixed and Default Attribute Specifications*

In XML the use of an attribute can be specified as prohibited, required, or optional. Also, Fixed and Default slots can be given any string values.

For example, in a Schema, we can have:

```
<xs:attribute name="example" type="xs:int"
  use="required" fixed="2" />
```

describes a situation where the example attribute is an integer type whose use is required and whose value is always "2."

Rule-based Approach

```
<xs:attribute name="example" type="xs:int"
  use="prohibited" fixed="2" />
```

describes a situation where the example attribute is an integer type whose use is prohibited but whose value would be "2" if its use were to be allowed.

Now, note that the *use* slot does not have to be present. For example,

```
<xs:attribute name="example" type="xs:int"
  default="2" />
```

Specifies that the default value is "2" for the example attribute but does not constrain the use.

By examining all combinations, one can conclude that XML places two constraint rules on the various combinations:

1. if default is present then use, if present, must be optional

So, for example, the following is not valid:

```
<xs:attribute name="example" type="xs:int"
use="required" default="2" />
```

2. fixed and default can't both be present

So, for example, the following is not valid:

```
<xs:attribute name="example" type="xs:int"
  default="2" fixed="3"/>
```

The apparent reason for this constraint is that a contradiction is possible if both the default and fixed values are specified and are different.

In the following discussion, we will use the SES and pruning constraints to see how we might specify various kinds of restrictions other than those chosen by the developers of XML.

We begin by representing the attribute structure by the SES in Figure 9.2. Note that each possibility for choice — use, fixed, and default — is represented as a component with an associated specialization.

Notes:

- Each such specialization has an alternative labeled NONE which, when selected, acts to remove the component from the

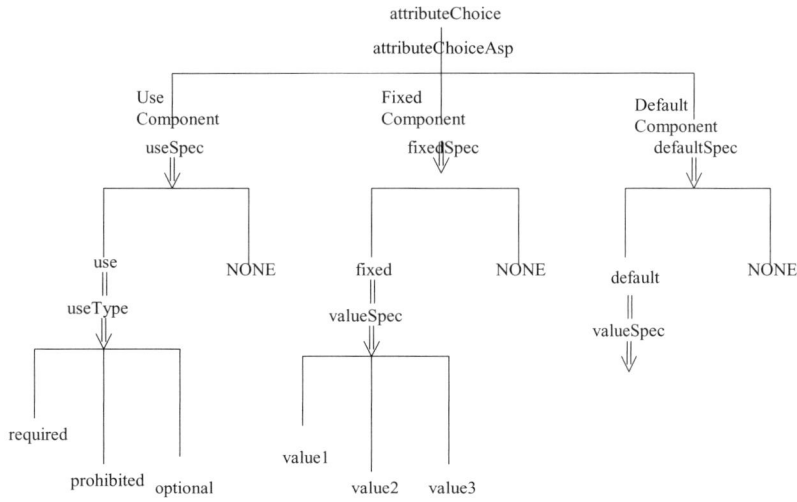

Figure 9.2

SES for Attribute Choice

structure. This models the fact that in XML, one need not state the *use* attribute at all, and similarly with the other two.
- The possible values for fixed and default are provided by the same specialization valueSpec. Using the uniformity axiom, this models the requirement to have a fixed or a default value coming from the same range set. For example, although this is not considered invalid by XML, we would not want to have the choice as an integer for one and string for the other.

Figure 9.3 shows a PES that is considered valid in XML. It corresponds to the XML specification

```
<xs:attribute name="example" use="optional"
  default="value2" />
```

Figure 9.4 shows two cases that XML considers invalid — case (a) because, with a default specified, the use is specified as "required" (it needs to be "optional"); case (b) because both fixed and default values are present.

To replicate the XML constraints on combinations that include a default specification, we can introduce a pruning rule of the form:

If select default from defaultSpec then must select use from useSpec and optional from useType

Rule-based Approach

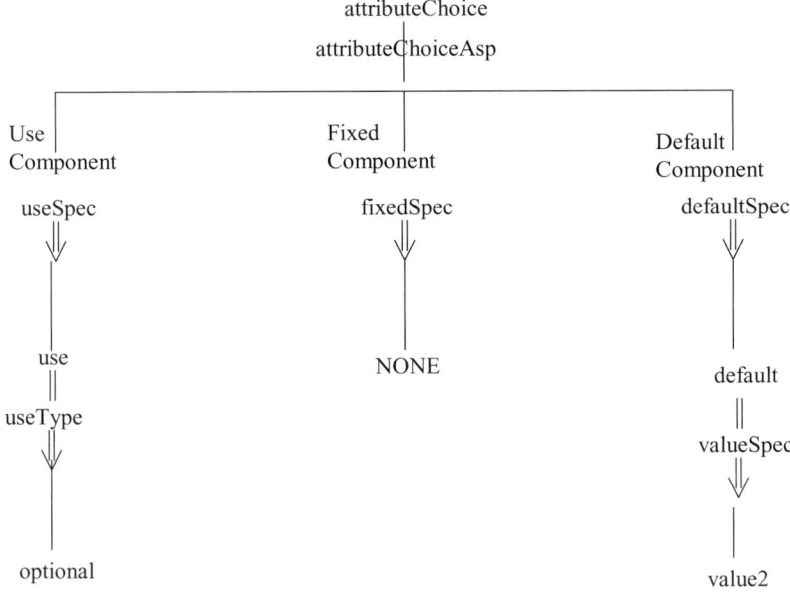

Figure 9.3

PES That Represents a Valid XML Attribute Choice

This rule will clearly distinguish Figure 9.2 from Figure 9.4(a).

On the other hand, the mutual exclusion constraint violated in Figure 9.4(b) is not as easy to enforce. It states that fixed and default specifications must not both be present in the same structure. Appendix A uses propositional logic to help you understand how to write rules with some assurance. In the current application, the mutual exclusion constraint can be stated as

$$\neg(p \wedge q)$$

which reads "not both p and q"

where:

- p = select fixed from fixedSpec
- q = select default from defaultSpec

An equivalent form is

$$(p \Rightarrow \neg q) \wedge (q \Rightarrow \neg p)$$

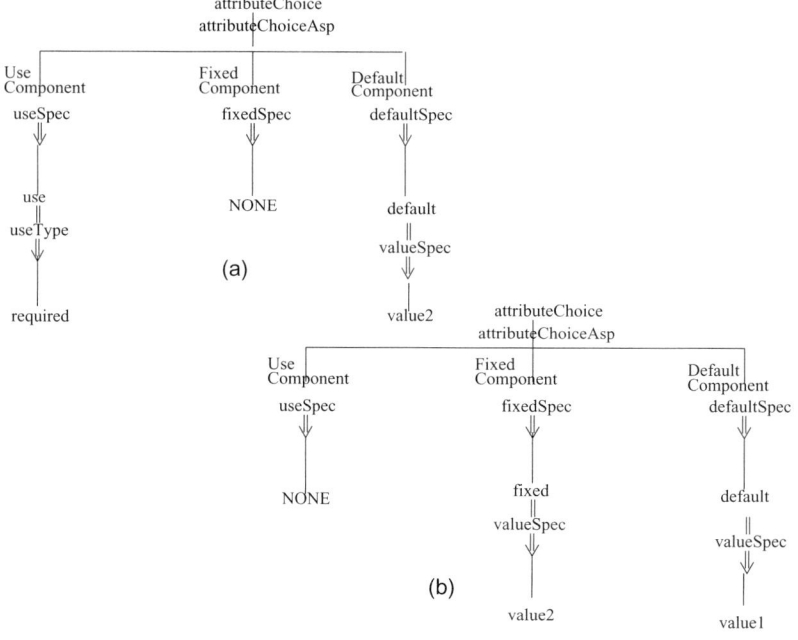

Figure 9.4

Two PESs That Represent Invalid Attributed Choices in XML

which leads to two selection rules:

If select default from defaultSpec then must **not** select fixed from fixedSpec

If select fixed from fixedSpec then must **not** select default from defaultSpec

We can convert the negation into a selection from the remaining choices, and write these rules in the form:

If select default from defaultSpec then must select NONE from fixedSpec

If select fixed from fixedSpec then must select NONE from defaultSpec

These rules will together distinguish the valid combination in Figure 9.2 from the invalid selection in Figure 9.4(b).

We conclude for the SES of Figure 9.2 that three rules are needed to enforce the constraints imposed by XML.

Rule-based Approach

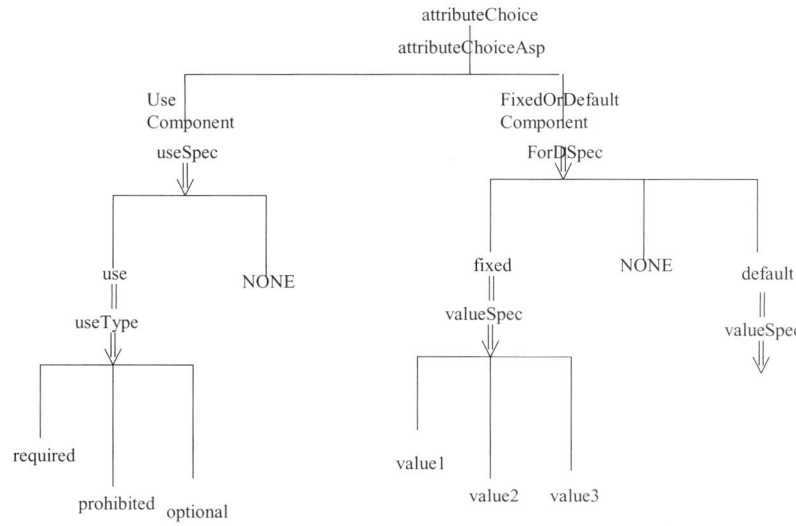

Figure 9.5

Another SES for Attribute Choice

An Alternative SES

Often there are several ways to represent the same behavior with different structures — with the implication that some may be better than others for particular criteria. In the case of attribute choice, Figure 9.5 presents an SES that requires *only one* selection rule. This is the first one discussed that restricts use to optional when default is present. The mutual exclusion constraint between fixed and default specifications is "built in" by having them both be under the same specialization. Indeed, any specialization requires that exactly one of its entities be selected. Note that in order to allow for the fact that *neither* default nor fixed specifications need to be present, we must add the entity NONE to the specialization. Such constraints implicit in the pruning of the SES should be kept in mind when constructing system entity structures, if the goal is to require as few selection rules as possible.

Exercises

1. For the SES of Figure 9.5,
    a. write the rule for "default" selection to enforce the requirement that if default is present then use must be optional.

b. write the negated equivalent form as a rule for "optional" selection and interpret the result.

c. write one or more rules that enforce the choice of fixed and default values to be identical if both specifications are present. *Hint:* assume that only three value entities are in the valueSpec (as shown), write the logical expression that describes the constraint, and then the selection rules that enforce this logical expression. How many rules are required?

2. Construct an SES and write pruning rules that implement the following informally stated rules for a client-server pair:

Rule 1:
If the client does include a requestID in its request, the server must use it in its subsequent responses.

Rule 2:
If the client does not include a requestID in its request and does include a ResponseHandler in its request (i.e., enables asynchronous mode), the server must generate a requestID and put it in the acknowledgment message (sent to the ResponseHandler).

Rule 3:
If the client does not include a requestID in its request and does not include a ResponseHandler in its request (i.e., enables synchronous mode) the server must not put a requestID in its reponses.

Relation-based Approach

Rather than writing a set of rules, we can implement constrained pruning by enabling the user to specify constraints in tabular or relational form and providing methods to enforce these constraints. This approach has the advantage that it provides a more compact form for constraint specification (tables vs. rules) and allows more of a global view of how multiple relations interact. However, on the negative side, the relations must be provided explicitly as pairs in a relation. Consequently, the relational approach cannot accept rules that are stated for infinite sets, as is possible in predicate logic.

Figure 9.6
SES for Ball

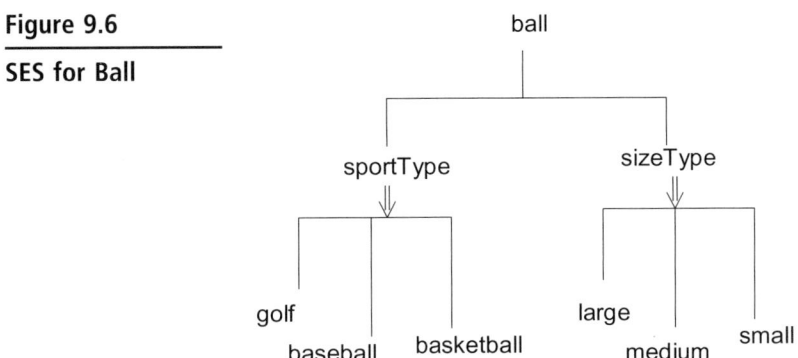

Table 9.2. Ball Size as a Function of Sport Type

sportType/sizeType	golf	baseball	basketball
small	x		
medium		x	
large			x

We start with an example:

Ball can be golf, baseball, or basketball in sportType
Ball can be large, medium, or small in sizeType

specifies the SES shown in Figure 9.6.

In this example the size of a ball is directly related to the sport in which it is used (Table 9.2).

The overall process of relation-based pruning is shown in Figure 9.7. We start with an SES expressed in XML, such as that for ball. The next step employs the generic pruning method (Chapter 8) to generate the Generic Pruned Entity Structure (GPES) for this SES. Finally, relation-based pruning is applied to the GPES to produce a completely Pruned Entity Structure (PES).

A table such as Table 9.2 is represented by a relation, such as:

$R_{sportType,sizeType} \subseteq entities_{sportType} \times entities_{sizeType} = \{(\text{golf},\text{small}),$
$(\text{baseball},\text{medium}), (\text{basketball},\text{large})\}$

Relation-based pruning has to enforce the restriction on pruning imposed by this relation.

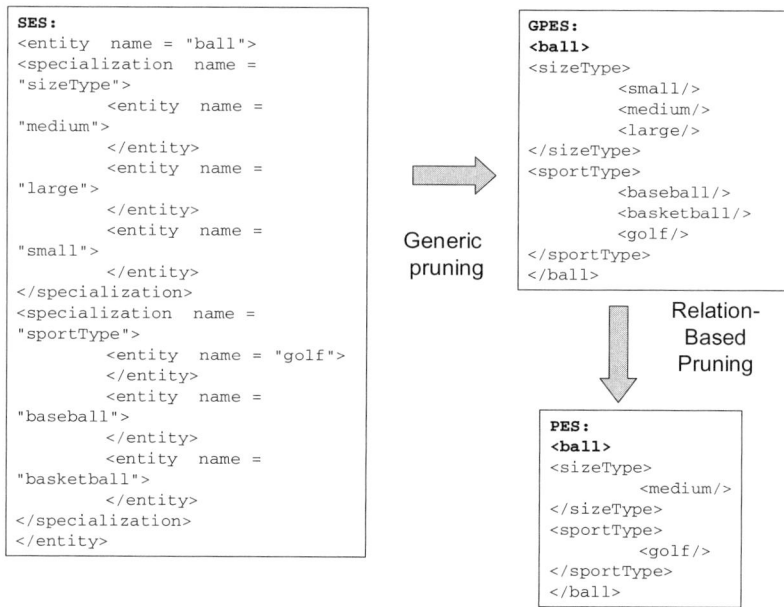

Figure 9.7

Overall Process of Relation-based Pruning

The restriction can be stated as:

$$(x,y) \in entities_{sportType} \times entities_{sizeType} \Rightarrow (x,y) \in R_{sportType,sizeType}$$

i.e., (x,y) is one of the pairs: {(golf,small), (baseball,medium), (basketball,large)}

As we have seen, one way to impose this restriction is through rules, stating, for example, that if golf is selected from sportType then small must be selected from sizeType.

To specify such relations we will attach a function to an SES that tells whether on not an ordered pair of specializations imposes a restrictive relation on pruning. Thus, for a given SES, we define:

restrictRelationFn: specializations × *specializations* → *Relations*

such that

$restrictRelationFn(spec_1, spec_2) = R_{spec_1,spec_2}$

where

$R_{spec_1,spec_2} \subseteq entities_{spec_1} \times entities_{spec_2}$

Implementation

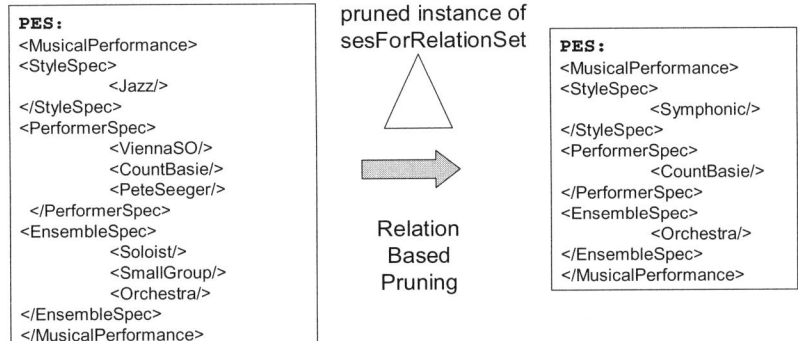

Figure 9.8

Attaching a Set of Restrictive Relations to Guide Pruning

In the ball example there is only one relation, so we have:

$restrictRelationFn(sportType, sizeType) = R_{sportType, sizeType}$

and otherwise

$restrictRelationFn(spec_1, spec_2) = \emptyset$

Implementation

Appendix B develops some theory that supports the implementation of pruning constrained by the restriction relations just discussed. Please refer to this appendix for more detail. We'll employ the musical performance relation in Table 9.1 to discuss the methods provided in SES/JAVA. See Figure 9.8.

First an SES called relationSet allows the user to specify a set of relations as an XML instance document. Then relation-based pruning works from this instance to impose the restrictions it specifies.

Defining Restrictive Relations

An SES for defining restrictive relations is shown in Figure 9.9(a). A relationSet consists of several relations, each of which is a set of pairs, each pair having a key and value. Each relation is identified by its domain and range specializations, domSpec and rangeSpec, which are the names of the corresponding specializations in an SES. An instance of this structure is illustrated in Figure 9.9(b) where the relations based on pairs (StyleSpec,

158 Chapter 9 **Constrained Pruning**

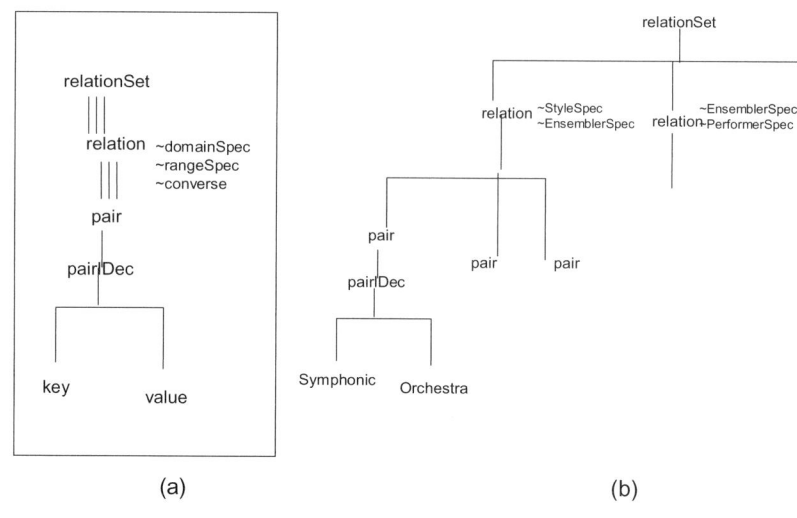

Figure 9.9

SES to Specify a Set of Restrictive Relations

EnsembleSpec) and (EnsembleSpec, PerformerSpec) are sketched for application to pruning the Musical Performance SES.

The converse of a relation can be specified by setting the *converse* slot in a relationSet instance to true. This will switch the domainSpec with the rangeSpec, thus changing the processing order, as well as interpreting the pairs in a converse manner (see Appendix B).

Figure 9.8 illustrates how instances of relationSet are employed to restrict the pruning process. The complete process is implemented in the relation-based pruning operator:

generatePrunings.pruneWRelation(
    folder +" relationSetInstance.xml",
    folder +"sesForMusicalPerformance.xml"
    folder + "MusicalPerformanceInstance.xml")

This method generates a pruning of the PES, "Musical PerformanceInstance.xml," that satisfies the restrictive relations (provided that the original state is compatible with these relations — see Appendix B). The PES should have been previously constructed by using the generic pruning method or in any of the ways discussed in Chapter 8. We note that the resulting PES is left in memory so that it can be written to a different xml file if desired. For example,

generatePrunings.writePruneSchemaDoc(folder+
  "pruned MusicalPerformanceInstance.xml",
  " MusicalPerformance.xsd")

writes the result of pruning to "prunedMusicalPerformance Instance.xml" without overwriting the original PES.

A one argument form of pruneWRelation takes a target SES as its argument:

generatePrunings.pruneWRelation(folder+
  " sesForMusicalPerformance.xml ").

In this case, pruneWRelation will generate a generic pruned entity structure to be pruned further. Since the restrictive relation set is empty by default, this represents ordinary pruning without restrictions. Thus, pruneWRelation provides a convenient way to create pruned entity structures with or without relation-based pruning.

Design of Pruning Relations and Processing Order

Choice of relations (or their converses) and processing order are important factors in determining the outcome of pruning. Experimentation is needed to determine whether the outcome is satisfactory. Figure 9.10 displays a process for performing this experimentation (see Appendix B for background).

Once a sequence of relations is specified, the outcome of pruning is obtained by propagating the restrictions imposed by the relations along this sequence. The propagation begins with a down-selection from the domain specialization at stage 1. At each subsequent stage n, assume that the domain specialization has been down-selected by processing earlier relations. Then the relation at stage n restricts the possible values of its range specialization to those that are consistent with the down-selected domain specialization. This smaller range specialization, in turn, down-selects the domain specializations with same name in subsequent stages, $n + 1, n + 2, \ldots$.

Root-based Pruning

We expand the possible bases for pruning to include awareness of the context provided by the root of the SES. In other words, a relation can include possible roots of SESs in its domain. In this way, when the relation is processed in the context of a particular SES, the root entity will constitute the restricted domain for starting a propagation sequence as just discussed.

160 Chapter 9 **Constrained Pruning**

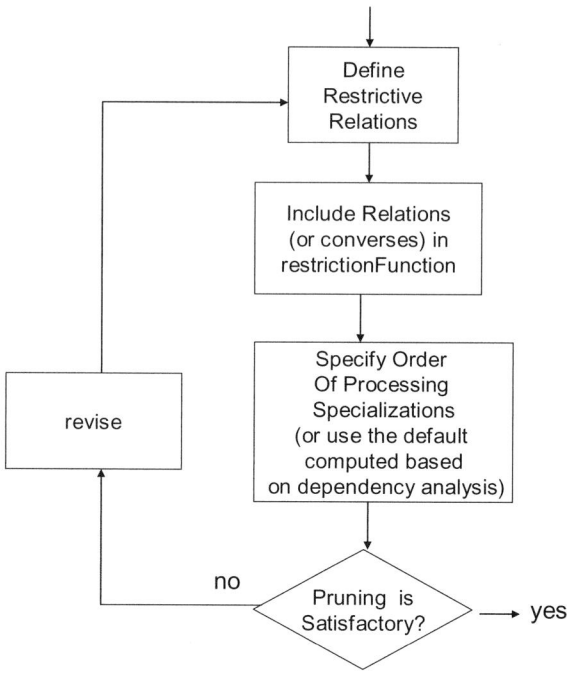

Figure 9.10

Flow Diagram for Developing a Set of Restrictive Relations

Relation-based Pruning for multiAspects

Recall that multiAspects offer the mechanism for expanding the grist for pruning (Chapter 8). They also offer a platform for investigating the possible PESs that result from a given set of restrictive relations. For example, we can enumerate the possible combinations of musical performances explicitly by considering a multiAspect with the generating entity MusicPerformance. In this vein, we can say:

From a multi perspective, musicPerformances are made of more than one musicPerformance

This gives rise to an SES similar to that in Figure 9.11.

Now by pruning the multiAspect, we can generate the different possibilities allowed for musicalPerformance and add them to the multiAspect, as illustrated in Figure 9.12.

Relation-based Pruning for multiAspects   161

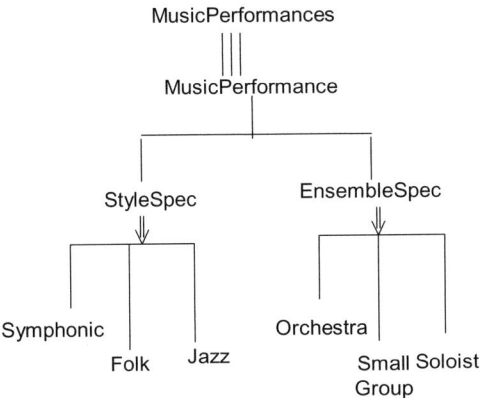

Figure 9.11

An SES with a multiAspect for Musical Performances

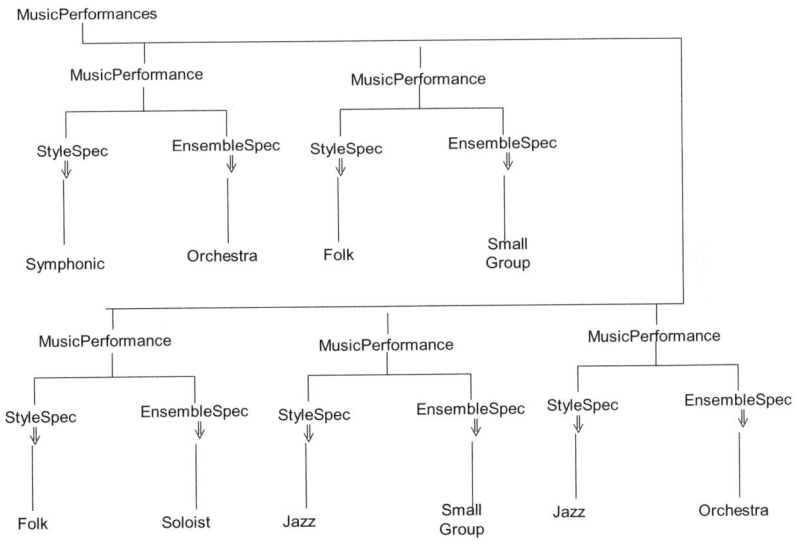

Figure 9.12

Pruned multiAspect for Musical Performances

Interaction among multiAspects

More interesting pruning situations arise where multiple multi-Aspects are present and pruning is constrained by relations among their subspecializations. A common pattern of this kind occurs when a system 1) has a structure in which there are choices for its components, and 2) it comes in particular forms that constrain the possibilities for selections. For example, a broadcast network might be described by the following:

From a structure perspective, a broadcastNetwork is made of channels
A broadcastNetwork can be CBS, NBC, or ABC in company
From a multiperspective, channels is made of more than one channel
A channel can be CH1, CH2, CH3, CH4, or CH5 in name

which specifies an SES as in Figure 9.13. We see here that any particular network is made up of stations and channels from which there are five possibilities to choose. Now, the network companies under consideration are ABC, NBC, and CBS. Since companies cannot broadcast on the same channel, there must be a relation between the companies and the channels.

Such a relation is shown in Table 9.3. There might have been a deliberate allocation of channels or it might have arisen through historical happenstance. In any case, Table 9.3 illustrates a situation in which ABC operates on channels CH1, CH2, and

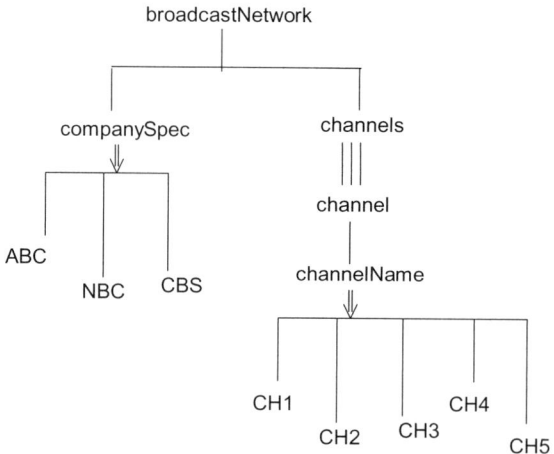

Figure 9.13

Broadcast Network SES

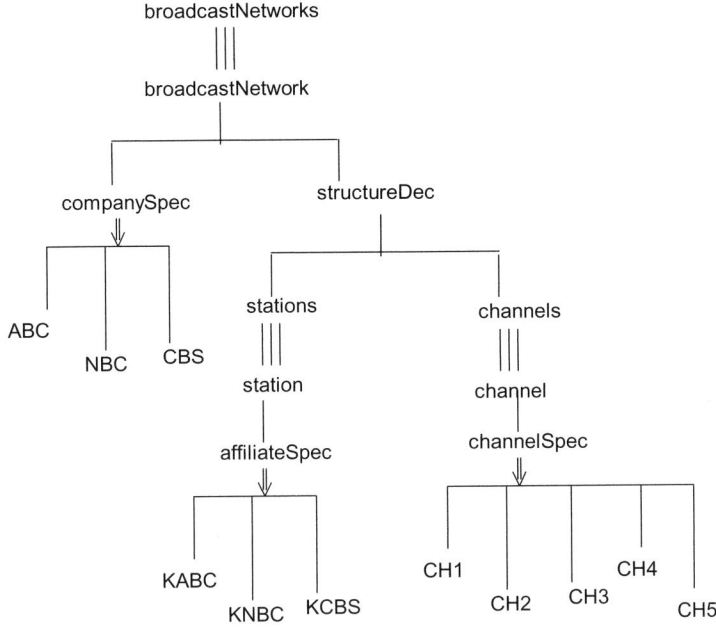

Figure 9.14

Expanded Broadcast Network SES

Table 9.3. Allocation of Channels to Companies

CompanySpec/ channelName	CH1	CH2	CH3	CH4	CH5
ABC	x	x			
NBC			x		x
CBS				x	

CH4; NBC on channels CH3 and CH5; and CBS is relegated to CH4.

Note that a specialization for a system might determine not just one, but many, possibly related, selections for its components at different locations in the SES. For example, in Figure 9.14, the broadcast network SES is expanded to include local station affiliates. Now, selecting ABC from the company specialization would not just constrain the channels it uses, but also the local affiliate, for example, KABC. Indeed, it might be the case that national companies are constrained to at most one local affiliate and that local affiliate is constrained to particular channels to broadcast.

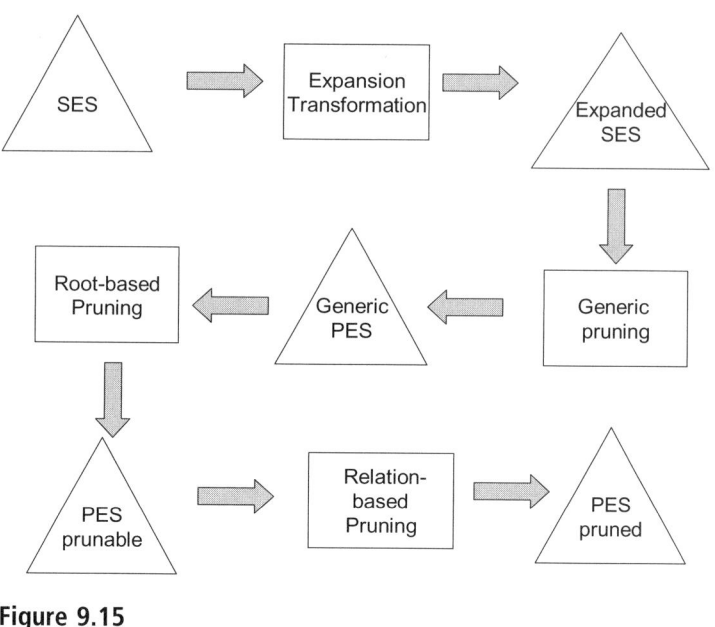

Figure 9.15

Relation-based Pruning of an SES with multiAspects

This shows that the SES is capable of representing the possible combinations of specialization selections that might be quite complex in nature. The question we now address is how it can be made to represent and enforce constraints on combinations such as those represented in Table 9.3. The answer is based on applying the multiAspect expansion restructuring (Chapter 7) to set up an SES with the right material to work on. For example, in Figure 9.14, based on the companySpec, broadcastNetworks is expanded to ABC_broadcastNetwork, NBC_broadcastNetwork, and CBS_broadcastNetwork. Now with the replica of the broadcastNetwork SES under each company, we can proceed to prune that SES in the context of that company. In the case of ABC, root-based pruning down-selects companySpec to ABC and then propagates this restriction through the relation involving companySpec and channelSpec, down-selecting the latter to channels CH1 and CH2. Similarly, if companySpec constrains the choices of stations under affiliateSpec, then after root-based pruning, affiliateSpec might be down-selected to KABC. Subsequently, relation-based pruning would constrain the choices under channelSpec to those owned by KABC. This process is summarized in Figure 9.15.

Summary

We said in Chapter 2 that handling of constraints are characteristic of ontologies that are more capable of expressing semantics and pragmatics than syntactic representations such as XML. In this chapter we have shown that by adding restrictive relations to the system entity structure we can constrain the pruning process to generate a desired restricted subset of world state descriptions. Chapters 13 and 18 will contain real-world applications of the relation-based pruning capability.

References

1. Zeigler, B. P. "Object-Oriented Simulation with Hierarchical, Modular Models: Intelligent Agents and Endomorphic Systems," Boston: Academic Press, 1990
2. Rozenblit, J. W., J. Hu, B. P., Zeigler, and T. G. Kim, "Knowledge-Based Design and Simulation Environment (KBDSE): Foundational Concepts and Implementation," *J. Operations Research Society,* Vol. 41, No. 6, pp. 475–489, 1990
3. Couretas, Jerry M., Bernard P. Zeigler, and George V. Mignon, "SEAE-SES enterprise alternative evaluator: design and implementation of a manufacturing enterprise alternative evaluation tool," Proceedings of SPIE — Volume 3696, Enabling Technology for Simulation Science III, Alex F. Sisti, Editor, pp. 136–146, 1999
4. Couretas, Jerry M., Bernard P. Zeigler, and Patel, U. "Automatic Generation of System Entity Structure Alternatives: Application to Initial Manufacturing Facility Design," *Transactions of the SCS,* 1999, 16(4), pp. 173–185

Appendix A: Logical Formulation of Selection Rules

A selection rule looks much like a logical material implication:

$$p \Rightarrow q$$

where

- p = select Symphonic from StyleSpec
- q = select Orchestra from EnsembleSpec
- and $\Rightarrow$ is interpreted as "must."

In fact, material implication does not convey all of the meaning of the selection rule, but it does offer some useful insight into the combinations of selections that follow from executing a rule.

For example, in propositional logic, it is known that

$$p \Rightarrow q \text{ is equivalent to } \neg q \Rightarrow \neg p$$

We can interpret this equivalence as transforming the original rule into a new rule of the form:

If select x from B-b then must select y from A-a

where B-b means set obtained by removing the element b. For example, with B = StyleSpec = {Symphonic, Folk, Jazz} and b = Symphonic, we have B-b = {Folk, Jazz}.

where we wrote {Folk, Jazz} as "Folk or Jazz" and where the "or" means "mutually exclusive or," that is, exactly one of the elements must be chosen.

In other words, if p = select Symphonic from StyleSpec then we interpret $\neg p$ = select Folk or Jazz StyleSpec.

So that if $p \Rightarrow q$ represents the Symphonic selection rule, then the negation $\neg q \Rightarrow \neg p$ takes the form:

If select Small Group or Soloist from EnsembleSpec then must select Folk or Jazz from StyleSpec

StyleSpec/EnsembleSpec	Small group	Soloist
Folk	x	x
Jazz	x	

This table actually requires a more restrictive rule for Soloist:

If select Soloist from EnsembleSpec then must select Folk from StyleSpec

Now in propositional logic,

$$p \Rightarrow q \text{ is } \mathbf{not} \text{ equivalent to } \neg p \Rightarrow \neg q$$

For example, if we use this transformation on the original rule, we get

If select Folk or Jazz from StyleSpec then must select Small Group or Soloist from EnsembleSpec

is not correct since Jazz does go with Orchestra.
We know that

$$q \Rightarrow p \text{ is equivalent to } \neg p \Rightarrow \neg q$$

So we also have

$$p \Rightarrow q \text{ is } \textbf{not} \text{ equivalent to } q \Rightarrow p$$

If select Orchestra from EnsembleSpec then must select Symphonic from StyleSpec

is not correct since an Orchestra can play Jazz as well as Symphonic music.

Suppose we remove Jazz as played by an Orchestra as in the following table.

StyleSpec/EnsembleSpec	Orchestra	Small group	Soloist
Symphonic	x		
Folk		x	x
Jazz		x	

Then Symphonic music is the only possibility that goes with Orchestra and in this case, the selection rule goes in both directions:

If select Orchestra from EnsembleSpec then must select Symphonic from StyleSpec
If select Symphonic from StyleSpec then must select Orchestra from EnsembleSpec

Appendix B: Theory Support for Relation-based Pruning

There is a technical requirement for consistency that needs to be addressed here. We will require that a pair of specializations has at most one non-empty relation associated with it. In other words,

$restrictRelationFn(spec_1, spec_2) = R_{spec_1,spec_2} \land R_{spec_1,spec_2} \neq \emptyset$
$\Rightarrow R_{spec_2,spec_1} = \emptyset$

For example, $R_{sportType,sizeType}$ is not empty, so $R_{sizeType,sportType}$ must be empty. Thus only one of the ordered pairs of the same unordered specialization pair can impose a restriction on pruning. We call the first specialization of such a pair the *key* specialization. In pruning, the key specialization will always be processed before the other specialization in the pair.

The *converse of* $R_{spec_1,spec_2}$ is $R_{spec_2,spec_1}$ where

$R_{spec_2,spec_1}(x,y) \Leftrightarrow R_{spec_1,spec_2}(y,x)$

The converse represents the same information available in the original relation. Note however, that the pair of specializations is listed in the reverse order and therefore the key specialization is different. This means that the order in which pruning examines specializations will be affected. So we have a distinct choice in deciding whether to include a relation or its converse in the restriction function.

Also, we consider only *distinct* pairs as capable of imposing restrictions. In other words,

$restrictRelationFn(spec_1, spec_2) = R_{spec_1,spec_2} \land R_{spec_1,spec_2} \neq \emptyset$
$\Rightarrow spec_1 \neq spec_2$

The domain of the restriction function is the set of ordered pairs of specializations that impose restrictions on pruning. Namely,

$domain(restrictRelationFn) = \{(spec_1, spec_2) \mid R_{spec_1,spec_2} \neq \emptyset\}$

The range of the restriction function is the set of relations that impose restrictions on pruning. Namely,

$range(restrictRelationFn) = \{R \mid R_{spec_1,spec_2} \neq \emptyset$ for some
    $(spec_1, spec_2) \in specializations \times specializations\}$

In terms of the ontological framework of Chapter 1, such a system of relations is an example of constraints placed on the world states or structures that can be described by the ontology. Often, there is not a particular step-by-step procedure that will unequiv-

Appendix B: Theory Support for Relation-based Pruning

ocally produce such a structure. So it is convenient to think of such structures as nonunique "solutions" to the constraints. In the present case, consider an assignment to each specialization of one of its specialized entities. This assignment is a solution to the restriction relation function if it satisfies each of the relations in its domain. Formally, let

prunedState : *specializations* → *Entities*

such that

prunedState (*spec*) ∈ *entities*$_{spec}$

Then

prunedState satisfies *restrictRelationFn* if

for all (*spec*$_1$,*spec*$_2$) ∈ *domain*(*restrictRelationFn*)

we have

(*prunedState*(*spec*$_1$), *prunedState*(*spec*$_2$)) ∈ $R_{spec_1,spec_2}$

We will also say that a pruned state is *compatible* with the restriction function if it satisfies this function.

In our example, there are three compatible pruned states:

(sportType = golf,sizeType = small),
(sportType = baseball,sizeType = medium),
(sportType = basketball,sizeType = large)

Pruned states represent pruned entity structures that are completely pruned, i.e., where a single element has been selected for every specialization. However, we also will want to deal with partially pruned structures where there are still non-singleton subsets of entities under specializations. We'll say that such a partial pruning is *compatible* with the restriction function if every specialized entity is a part of at least one compatible pruned state that can be formed from the pruning.

StyleSpec/EnsembleSpec	Orchestra	Small group	Soloist
Symphonic	x		
Folk		x	x
Jazz	x	x	

In Figure 9B.1 the pruning at the bottom left is not compatible since Soloist cannot be paired with either Symphonic or Jazz to

Figure 9B.1

Illustrating Partial Pruning Compatibility

constitute a compatible pruned state. Thus compatibility of a partial pruning means that there is at least one possible completely pruned entity structure that can result from further pruning.

Exercise

Imagine a game called the pruning game. At each turn, players remove one or more entities from specializations in the generic PES of an SES with a restriction function. Of course, once a specialization has exactly one entity it cannot be pruned anymore. A player can bet that the selected entity is not compatible with any completely pruned state. In subsequent turns, players attempt to reach pruned states that are compatible with the restriction function. Any player succeeding in this endeavor wins. The betting player wins if no other player can claim victory.

 **Example**

Given

$R_{styleSpec,EnsembleSpec} \subseteq entities_{StyleSpec} \times entities_{EnsembleSpec}$
= {(Symphonic,Orchestra),(Folk,SmallGroup),(Folk,Soloist), (Jazz,Orchestra),(Jazz,SmallGroup)}

with

$restrictRelationFn(StyleSpec,EnsembleSpec) = R_{StyleSpec,EnsembleSpec}$

and otherwise

$restrictRelationFn(spec_1, spec_2) = \emptyset$

 Exercise

Enumerate the pruned states that are compatible with this restriction.

Multiple Restriction Relations

DIFFICULTIES While pruning supported by several relations is a very powerful aid to achieving desired restrictions of pruning outcomes, there are two main difficulties that must be dealt with.

1. The outcomes may be dependent on the nature of the relations and the order in which they are processed and take effect.
2. The relations may form dependency cycles which do not allow local step-by-step selection.

We'll discuss these problems and then offer a design approach to supporting relation-based pruning.

ORDER DEPENDENCE Consider an SES for musical performance, similar to Figure 9B.2, used by the event planner for a music hall who has the choice of several performers, each of which has a particular style and ensemble composition (Table 9B.1).

$restrictRelationFn(StyleSpec,EnsembleSpec) = R_{StyleSpec,EnsembleSpec}$
$restrictRelationFn(PerformerSpec,EnsembleSpec) =$
    $R_{performerSpec,EnsembleSpec}$
$restrictRelationFn(PerformerSpec,StyleSpec) = R_{performerSpec,StyleSpec}$

If we use a sequence starting with a choice of performer, this will determine the style and ensemble type. For example the sequence: *PerformerSpec, EnsembleSpec, StyleSpec.*

172   Chapter 9   Constrained Pruning

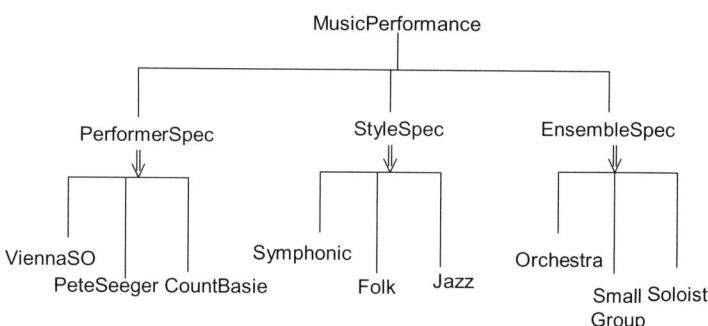

Figure 9B.2

Extended SES for Musical Performance

Table 9B.1.   Relations for Extended Musical Performance

PerformerSpec/ EnsembleSpec	Orchestra	Small group	Soloist
ViennaSO	x		
PeteSeeger			x
CountBasie	x		
ViennaSO (Vienna Symphony Orchestra)	x		
PeteSeeger		x	
CountBasie			x

This gives rise to a prunedState:

(PerformerSpec = PeteSeeger, StyleSpec = Folk,
   EnsembleSpec = Soloist)

which is compatible with the restriction function.

Now consider the sequence: *StyleSpec, EnsembleSpec, PerformerSpec*

Now select StyleSpec = Folk. This restricts EnsembleSpec = {SmallGroup, Soloist}

Selecting SmallGroup, results in:

(StyleSpec = Folk, EnsembleSpec = SmallGroup,
   PerformerSpec = ?)

Appendix B: Theory Support for Relation-based Pruning

which is incompatible with the restriction function since no available performer fits the bill.

On the other hand, selecting StyleSpec=Folk restricts

EnsembleSpec = {SmallGroup, Soloist)

Selecting Soloist, results in:

(StyleSpec=Folk, EnsembleSpec = Soloist,
  PerformerSpec = PeteSeeger)

which is compatible with the restriction function.

Exercise

Select StyleSpec=Jazz. This restricts

EnsembleSpec = {SmallGroup, Orchestra)

Selecting Orchestra, results in:

(StyleSpec=Jazz, EnsembleSpec = Orchestra,
  PerformerSpec = ?)

which is compatible with the restriction function.

Show that both desired and undesired states are compatible with the restrictive relations.

Dependency-based Sequences and Cycles

Each relation can be viewed as a component with an input port, viz., its domain specialization and an output port, viz., its range specialization. Dependency analysis (Chapter 17) orders these components according to their input/output dependencies, i.e., a component must precede another if the output port of the first matches the input port of the second.

$$R_{spec_1, spec_2} \text{ precedes } R_{spec_2, spec_3}$$

For example,

$$R_{performerSpec, StyleSpec} \text{ precedes } R_{StyleSpec, EnsembleSpec}$$

A *computation sequence* is a sequence of relations in which the precedence relation is respected. In other words,

$R_1, R_2, \ldots, R_i, R_{i+1}, \ldots, R_n$ is a computation sequence if

R_i precedes $R_j \Rightarrow i < j$

A computation sequence provides an efficient means of generating pruned states that are compatible with the relations. This is so since only one pass is required to propagate the selections made from one relation to the next and to obtain a global solution. Let's examine how selections are propagated in such a sequence.

A sequence of restrictive relations, $R_1, R_2, \ldots, R_i, R_{i+1}, \ldots, R_n$ gives rise to sequence of specializations:

$$spec_{1d}, spec_{1r}, spec_{2d}, spec_{2r} \ldots spec_{n,d}, spec_{n,r}$$

where $spec_{i,d}, spec_{i,r}$ is the pair of domain and range specializations associated with R_i.

Then processing of specializations proceeds in the order given by the above sequence. That is,

- (Basis step) we make a choice selection from the first specialization — this means that we reduce the set of entity children of this specialization. The pruning state is now represented by

$spec_{1,d} = entities'$ where
$entities' \subseteq entities_{spec_1}$ (the initial children of $spec_1$)

Of course if we select a specific specialized child then the resulting set of entities is a singleton. We allow for non-singleton selections, since this is the usual situation as selections propagate along the sequence.

Now, propagation occurs as follows:

- (Inductive Step)

Let $entities_i'$ be the current subset of entity children of the ith specialization in the sequence.
Let $influences_i = \{spec_j \mid (spec_i, spec_j) \in domain(restrict\ RelationFn)$.

The influences of a specialization are the specializations that are immediately dependent on it.

Appendix B: Theory Support for Relation-based Pruning

Then R_i restricts the entity children of each $spec_j \in influences_i$ as follows:

$$entities_j' = \{y \mid (x, y) \in R_i \land x \in entities_i'\}$$
$$= \bigcup_{x \in entities_i'} R_i(x)$$

where $R_i(x) = \{y \mid R_i(x,y)\}$

In other words, when a specialization's child entities become restricted to a subset, the child entities of each of its influences become restricted to a subset that is compatible with the connecting relation.

In a computation sequence there is never a situation where a selection is affected by choices made in processing a relation later in the sequence. However, finding a computation sequence is not possible if there is a cycle in the dependency graph.

For example, suppose we have three relations,

$$R_{PerformerSpec,StyleSpec}, \quad R_{StyleSpec,EnsembleSpec}, \quad R_{EnsembleSpec,PerformerSpec}$$

and therefore

$R_{PerformerSpec,StyleSpec}$ precedes $R_{StyleSpec,EnsembleSpec}$
$R_{StyleSpec,EnsembleSpec}$ precedes $R_{EnsembleSpec,PerformerSpec}$
$R_{EnsembleSpec,PerformerSpec}$ precedes $R_{PerformerSpec,StyleSpec}$

These relations form a cycle of dependencies and cannot be placed into a computation sequence.

Exercise

Suppose you select a performer, e.g., CountBasie. Trace the successive restrictions placed on styles, ensembles, and, finally, performers.

10

Pruned Entity Structures: Data Extraction and Change-based Information Exchange

The axiomatic structure imposed by the SES allows us to tackle certain problems that arise in the processing of the XML instances. Solutions of these problems are based on the properties that prunable entity structures (PES) derive from their underlying SESs. One problem to be addressed is how to provide the minimum context needed to disambiguate multiple occurrences of an item. This is needed for efficient extraction of the data from the XML document received by the consumer. An algorithm for determining the minimum context and associated algorithm for extracting data based on it underlie the processing chain concepts discussed in Chapter 15. We'll discuss such an algorithm before going on to consider the more advanced form of information exchange in which changes in world state are the basis for information exchange rather than the before and after states themselves.

Extracting Data from Pruned Entity Structures

Recall that a PES is a tree-like structure whose nodes are items with labels. In the XML representation, these labels are represented by tag names. In processing such XML schema instances, an automated process will have to access such items, for example, to do pruning operations, or to set/get values of associated attributes. We can use a tag name to access elements in the DOM representation of a PES by using the method getElementsByTagName. The problem is that there may be many elements with the same tag name. How do we identify a specific one?

As an example, consider the XML instance first discussed in Chapter 2:

```
<?xml version="1.0" encoding="UTF-8"?>
<!DOCTYPE notice DTDForRegistration.dtd">
<notice>
        <buyer firstName="Bernard"
          lastName="Zeigler" SSN="000-00-0000"/>
        <vehicle make="Toyota" model="Prius"
          VIN="00000-00000-00000-00000"/>
</notice>
```

When this instance is received it is converted to an instance of Document, the JAVA class implementation of the DOM interface. We can use getElementsByTagName within the following method to get an element with a given tag name:

```
public static Element getElementOccurrence(Document doc,
   String tag) {
     NodeList nl = doc.getElementsByTagName(tag);
     return (Element)nl.item(0);
}
```

Having an element, we can use the following methods to get the value of one of its attributes:

```
public static String getAttrVal(Document doc, String tag, String
   attr) {
     Element el = getElementOccurrence(doc, tag);
     if (el == null) {
       return " ";
```

}
return el.getAttribute(attr);
}

For example, using doc as the Document instance, we can write:

String buyerFirstName = getAttrVal(doc, "buyer", "firstName");

Exercise

Write a statement to get the model of the vehicle.

The problem mentioned above arises when there are multiple occurrences of elements such as buyers. For example, Figure 10.1(a) gives an SES that generates a DTD for multiple notices. Figure 10.1(b) shows a PES having two notices that conform to this SES (Chapter 8).

The following fragment shows an example XML document that implements the PES in Figure 10.1(b).

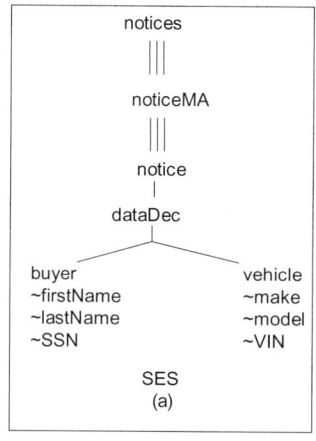
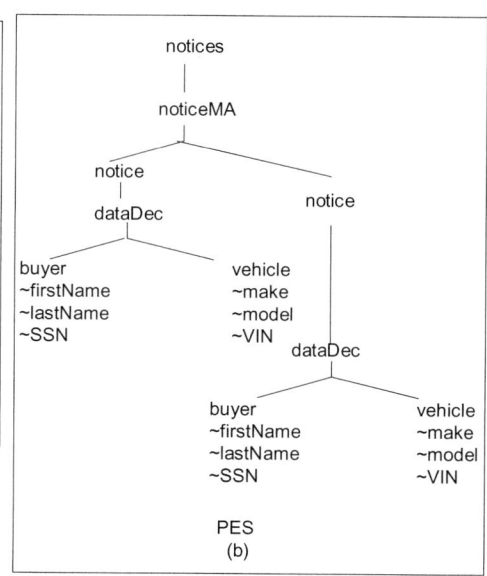

Figure 10.1

SES for Multiple Notices and a Conforming PES

```xml
<?xml version="1.0" encoding="UTF-8"?>
<!DOCTYPE notice DTDForRegistration.dtd">
<notices>
...
<notice>
...
        <buyer firstName="Bernard"
          lastName="Zeigler" SSN="000-00-0000"/>
        <vehicle make="Toyota" model="Prius"
          VIN="00000-00000-00000-00000"/>
...
</notice>
<notice>
...
        <buyer firstName="John" lastName="Public"
          SSN="000-00-0001"/>
        <vehicle make="Ford" model="Taurus"
          VIN="00000-00000-00000-00001"/>
...
</notice>
...
</notices>
```

How do we access the data for a specific buyer in this document? The problem is that getElementOccurrence("buyer") will return only the first buyer occurrence (with name "Bernard"). Looking inside this method, we see that the method getElementsByTagName("buyer") will return the list of the two elements having the tag, "buyer." How can we distinguish these elements? In this particular case we can use the SSN as a unique identifier. However, in general we are not guaranteed to have such an externally given unique identifier. So we need an approach that makes no specific assumption about the PES under consideration.

Using Context to Shorten Path Identifiers

To advance toward a solution, recall that there is a unique labeled path from the root to any item in an SES (Chapter 5). It is almost immediate that the same property is inherited by a PES that conforms to an SES without multiAspects. Recall that, as illustrated in Figure 10.1, a multiAspect allows multiple copies of the same SES to co-exist in a PES. It follows that to retain the unique path labeling property in such a PES we need to distinguish root entities of the SESs introduced in the PES. For example, in Figure 10.1(b), we might add distinct suffixes such as in notice[1], notice[2], etc. These suffixes can be generated automatically if there are agreed-upon distinguishing variables attached to the generating entities of multiAspects. For example, in Figure 10.1(a), the notice entity would have an integer variable that can be incremented automatically when a new copy of the subSES for notice is created. *From here on we will assume that variables named id or number are attached to multiAspects for this purpose.* Under this assumption, and the implementation of automatic suffixing, it follows that the unique labeling property holds for all PESs. For example, paths:

buyer, dataDec, notice[1]

and

buyer, dataDec, notice[2]

distinguish the two occurrences of buyer in Figure 10.1.

Algorithm for Ancestor Contexts

The problem with this solution is that the uniquely labeled paths it uses may be very long for deep SESs (for example see the SES discussed in Chapter 14). We need a way to shorten the information required to uniquely access items in the PES.

Our inspiration for such a solution starts from the way books are broken down into chapters and sections. Using suffix identifiers as illustrated in Figure 10.2(b), we can locate a section with a chapter, for example, section[1] in chapter[2]. This means that chapter and section root entities provide enough *context* for locating any occurrence in a book PES. For example, suppose that the SES for section does not contain any multiAspects. Then an item inside any one of the four pruned copies in Figure 10.2(b) can be identified using its unique identifier within the SES and the section within chapter context. For example, a path of the form, exercise ... section[2].chapter[2] is enough to distinguish a particular exercise.

Exercise

Suppose that a section is specialized into introduction, exposition, summary, or exercise. Show that this subSES satisfies the

Using Context to Shorten Path Identifiers 181

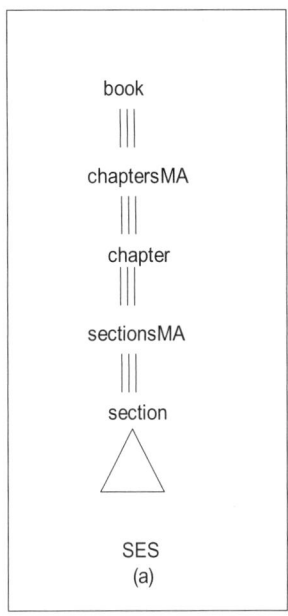 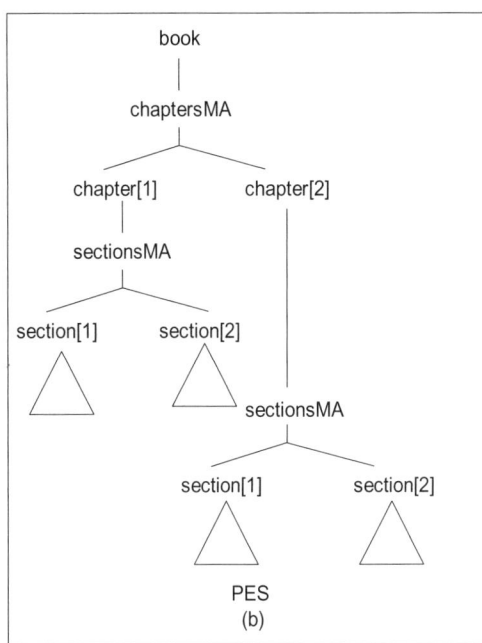

Figure 10.2

SES for Book with Chapters and Sections and Conforming PES

requirements just discussed and provide the unique path to any section in the book.

In this approach, the root entities of multiAspects provide the contexts that allow shortening path identifiers by skipping intermediate subpaths. The question now is how to characterize such contexts in general and then design an algorithm to find them when given any PES. We present such a characterization and the algorithm in Appendix A. Figure 10.3 briefly summarizes the solution. Starting from the tag with multiple occurrences we iterate a process to find successive top ancestors until all occurrences are uniquely identified by sequences of such ancestors.

Directly accessing an element using its tag and context information can now be achieved with:

```
public static Element getElementOccurrence(Document doc,
    String tag, String[ ] topSequence) {
      NodeList nl = doc.getElementsByTagName(tag);
      return (getUniqueElement(nl,topSequence));
}
```

182 Chapter 10 Data Extraction and Change-based Information Exchange

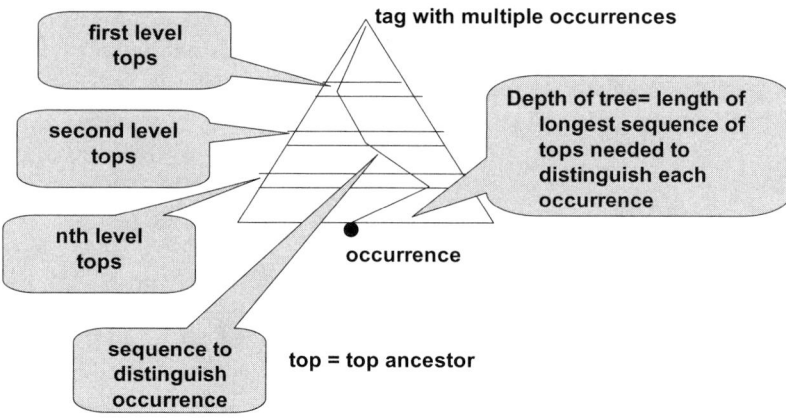

Figure 10.3

Reducing Path Identifiers Using Top Ancestors

For example, to get the exercise in the first section of the second chapter, we use:

Element exercise = getElementOccurrence(doc, "exercise", new String[ ]{"section[1]","chapter[2]"});

Knowing the path from the root entity to an Element we can get its context information:

String[ ] topSequence = public static String[ ] getTopSequence(doc, el, ,pathFromRoot);

This method can be used when traversing the PES to extract its data, as illustrated in Algorithm 10.1.

ALGORITHM 10.1

Extract Data from PES

//initialize
traverseEntity(rootEntity," ")
//
traverseEntity(entity, path){
specializations = entityHasSpecializations(entity)
for each spec ∈ specializations

```
    traverseSpecialization(spec, path + spec)
}

traverseSpecialization(spec, path){
    entity = specializationHasEntity(spec)//assume complete
       pruning
    report spec selection is entity at getTopSequence(entity,
       path)
    traverseEntity(entity, path + entity)}
}
//similarly for variable,aspects,multiaspects,
```

Algorithm 10.1 proceeds by traversing the PES while keeping track of the path. When a data element of the PES is found, the value is reported along with the context information needed to access the data element. Traversal continues until all nodes have been visited.

Change-based Information Exchange

Instead of sending world state descriptions as in Figure 8.2 we can send the change in such descriptions. As illustrated in Figure 10.4, one approach is to compare the PES that describes the new world state after an event with the PES that pertained before this event. The difference resulting from this comparison is sent in a message and interpreted by the receiver by adding the difference to its stored PES. The PES that results from this reconstruction becomes the stored PES for the next update. Such an update process is considered in a report by the ISO committee on Geographic information.[1]

Comparing Two Prunings of the Same SES

The essence of the approach in Figure 10.4 is comparing two prunings of the same SES. Our approach is to follow the recursive pattern introduced earlier (Chapter 6). However, as illustrated in Figure 10.5, since two structures are involved, we must include a means of placing their items into correspondence at each level before continuing with the analysis and recursion. Indeed, the key to detecting differences at each level lies in attempting to

Chapter 10 Data Extraction and Change-based Information Exchange

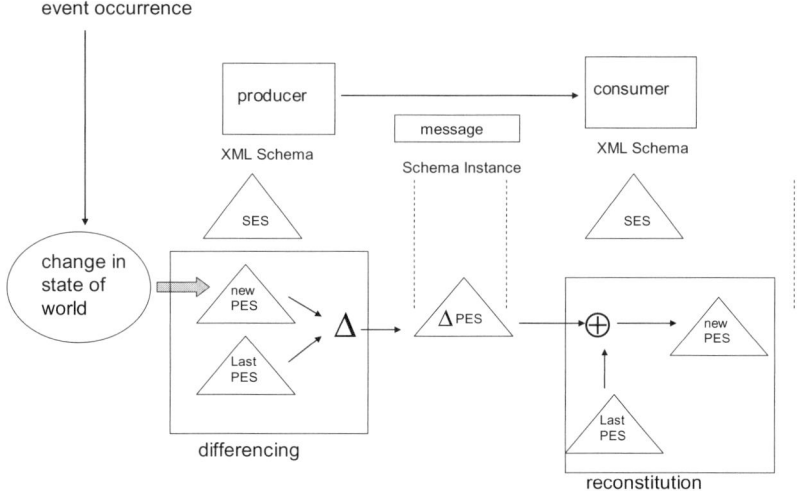

Figure 10.4

Restating the Information Exchange Framework in the SES Context

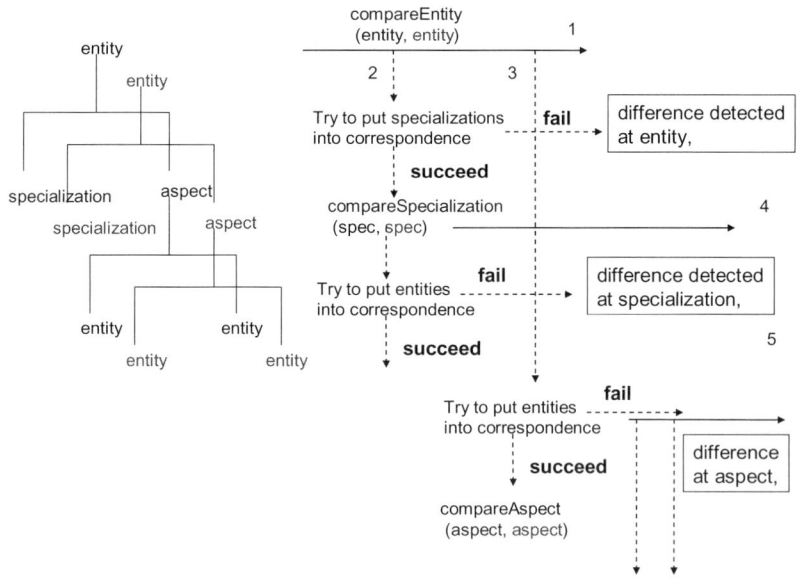

Figure 10.5

Recursive Pattern Applied to Comparison of PESs

make such correspondences. If we succeed in making a one-to-one match, then no difference is reported and we proceed to employ the correspondence to continue with the recursion to next lower levels. On the other hand, a failure to make such a correspondence indicates that a difference exists, and the new element is reported along with information about the path to the item where the difference exists. Here again we use the context to minimize the amount of information that has to be sent. When received by the consumer, the path information allows the reconstruction process to replace the element in the stored PES with the element just received.

At each level, we must perform this comparison:

1. *variables attached to paired entities* — put the variables into correspondence by name and check whether the corresponding variables have the same assigned values; if not, report difference at the entities
2. *specializations under paired entities* — put the specializations into correspondence by name and for each pair of specializations, go to 5
3. *aspects under paired entities* — attempt to put the aspects into correspondence by name; if fail, report difference at the entities, otherwise, for each pair of aspects, go to 5
4. *multiAspects under paired entities* — attempt to put the multiAspects into correspondence by name; if fail, report difference at the entities, otherwise, for each pair of multiAspects, go to 7
5. *entities under paired specializations* — attempt to put the entities into correspondence by name; if fail, report difference at the specializations, otherwise, for each pair of entities, recursively apply this comparison
6. *entities under paired aspects* — put the entities into correspondence by name and for each pair of entities, recursively apply this comparison
7. *entities under paired multiAspects* — attempt to put the entities into correspondence by identification; if fail, report difference, otherwise, for each pair of entities, recursively apply this comparison

More detail is given in Algorithm 10.2 elaborated for comparison of specializations.

ALGORITHM 10.2

> **Compare Two PESs for Differences**
>
> //initialize
> $compareEntity(rootEntity1, rootEntity2, " ")$
> //
> $compareEntity(entity1, entity2, path)$ {
> $specializations1 = entityHasSpecializations(entity1)$
> $specializations2 = entityHasSpecializations(entity2)$
> $f = match(specializations1, specializations2)$
>   for each $pair \in f$, $pair = (spec1, spec2)$
> $compareSpecialization(spec1, spec2, path + spec1)$
> }
> $compareSpecialization(spec1, spec2, path)$ {
> $entitys1 = specializationHasEntity(spec1)$
> $entitys2 = specializationHasEntity(spec2)$
> $f = match(entitys1, entitys1)$
>   if($entitys1.size()\ != entitys2.size() \lor f.size()\ != entitys1.size()$){
>   report difference at $getTopSequence(spec1, path)$
>   }
>   else {
>   for each $pair \in f, pair = (entity1, entity2)$
> $compareEntity(entity1, entity2, path + entity1)$}
> }
> //similarly for variable, aspects, multiaspects,

Figure 10.6 illustrates application of Algorithm 10.2 to two prunings of an SES for book. By traversing each PES in synchronization, we find a difference in the color selected for frontCover, and that a new page has been added under the multiAspect.

Summary

Using the properties that PES derive from their underlying SESs, we presented an approach to provide the minimum context needed to disambiguate multiple occurrences of an item in a PES, and its realization as an XML instance. We also considered use of changes in world state as the basis for sending information, rather than world state descriptions themselves. When there is relatively little change from one event to the next in the world

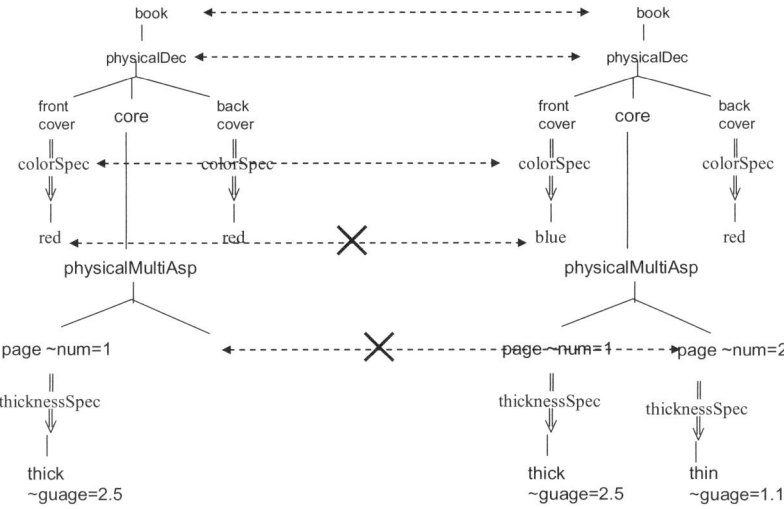

Figure 10.6

Example of Pruning Comparison

state of interest to the pragmatic frame, an ontology that supports sending changes can result in more efficient and effective net-centric information exchanges. Quantization, to be discussed in Chapter 11, is a well-established example of such change-based representations.

Reference

1. International Standards Organization, "Geographic information — Encoding," sent to the ISO Central Secretariat for publication, ISO/TC 211 Geographic information/Geomatics ISO reference number: 19118, 2006

Appendix A: Top Ancestor as Context Information

We discuss an algorithm for discovering top ancestors. See Figure 10A.1.

We start with all elements with the given tag name. For all elements, we trace back through their ancestors, grouping them into equivalence classes that are not distinguished by the paths so generated. This refinement process stops when all elements

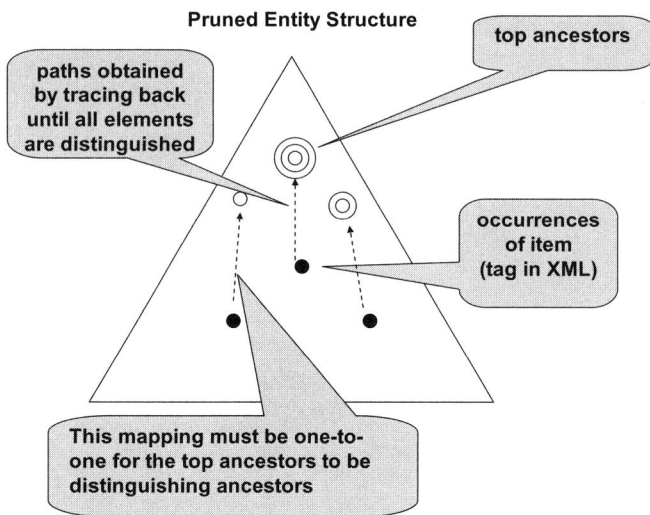

Figure 10A.1

Illustrating Top Ancestors

are distinguished, i.e., placed into singleton equivalence classes. The top ancestors at this stage are the last labels of the paths. If the labels alone are sufficient to distinguish all elements then we terminate with these ancestors. If the labels are not sufficiently distinguishing then we apply the refinement process to each group of original elements that terminate in the same labels. The iteration continues until paths cannot be further reduced. In this way, a tree is established that yields successive top ancestors to form a minimal context path.

EXAMPLE: The book PES in Figure 10A.2 has pruned entity red from all specializations for illustration purposes.

1. Each of the elements having the label red are placed into an initial relation with empty paths. There is one equivalence class [1, 2, 3, 4].

   1 " "
   2 " "
   3 " "
   4 " "

2. The first stage refinement process begins. The parent labels of each element are concatenated to each path, with the result that elements of the equivalence classes are [1, 3] and [2, 4].

Appendix A: Top Ancestor as Context Information

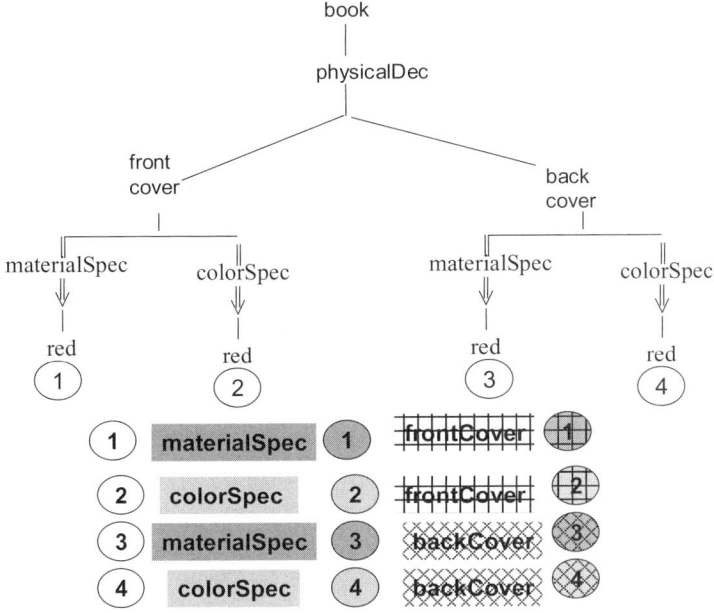

Figure 10A.2

Example Book PES

1 materialSpec
2 colorSpec
3 materialSpec
4 colorSpec

3. Continuing, the next level ancestor labels are concatenated to each path, with the result that elements of the equivalence classes are [1], [3], [2], and [4], thus terminating this iteration:

1 materialSpec.frontcover
2 colorSpec.frontcover
3 materialSpec.backcover
4 colorSpec.backcover

4. The top labels after the first iteration are {frontCover, backCover}. Using these labels is not sufficient to distinguish all elements. They are grouped into two sets:

frontCover   1, 2
backCover   3, 4

5. The second stage refinement process is carried out for each group. For the group under frontCover, we start with the relation:

   1   " "
   2   " "

6. After one step, we have

   1   materialSpec
   2   colorSpec

which yields a distinguishing set of tops = {materialSpec, colorSpec}.

Similarly, for the group under backCover we find the same set of tops. See Figure 10A.3.

In this case, the sequences of tops replicate the original unique identifiers. However, in general, this is not the case.

Exercise

Apply the top ancestor discovery process above to disambiguate the label, thick. Show that a single level of tops is needed. Now create an SES for a library that has a multiAspect with generating entity, book. Apply the discovery process to the library and show that a second level of tops is needed to disambiguate the label, thick. See Figure 10A.4.

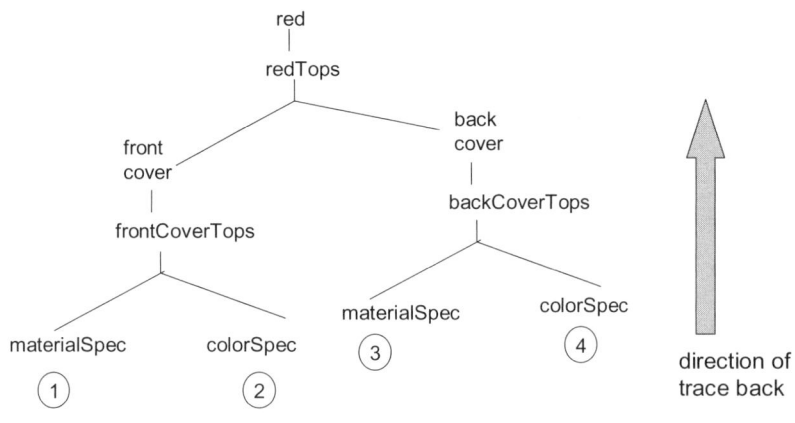

Figure 10A.3

Tree of Top Ancestors

Figure 10A.4

Example of PES

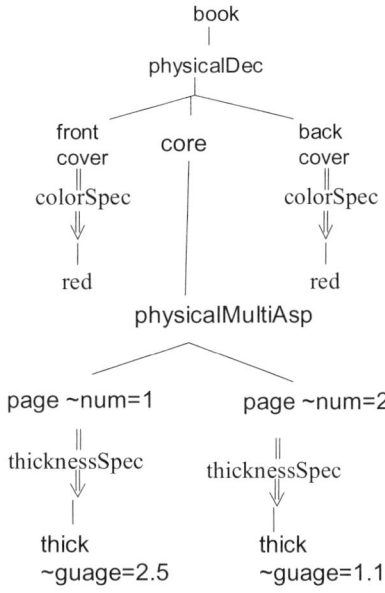

Exercise

Find the flaw in the following proof of the claim that the top ancestor is always a distinguishing ancestor.

Proof: Suppose that two unique identifier paths converge on the same node. If the paths are the same, then the occurrences must be the same. Since they are different, the paths must be different and there is a first label in the direction toward the root where they differ. But now the paths from occurrences to this label (inclusive) are shorter unique identifiers, contradicting the shortness claim of the unique identifier paths.

Hint: what are the unique identifiers of the four occurrences of red in the PES of Figure 10A.2? What are the top ancestors? Does the number of top ancestors equal to 4?

Part III

Modeling and Simulation and Data Engineering

11
Hierarchical Systems, Models, and Simulations: The SES Ontology

This chapter connects the SES with the domain of simulation modeling. First, we motivate the need for hierarchical, modular structure with an example specification, that of an existing tactical data standard, in which it is only partly apparent. Then the SES, as characterized by the axioms introduced in Chapter 5, is interpreted as an ontology for the domain of hierarchical, modular simulation models. After showing the workings of this ontology, we review some basic concepts in systems theory with particular focus on discrete event systems. The chapter closes with an example of how the SES can be used to support a web service that supplies weather simulation models to satisfy consumer requests. This chapter sets up the basics to understand how the SES and its simulation models are employed for multilevel testing of net-centric systems (Chapters 17 and 18).

Chapter 11 Hierarchical Systems, Models, and Simulations

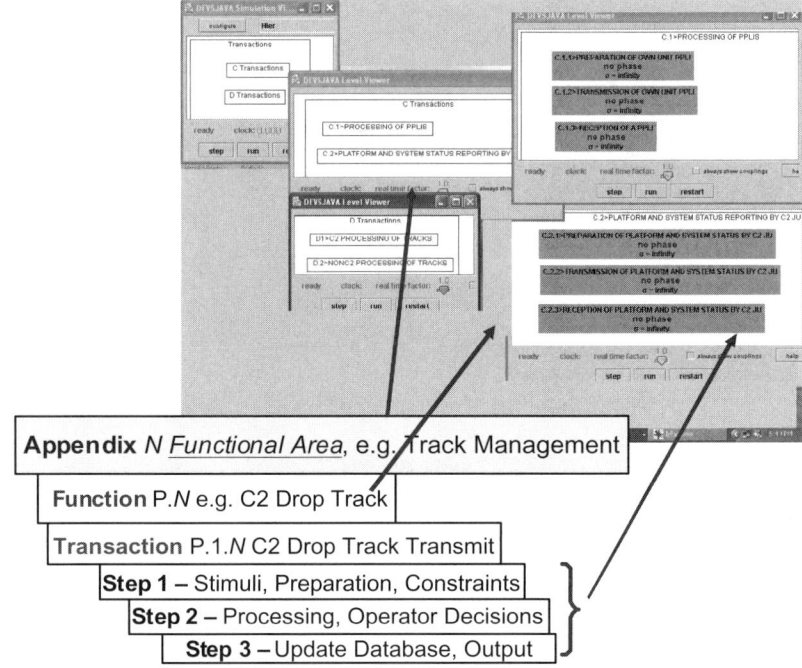

Figure 11.1

The Natural Hierarchy That Underlies the Link16 Standard

Why Hierarchical Structure?

Probably due to human preferences for hierarchy to help manage otherwise overwhelming complexity, technological systems often appear to have a nominal hierarchical structure with interactions that at least partially conform to that structure.[1] One of our main tasks is to analyze the system to model this "natural hierarchy" in a similar structure as a point of departure for better understanding it, modifying it, or imposing on it a higher level of interoperability and harmony.

An example is Link16, an important standard for tactical data communication, command, and control.[1] The specification document is hierarchically organized into a breakdown of Appendices, Functions, Transactions, and Processing steps as illustrated in Figure 11.1.

[1] Limitations on human working memory capacity[6] appear to necessitate chunking of information into, at most, nine components,[7] leading to hierarchical structures to organize thought.[8]

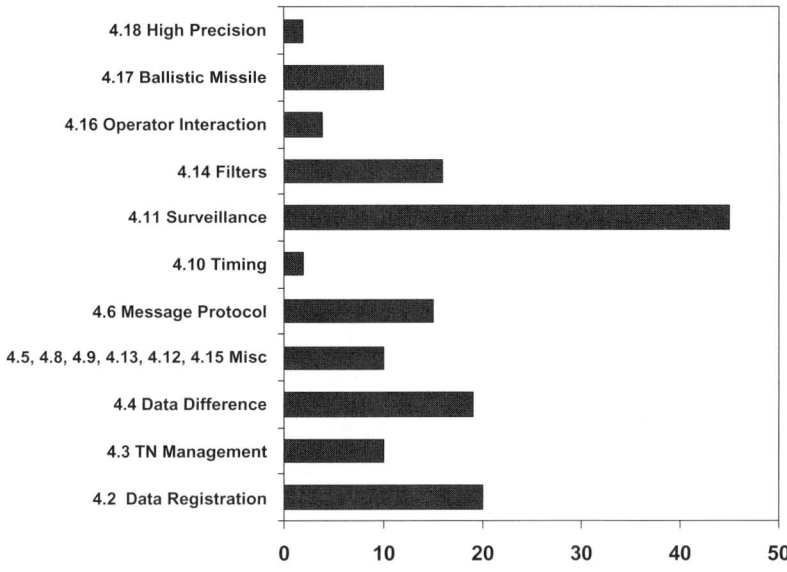

Figure 11.2

Appendices of Link-16 Specification

The Appendices roughly correspond to functional areas that are depicted in Figure 11.2 along with a measure of the relative proportion of detail devoted to each one.

Functional areas are collections of closely related functions, in which the latter consist of transactions made up of steps to execute in somewhat sequential manner. This hierarchy is a loose one, since to carry out a task usually requires the collaboration of a number of functions whose interactions often cross organization boundaries and invoke each other at the transactional or processing step levels. Nevertheless, such a hierarchical structure is a useful organizing principle since, without it, the specification document would appear as entangled as the proverbial bowl of spaghetti.

Hierarchical Modular Simulation Models — Universe of Discourse for the SES Ontology

In its original, and most expressive, interpretation, a System Entity Structure specifies a family of hierarchical, modular simulation models, each of which corresponds to a complete pruning of the SES. Thus, the SES formalism can be viewed as an ontology with the set of all simulation models as its domain of discourse. The mapping from SES to the System's formalism,

Table 11.1. Mapping SES Elements to Simulation Model Elements

SES element	Maps to
Entity	Component in a composite model
Aspect	Decomposition of a composite model into components corresponding to the aspect's children
multiAspect	Decomposition of a composite model into components, each of which is derived from the aspect's single child entity
Coupling of an aspect	Specifies the routing paths for information flow among the components corresponding to the aspect's children
Specialization	A family of alternative "plug-ins" for a component corresponding to the parent entity
Variables of an entity	Variables, including state variables and parameters, of the component corresponding to the entity

particularly to the DEVS formalism, is discussed in Ref. 2. Simulation models include both static and dynamic elements, and we can view the interpretations of the SES discussed up to now as restricting models to their static elements only.

Table 11.1 summarizes the mapping of SES elements to simulation model elements.

This mapping is illustrated in Figures 11.3–11.6.

Figure 11.3 illustrates how an aspect is interpreted as a recipe for constructing a composite model. The components of the composite model correspond to the entity children of the aspect. A coupling slot is associated with the aspect which specifies the

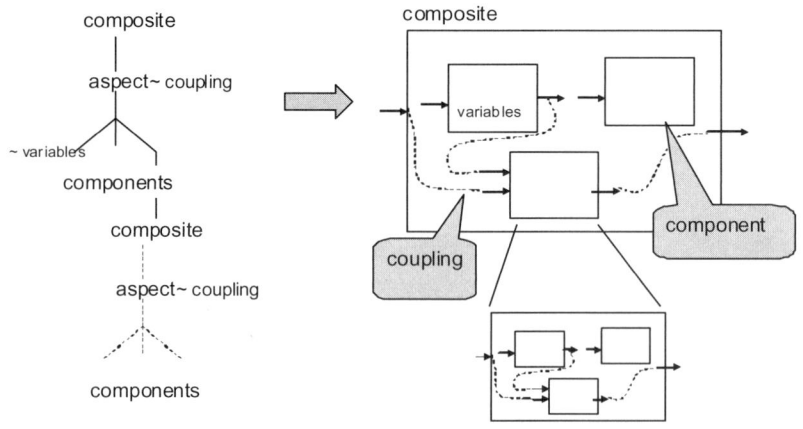

Figure 11.3

Mapping Aspects to Composite Models

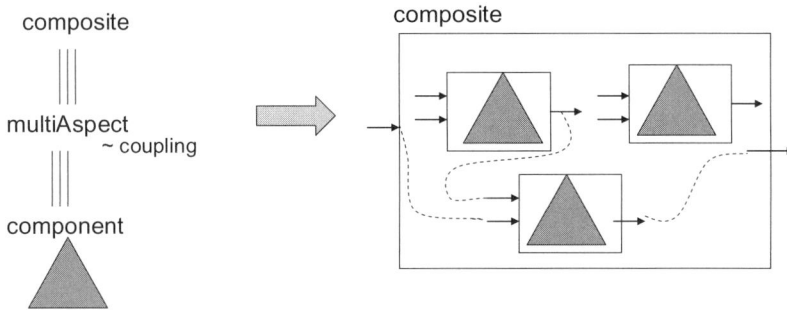

Figure 11.4

Mapping multiAspects to Composite Models

routing paths for information flow among the components corresponding to the aspect's children. In the DEVS interpretation, these paths are specified as connections of input and output ports and the composite is called a coupled model. The construction is hierarchical and modular as enabled by a proof of closure under coupling.[2] As illustrated, if an entity of an aspect itself has an aspect, this leads to replacing the corresponding component with the composite model specified by the second aspect.

The mapping of a multiAspect is illustrated in Figure 11.4. The mapping is similar to that for an aspect except that all components correspond to the entities generated by the multiAspect's generating entity.

The entities of a specialization provide a family of alternatives for a component corresponding to the parent entity. As depicted in Figure 11.5, each of the entities of a specialization map to a component that can be plugged into the slot of the parent entity of the specialization.

Figure 11.6 portrays the meaning of more than one aspect under the same entity. These aspects offer alternative decompositions that can be employed to construct a composite model corresponding to the parent entity.

The Levels of System Specifications

System theory deals with a hierarchy of system specification which defines levels at which a system may be known or specified. Table 11.2 shows Levels of System Specification (in simplified form, see Ref. 2 for an in-depth discussion).

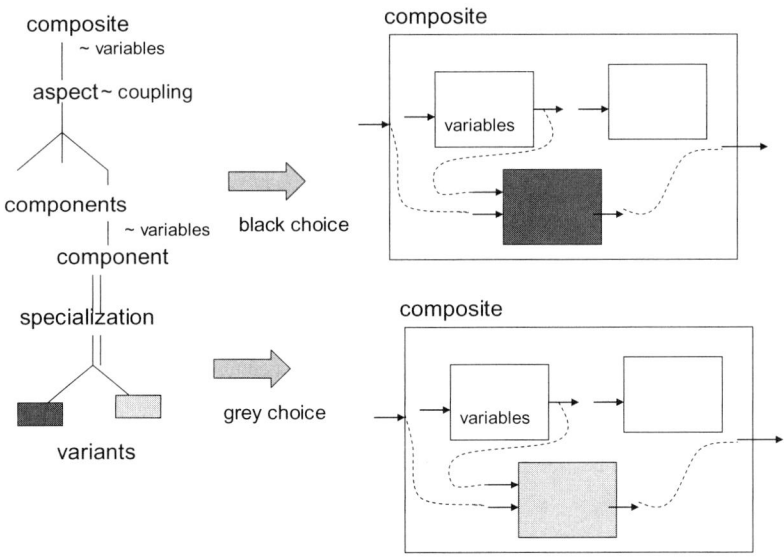

Figure 11.5

Mapping multiAspects to Composite Models

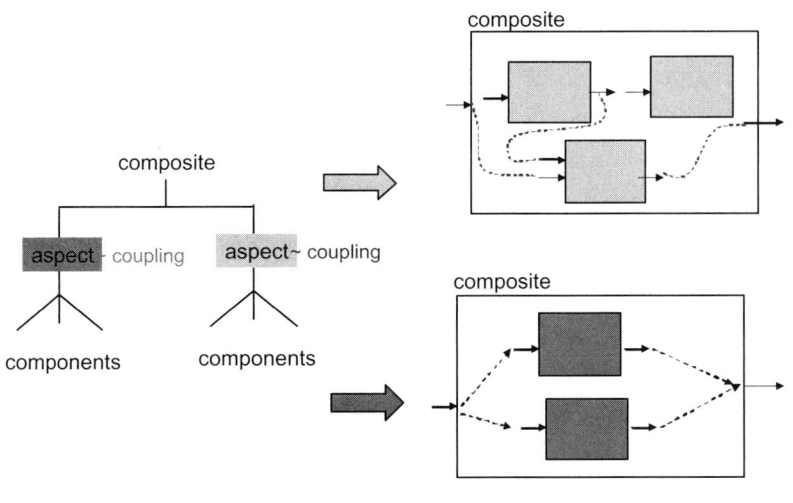

Figure 11.6

Interpreting Multiple Aspects as Alternative Decompositions

Dynamic Data Representations

Table 11.2. Levels of System Specification

Level	Name	What we specify at this level
4	Coupled systems	System built up by several component systems which are coupled together.
3	I/O system	System with state and state transitions to generate the behavior.
2	I/O function	Collection of input/output pairs constituting the allowed behavior partitioned according to the initial state the system is in when the input is applied. The collection of I/O functions is infinite in principle because, typically, there are numerous states to start from and the inputs can be extended indefinitely.
1	I/O behavior	Collection of input/output pairs constituting the allowed behavior of the system from an external Black Box view.
0	I/O frame	Input and output variables and ports together with allowed values.

Dynamic Data Representations

The levels of system specification provide a framework to discuss representation of time-indexed data in SES and XML, especially focusing on change-based representations. Figure 11.7(a) depicts a time segment in which the domain is an interval of time and the range is an arbitrary set. Such a time segment can play the role of an input segment or an output segment at Level 1 in Table 11.2. Here we are interested in the case in which the segment is an output segment of a system called a *generator*. Although shown as if the time interval were continuous, digital computation requires that only a finite number of points, or event times, represent the interval, as in Figure 11.7(b). This finite number allows us to represent the segment as a list of pairs called an *event set*. A generic event set can be described by an SES as in Figure 11.7(c). Any particular event set can be represented as a pruning of this SES. Finally, such a pruning, or XML instance, can be given to a generic generator depicted in Figure 11.7(d). The generic generator is a DEVS model specified at Level 3 in Table 11.2.

An example of a time segment representation by event sets is illustrated in Figure 11.8. A plot of the daily normal temperature taken from a U.S. weather service database is shown in Figure 11.8(a). It provides a temperature for each of the 31 days in

Chapter 11 Hierarchical Systems, Models, and Simulations

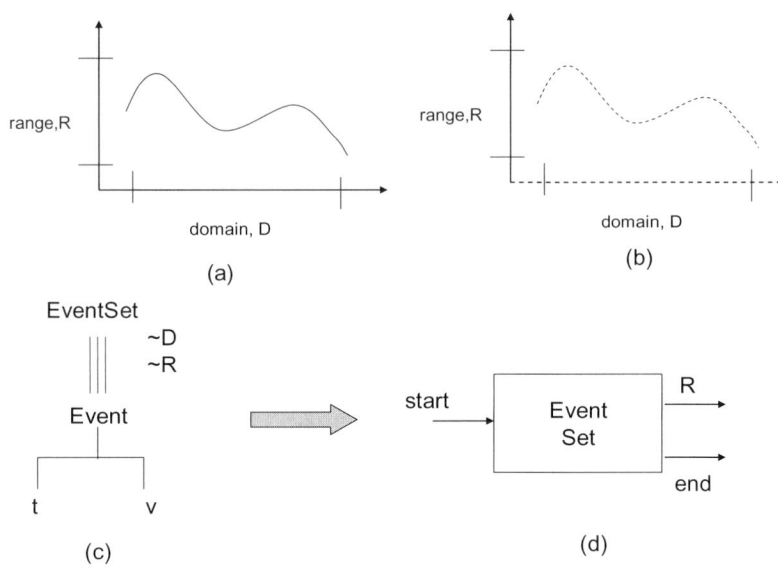

Figure 11.7

Representation of Time Segments by Event Sets and Event Set Generator

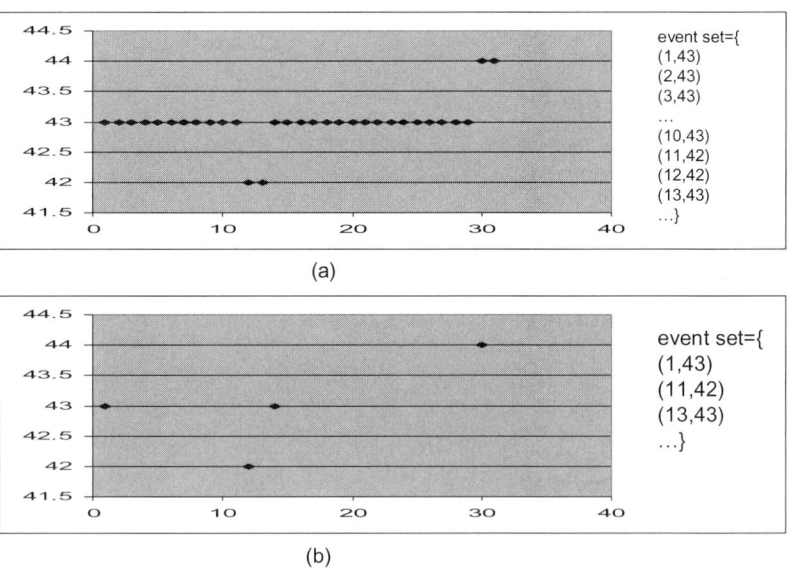

Figure 11.8

Representation of Time Segments by Quantized Event Sets

January at some particular weather station. The corresponding event set is shown to its right. In Figure 11.8(b), the same data is presented in a so-called *quantized* form in which only times and values for which there is a significant change (more than a quantum, here 1 degree) are retained. To transform to quantized form, we examine each successive value and compare it with the last saved value. When we come to a value that differs from the saved value by more than a quantum, it becomes the saved value and the time at which this happened is noted. It's apparent that the quantized representation can reduce the number of events needed to represent the same time segment. We can go one step further and replace the absolute times of events with incremental differences, so that pairs are of the form (t,v) where t is the difference between the current time and the last event time. For example, the event set in Figure 11.8(b) would be represented as $(1,43), (11,42), \ldots$. In this case, the order of the events must be retained to regenerate the original time segment.

Exercise

a) Write an algorithm to quantize a given event set (such as the one originally obtained from the weather service) and represent the result in incremental form.

b) Write an algorithm to reconstruct the original event set from the quantized, incremental form.

Figure 11.9 illustrates some basic DEVS models. Figure 11.9(a) is a simple delay. When it receives a start message while in phase passive, it transitions to phase active. It remains there for a delay time, τ, before returning to passive and emitting an end message. Figure 11.9(b) is an event set generator. It assumes event pairs are in the form (t,v) where t is the incremental time and v is the value to be emitted. When receiving a start message, it reads successive pairs (e.g., from an XML schema instance) until none are left and then returns to passive, emitting an end message. After reading a pair (t,v) it holds for a time t before outputting the value v. In this way, the event set values are streamed out in their proper order and timing.

Operations on Event Sets and DEVS Model Realizations

To build up large data sets, and likewise, large simulation generators for them, we discuss some fundamental operations on event

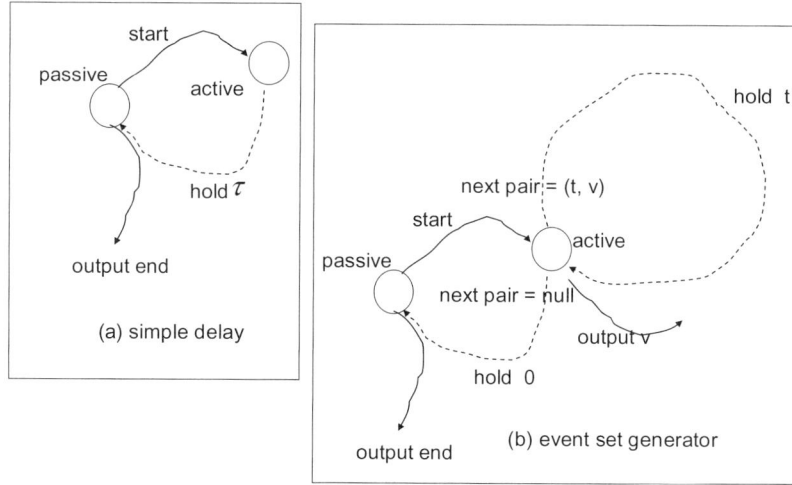

Figure 11.9

DEVS Models for Simple Delay and Event Set Generator

sets: *translation, concatenation,* and *cross product*. The translation operator, illustrated in Figure 11.10(a), moves a time segment to start at a different time without disturbing the relative sequence and timing of events within it. For example, a time series of average daily temperature in January might be translated to start in July while retaining the same pattern of temperature values. Figure 11.10(b) shows how, given an event set generator, we can build a generator for a translated version of its event set. In this case, the generator is a coupled model composed of a simple delay and the given event set generator with a coupling from the end port of the former to the start port of the latter. When the coupled model receives a start message, it passes it on to the delay which, after the prescribed duration has elapsed, triggers the start of the given generator. The values are produced at the port labeled R. The end messages are transmitted to the end port of the coupled model. In this way, the coupled model serves as a generator for the translated event set.

Concatenation of two time segments creates a new segment whose domain interval is the union of those of the original segments (Figure 11.11(a)). To be well-defined, the operation requires that the intervals be contiguous. Further, all values and timings of each of the components are retained in the concatenation. Given event set representations and their realizations by generators, we construct a realization of their concatenation using the coupled model of Figure 11.11(b).

Operations on Event Sets and DEVS Model Realizations 205

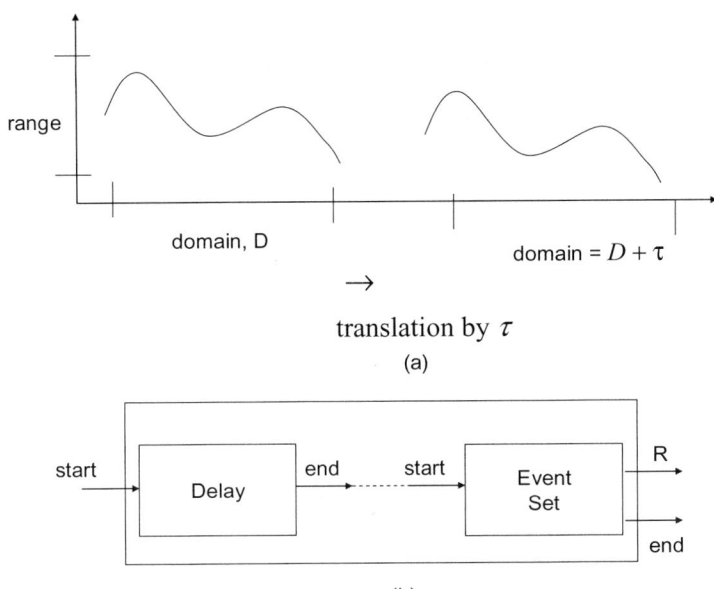

Figure 11.10

The Translation Operator and Its Realization by a Coupled Model

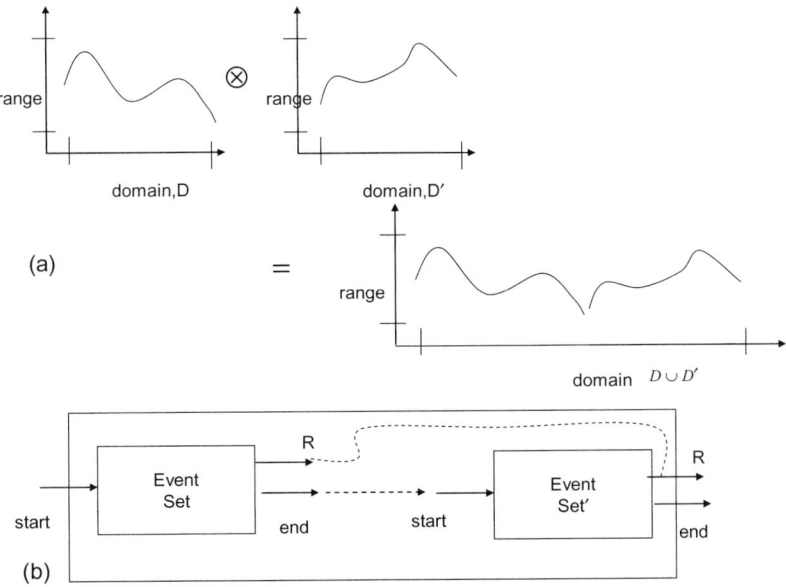

Figure 11.11

The Concatenation Operator and Its Realization by a Coupled Model

Exercise

Describe the operation of the coupled model in Figure 11.11(b). Show how it realizes the concatenation of the event sets.

Finally, the cross product operation is illustrated in Figure 11.12(a). It results in a time segment domain and is the same as both originals whose range is the cross product of the original ranges. To be well-defined, the domain intervals of the components must be equal. For example, given a time series of daily lows and one of daily highs for the same period, we can juxtapose the two by placing them on the same time axis such that for any day we have both the high and low for that day. Figure 11.12(b) shows how to construct a coupled model that generates the event set corresponding to the cross product of event sets. Such a composition can also be referred to as a parallel or concurrent composition since it amounts to starting the generators of the original components at the same time and having their values appear concurrently on distinct output ports.

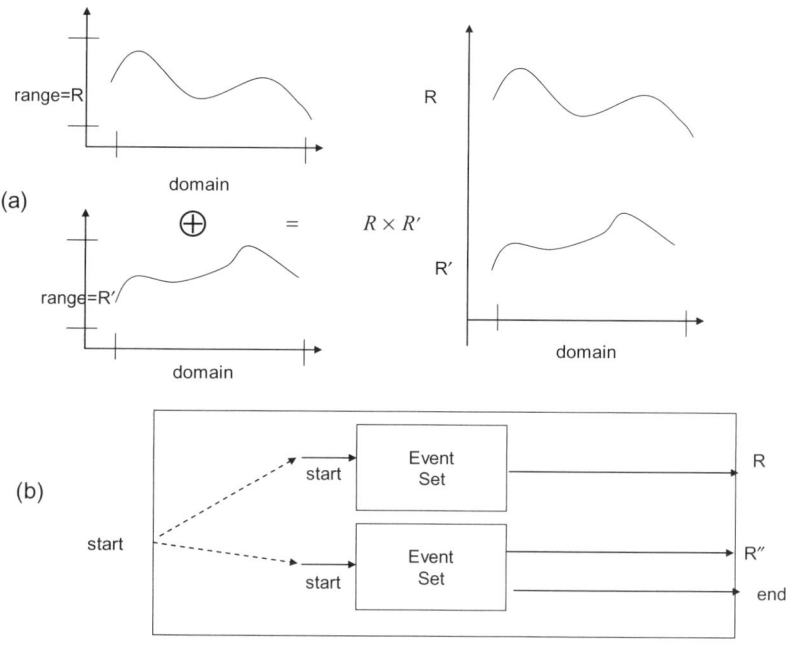

Figure 11.12

The Cross Product Operator and Its Realization by a Coupled Model

Exercise

Describe the operation of the coupled model in Figure 11.12(b). Show how it realizes the cross product of the given event sets.

The coupled model constructions just discussed illustrate the general concepts of coupled models introduced earlier in the chapter. For a detailed discussion of these concepts and the related simulation engine constructs, see Ref. 2.

Constructing Hierarchical Event Sets

Large hierarchical event sets and corresponding simulation generators can be specified and automatically synthesized using the ideas just discussed. The SES is used to specify a composite event set and its interpretation as a simulation model is used to construct the corresponding generator. The SES in Figure 11.13 includes a coupling specification at each aspect, labeled concat for concatenation or crossp for cross product. The couplings so specified are multi-argument generalizations of the binary operators discussed above. Thus, we concatenate event sets for December, January, and February for different variables (e.g., High,

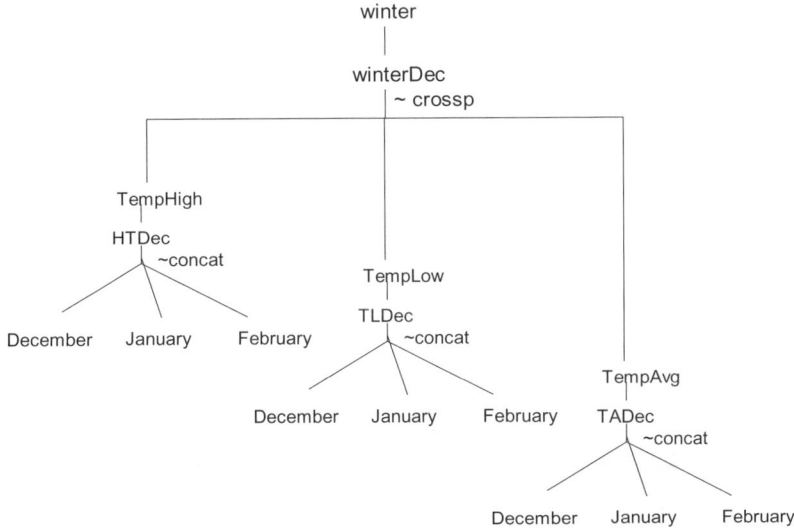

Figure 11.13

Synthesizing a Hierarchical Composite Event Set

Low, and Average temperatures), and then put them into a parallel composition to synthesize a generator for winter temperature profiles.

While we have used temperature as an example, there is no restriction in the approach to any particular variable or measurement. For example, weather variables such as precipitation and humidity can be concurrently generated with temperature. The precondition for application to any domain is that there is a large repository of raw time-indexed data expressed in a multiplicity of dimensions. The approach allows flexible combinations through concatenation and cross products oriented toward different "cuts" of the data. The web site www.devsworld.org describes a web service that uses the approach to organize a large collection of weather data. The service packages a subset of the data into a simulation weather generator and returns the resulting model in response to a consumer query.[9] An SES provides the framework for such queries as will be discussed in the next chapter.

Summary

The SES in its most expressive form constitutes an ontology for the domain of hierarchical, modular simulation models. An SES specifies a family of simulation models in which successive levels of coupled models are constructed in a stage-wise manner. Ultimately, these levels use the coupling specified at each level in the SES to connect components that reside in a model repository (see Ref. 3–5 for details). Pruning of the SES selects a particular member of the family of models to be synthesized. We showed how the SES can be used to support a web service that can respond to a variety of requests to synthesize weather simulation models from a large database. These concepts will set the foundation for understanding how the SES and its simulation models are employed for multilevel testing of net-centric systems (Chapter 17).

References

1. Single Link Interface Reference Specification for Link-16 (JSLIRS-16), 20003, http://www.stasys.co.uk/defence/datalinks/link_16.htm (accessed Nov. 2006)

2. Zeigler, B. P., T. G. Kim, and H. Praehofer, "Theory of Modeling and Simulation: Integrating Discrete Event and Continuous Complex Dynamic Systems," 2d ed. Boston: Academic Press, 2000
3. Kim, T. G., C. Lee, B. P. Zeigler, and E. R. Christensen, "System Entity Structuring and Model Base Management," *IEEE Trans. Sys. Man & Cyber.*, Vol. 20, No. 5, pp. 1013–1024, 1990
4. Luh, C., and B. P. Zeigler, "Model Base Management for Multifaceted Systems," *ACM Trans. on Modeling and Comp. Sim.*, Vol. 1, No. 3, pp. 195–218, 1991
5. Lee, Jong-Keun, Ye-Hwan Lim, and Sung-Do Chi, "Hierarchical Modeling and Simulation Environment for Intelligent Transportation Systems," SIMULATION, Vol. 80, No. 2, 61–76, 2004
6. Cowan, Nelson, *Working Memory Capacity*, New York: Psychology Press, 2005
7. C&A Software, chunking principle, http://www.chambers.com.au/glossary/chunk.htm, accessed Nov. 2006
8. Simon, Herbert, *The Sciences of the Artificial*, 3rd ed. Cambridge, MA: MIT Press, 1996
9. Cheon, S., "Experimental Frame Structuring for Automated Model Construction: Application to Simulated Weather Generation," Doct. Diss. Dept. of ECE, U. Arizona, Tucson, AZ, 2007.

12

Managing System Entity Structures: Composing Large Systems

We have seen that an SES specifies a family of hierarchical modular structures that are constructed in a stage-wise manner. Once constructed, such structures may be treated as components to be used in creating larger composites. In this chapter, we discuss how to merge SESs and their PESs to compose larger hierarchical structures from components. Constructing large complex structures in this way supports human preferences for hierarchy to help manage otherwise overwhelming complexity (Chapter 11).

Composition of SESs to create larger structures is supported by the merge operation illustrated in Figure 12.1.

Merging SESs

Consider the pair of SESs in Figure 12.1(a):

$SES^1 = \langle Entities^1, Aspects^1, Specializations^1,$
$rootEntity^1,$
$entityHasAspect^1, entityHasMultiAspect^1,$
$\quad entityHasSpecialization^1,$

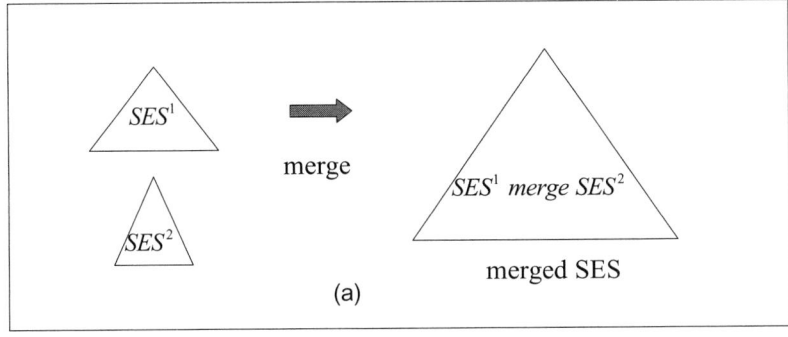
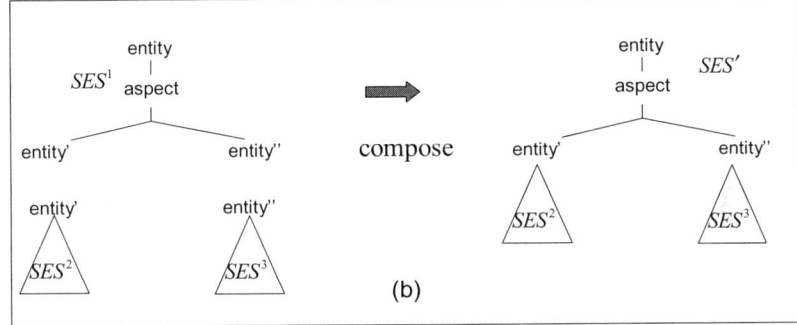

Figure 12.1

Merging and Composition of SESs

$$aspectHasEntity^1, multiAspectHasEntity^1,$$
$$specializationHasEntity^1\rangle$$

and

$SES^2 = \langle Entities^2, Aspects^2, Specializations^2,$
$rootEntity^2$
$entityHasAspect^2, entityHasMultiAspect^2,$
$entityHasSpecialization^2, aspectHasEntity^2,$
$multiAspectHasEntity^2, specializationHasEntity^2\rangle$

The merge of the pair is defined as:

SES^1 merge $SES^2 = \langle Entities^1 \cup Entities^2, Aspects^1 \cup Aspects^2,$
$Specializations^1 \cup Specializations^2,$
$rootEntity^1,$
$entityHasAspect^1 \cup entityHasAspect^1,$
$entityHasMultiAspect^1 \cup entityHasMultiAspect^2,$

$entityHasSpecialization^1 \cup entityHasSpecialization^2,$
$aspectHasEntity^1 \cup aspectHasEntity^2,$
$multiAspectHasEntity^1 \cup multiAspectHasEntity^2,$
$specializationHasEntity^1 \cup specializationHasEntity^2\rangle$

Note that the merge is basically a union of all the various sets and functions of the pair of SESs. However, the new root is that of the first SES.

Chapter 16 discusses an example of merging in the context of enhancing SESs with metadata slots.

Composition

An important special case of the merge operation is shown in Figure 12.1(b). Here one SES contains an aspect whose entities can be composed using this aspect, i.e., for each entity there is an SES whose root has the name of the entity. The result of merging is called the *composition* of the SESs using the aspect of the first SES. Since the composition can be shown to be associative, we can perform the overall operation in steps.

EXAMPLE: Consider SESs for book and for frontCover as illustrated in Figure 12.2. Note that frontCover is common to both sets of entities and that the root entity of the composed structure will be book. When merged the new structure is a valid SES and the

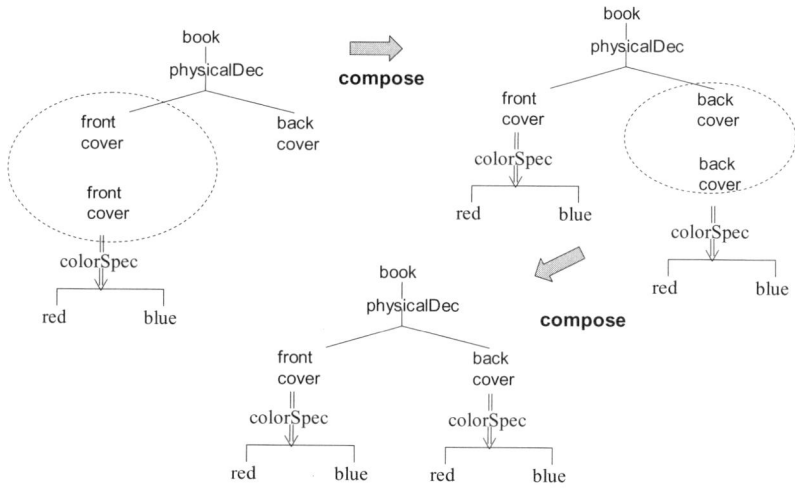

Figure 12.2

Illustrating the Merge Operation in a Book SES

Managing System Entity Structures: Composing Large Systems 213

common item, frontCover, serves to connect the two original components. So the end result is like pasting the SES for front-Cover into the SES for book. Continuing, we compose the result of the first composition with the SES for backCover to obtain the final result.

An as operation on SESs in relational form we have the pair of SESs to be merged:

$SesForBook = \langle$
$Entities = \{book\}$
$Aspects = \{physicalDec\}$
$Specializations = \{\}$
$rootEntity = book$
$entityHasAspect = \{(book, physicalDec)\}$
$entityHasMultiAspect = \{\}$
$entityHasSpecialization = \{\}$
$aspectHasEntity = \{(physicalDec, frontCover), (physicalDec, backCover)\}$
$multiAspectHasEntity = \{\}$
$specializationHasEntity = \{\}$
$\rangle$

$SesForFrontCover = \langle$
$Entities = \{frontCover, red, blue\}$
$Aspects = \{\}$
$Specializations = \{colorSpec\}$
$rootEntity = frontCover$
$entityHasAspect = \{\}$
$entityHasSpecialization = \{(frontCover, colorSpec)\}$
$aspectHasEntity = \{\}$
$entityHasMultiAspect = \{\}$
$specializationHasEntity = \{(colorSpec, red), (colorSpec, blue)\}$
$aspectHasEntity = \{(physicalDec, frontCover), (physicalDec, backCover)\}$
$multiAspectHasEntity = \{\}$
$\rangle$

Exercise

Merge the SES for frontCover into that for book using the union operations as required. Then create an SES for backCover similar to that for frontCover and merge it into the first composition. Use the natural language tool (www.devsworld.org) to specify the individual SESs and perform the merge.

Notes:

- Merging is not commutative — the order of the component SESs counts because the root entity of the resultant is that of the first SES.
- Merging is associative — merging three SESs in a particular order does not affect which pair is worked on first.

The merge is not necessarily a valid SES because it must satisfy the constraints that are imposed on SESs. First, the root entity of the merge must lie with the merged entities set. Although the items in the basic sets (entities, aspects, specializations, etc.) are all mutually disjoint in the components, they may not be disjoint in the composition. So we need to have this condition checked as a task of a validation tool. Further, the strict hierarchy axiom may not hold for the composition although it holds for the components.

Exercise

By providing a counterexample, verify that commutativity does not hold.

Prove that associativity does hold. *Hint*: Consider the associativity properties of the union operation.

Composing Hierarchical Systems and Models

As illustrated in Figure 12.3, we can exploit the hierarchical structure of a system, whether derived or built in, to model it by an SES, itself composed of smaller SES units. Sub-SESs can be merged, or plugged into, larger ones to create successively larger composites. Similarly, corresponding pruned entity structures can be composed in a hierarchical manner respecting that of the governing SESs.

EXAMPLE: *Weather Generator Simulation Model*

Chapter 11 introduced a web service for creating weather generating simulation models in response to consumer queries. The queries are created as pruned entity structures of a large SES that is created by merging smaller SESs as illustrated in Figure 12.4. The top entities in overall SES are spaces, time periods, and variables to enable the user to request weather generators for a

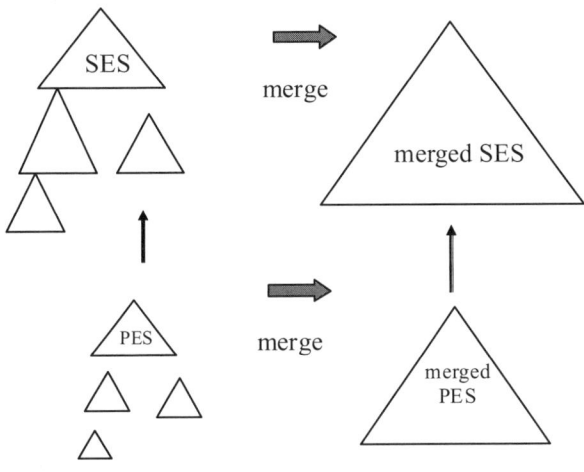

Figure 12.3
Composing SESs and Their PESs

particular space on the Earth, during a given period, for particular variables such as temperature or humidity. The user prunes the SES to create the query of interest. Individual SESs are maintained separately and are composed in the phased hierarchical manner just described to create the overall SES. To synthesize a weather generator, each component SES is pruned, and the individual PESs are likewise composed to create the overall PES. The latter is transformed into a simulation model that responds to the query. The individual SESs use the constructs of aspects, specializations, multiAspects, and variables to provide a wide range of possible ways of specifying combinations of spaces, time periods, and weather variables. See the web site www.devsworld.org for details.

Exercise

Using the discussion of hierarchical structure in Chapter 11 as background construct an SES with the following definitions:

From the structure perspective, a transaction is made of more than one rule
A rule has an ID
The range of rules's ID is string
From the structure perspective, a rule is made of a condition and an action

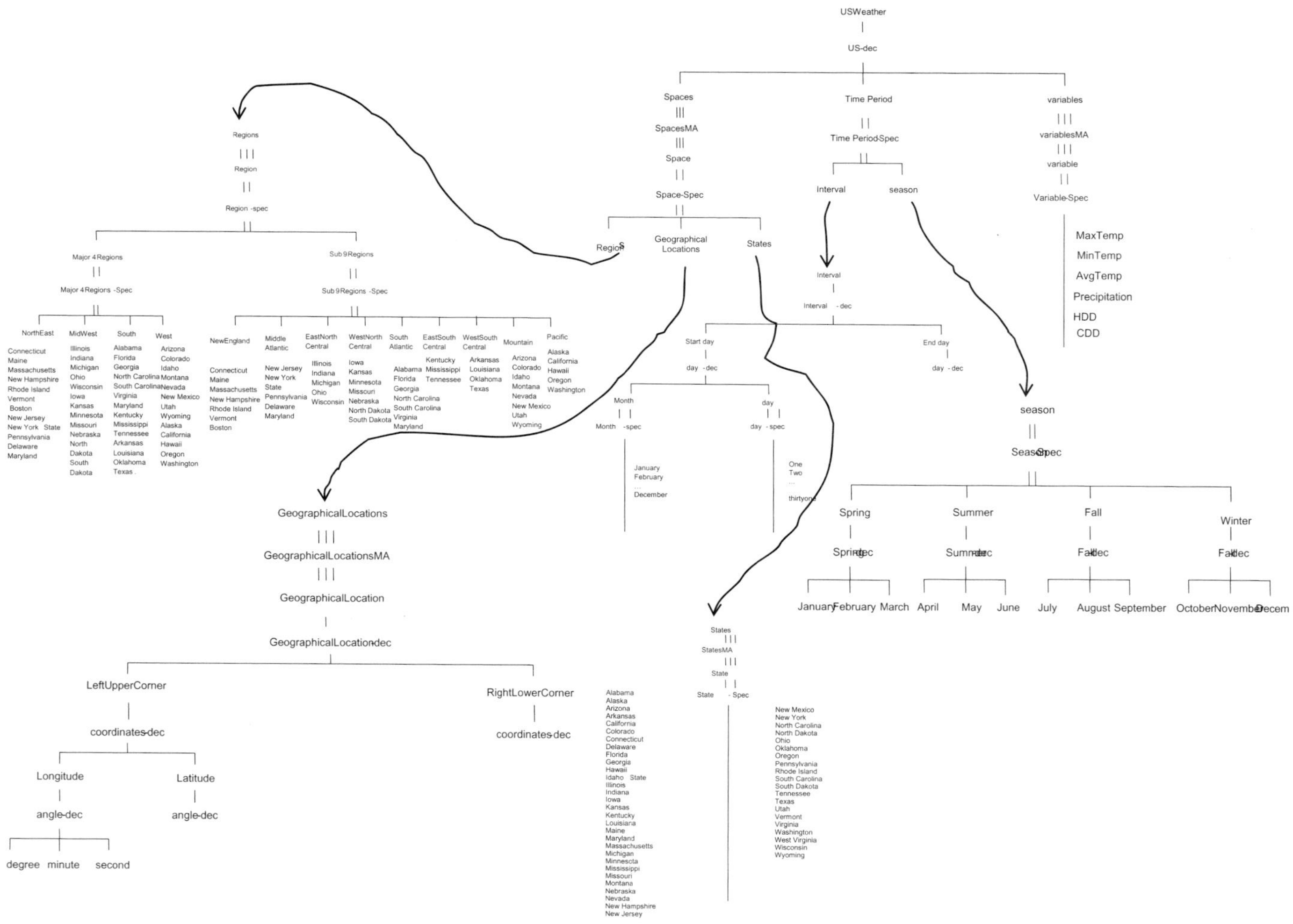

USWeather
US-dec
Spaces
SpacesMA
Space
Space-Spec
Regions
Geographical Locations
States
Time Period
Time PeriodSpec
Interval
season
variables
variablesMA
variable
Variable-Spec
MaxTemp
MinTemp
AvgTemp
Precipitation
HDD
CDD
Regions
Region
Region -spec
Major 4 Regions
Major 4 Regions -Spec
NorthEast
Connecticut Maine Massachusetts New Hampshire Rhode Island Vermont Boston New Jersey New York State Pennsylvania Delaware Maryland
MidWest
Illinois Indiana Michigan Ohio Wisconsin Iowa Kansas Minnesota Missouri Nebraska North Dakota South Dakota
South
Alabama Florida Georgia North Carolina South Carolina Virginia Maryland Kentucky Mississippi Tennessee Arkansas Louisiana Oklahoma Texas
West
Arizona Colorado Idaho Montana Nevada New Mexico Utah Wyoming Alaska California Hawaii Oregon Washington
Sub 9 Regions
Sub 9 Regions -Spec
NewEngland
Connecticut Maine Massachusetts New Hampshire Rhode Island Vermont Boston
Middle Atlantic
New Jersey New York State Pennsylvania Delaware Maryland
EastNorth Central
Illinois Indiana Michigan Ohio Wisconsin
WestNorth Central
Iowa Kansas Minnesota Missouri Nebraska North Dakota South Dakota
South Atlantic
Alabama Florida Georgia North Carolina South Carolina Virginia Maryland
EastSouth Central
Kentucky Mississippi Tennessee
WestSouth Central
Arkansas Louisiana Oklahoma Texas
Mountain
Arizona Colorado Idaho Montana Nevada New Mexico Utah Wyoming
Pacific
Alaska California Hawaii Oregon Washington
Interval
Interval -dec
Start day
day -dec
End day
day -dec
Month
Month -spec
January February ... December
day
day -spec
One Two ... thirtyone
season
SeasonSpec
Spring
Springdec
Summer
Summerdec
Fall
Falldec
Winter
Falldec
January February March April May June July August September October November December
GeographicalLocations
GeographicalLocationsMA
GeographicalLocation
GeographicalLocationdec
LeftUpperCorner
coordinates-dec
Longitude
angle-dec
degree minute second
Latitude
angle-dec
RightLowerCorner
coordinates-dec
States
StatesMA
State
State -Spec
Alabama Alaska Arizona Arkansas California Colorado Connecticut Delaware Florida Georgia Hawaii Idaho State Illinois Indiana Iowa Kansas Kentucky Louisiana Maine Maryland Massachusetts Michigan Minnesota Mississippi Missouri Montana Nebraska Nevada New Hampshire New Jersey
New Mexico New York North Carolina North Dakota Ohio Oklahoma Oregon Pennsylvania Rhode Island South Carolina South Dakota Tennessee Texas Utah Vermont Virginia Washington West Virginia Wisconsin Wyoming

Figure 12.4

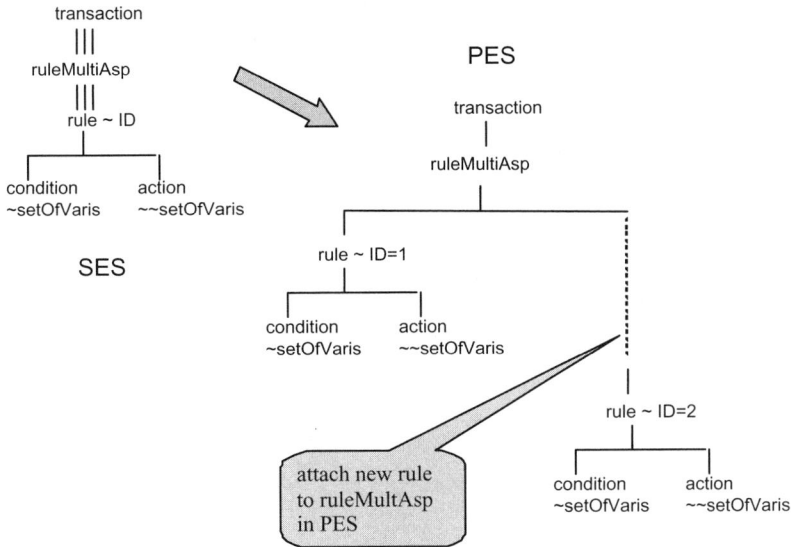

Figure 12.5

Attaching New Rules as They Are Constructed

A condition has a setOfVariables
The range of condition's setOfVariables is stringSet
An action has a setOfVariables
The range of action's setOfVariables is stringSet

Working at the same web site (www.devsworld.org), transform the SES into a PES as illustrated in Figure 12.5. As new rules are constructed, paste them into the appropriate slot of the PES.

Targeted Access to Substructures

We can extract substructures that satisfy given name patterns. For example,

SESOps.extractSesFromAllEntitiesWPrefix(string,folder)

will extract the sub-SESs from all entities whose names have the prefix given by string and place them in the folder.

Decomposition

To decompose an SES into its top level components, we use the method:

SESOps.extractSesFromFirstEntities(folder)

which will write the first level component SESs into the specified folder.

To generate the natural language description of top level components, we use the method:

SESOps.printSesTextsOfFirstEntities(folder);

which will write the natural language descriptions of the first level component SESs into the specified folder.

To compose the SESs, in arrayOfSESs, that are entities of an SES called topSES, we use the method:

mergeOps.mergeMultipleSes(topSES, arrayOfSESs,folder);

Summary

The ability to compose and decompose hierarchical modular structures is key to managing the complexity of data representations. We will see an application to a large standard for sensor metadata in Chapter 14.

13

Harmonizing Data Representations and Ontologies within Pragmatic Frames

> Data Engineering is concerned with the role of data in the design, development, management and utilization of information systems...
>
> Technical Committee on Data Engineering of the
> IEEE Computer Society[1]

Recall that in the information exchange framework (Chapter 1) we are concerned with data carried in messages from producer to consumer. For effective exchange, producer and consumer must agree upon the format of the data, how it is to be interpreted, and the uses to which it will be put. An ontology provides the framework for describing world states carried in the messages, so it follows that producer and consumer must agree on the ontology to be used as a necessary precondition for effective exchange (the condition is not sufficient in view of the pragmatic frame requirements to be considered below). Unfortunately, in the real world, agreement on the ontology to use is not easy to achieve. As we have seen in Chapter 3, data formats tend to be developed in a "stove piped" manner with little prior thought to interoperability. So a challenge emerges — finding a common ground for, or otherwise reconciling, the different data formats preferred by different producers and consumers. This aspect of data engineering is not explicit in the IEEE definition. When cast in an ontological framework, the problem of data format reconciliation becomes upgraded to ontology integration and the overall field in which solutions are sought may be called ontology engineering.[8-12]

It is helpful to distinguish two kinds of problems that are addressed in ontology integration. As suggested by Figure 13.1(a), ontologies may be geared toward the same real world aspect and yet provide different ways of describing its states. We'll refer to the problem of reconciling such ontologies as *harmonization*. Alternatively, as in Figure 13.1(b), ontologies may be geared toward describing different aspects of reality. The second case is of interest to Semantic Web researchers with the goal of developing intelligent agents that can synthesize new knowledge from ontologies for a profusion of subject matter accessible over the web. We'll refer to this problem as one of *merging* of ontologies. Of course, where the aspects addressed by ontologies overlap, merging must also involve harmonization. Note that we can

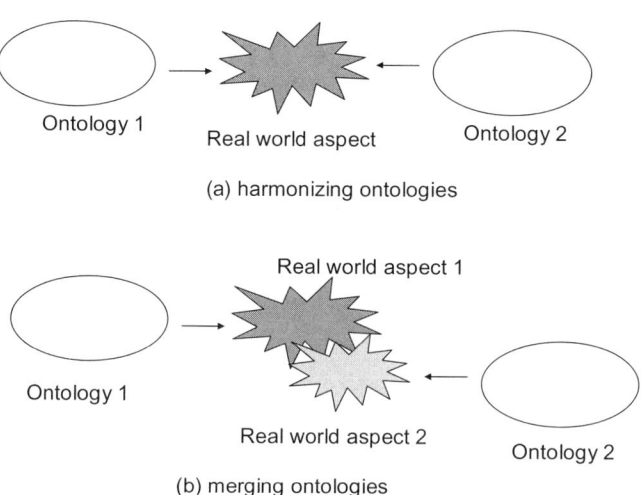

Figure 13.1

Integration of Ontologies: Harmonizing and Merging

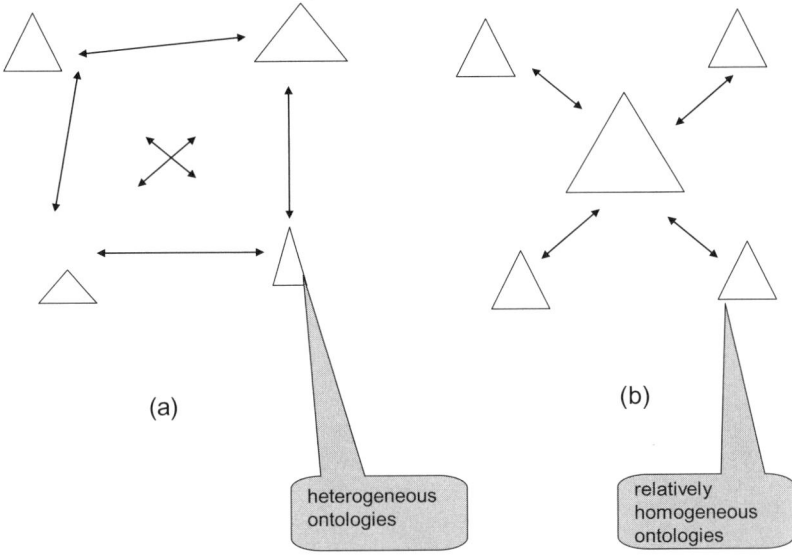

Figure 13.2

Two Approaches to Harmonization: (a) many-to-many alignment, (b) many-to-one alignment

contrast data engineering with ontology engineering from this perspective. Data engineering is concerned only with schema integration at the syntactic level, whereas ontology integration tries to reconcile the inherent meanings of concepts (the semantics) as well as the formats (the syntax).

In this chapter we are concerned with harmonization of multiple ontologies that address the same, or nearly the same, aspect of reality. The two general approaches to harmonization are illustrated in Figure 13.2. In the many-to-many approach, the various ontologies are individually harmonized with each other. In the many-to-one approach (also called "hub and spokes"), each individual ontology is harmonized with a common ontology.[2-5] Consequently in the latter approach, interoperation among ontologies is mediated by the common ontology.

In general the many-to-one approach is preferred because of its elegance — a common core evidences a deeper understanding than a series of ad hoc translations and scalability. To integrate a new ontology requires that it only be harmonized with one core, rather than all peer, ontologies. However, its two prerequisites may not be easily satisfied. These requirements are 1) a common reference ontology is available and 2) the legacy ontologies to be harmonized are sufficiently homogeneous that harmonization with a common core is feasible. Chapter 18 discusses an example in which the prerequisites appear to be satisfied in the case of a common information exchange model for command and control systems. If a common reference model is not available, then it must be constructed and the homogeneity of the legacy models, as well as the available subject matter expertise, will determine how straightforward this task is. Chapter 14 discusses an example in which a common reference was not available, but the domain of application was sufficiently well understood and circumscribed (synthetic aperture sensor systems) that it could be successfully developed. On the other hand, constructing a common ontology for all the communities of interest (COI) that might interact on a global information grid might well be infeasible.[13] This realization might underlie the data strategy policy of the Department of Defense which requires bottom-up development of COI-specific ontologies and their integration using many-to-many mediation once formed.[14]

Harmonization: Pragmatic Equivalence

The approach advocated in this book takes as a given that ontology and subject matter experts participate in an interactive knowledge-engineering process[6,7,11,15] (for example, see Chapter 14). In this approach, methods for harmonizing ontologies suggest to humans what manipulations might be appropriate and rely on human intuition to judge the validity of the processes and their results. However, we go beyond current methodology in viewing harmonization as a process by which ontologies are brought into an equivalence to be called pragmatic equivalence. The information exchange framework of Chapter 2 helps to throw some light on the right concept of equivalence to use. Recall that a pragmatic frame expresses how the world states generated by the ontology — expressed in a transmitted message from producer to consumer — will be used in downstream processing. This suggests that, as in Figure 13.3, the ultimate test of equivalence should be determined by whether the world states generated by the ontologies ultimately lead to the same downstream processing results. It also suggests that pragmatic equivalence is dependent on the frame of interest. Ontologies may be equivalent within one frame but not within another. Thus we have the definition:

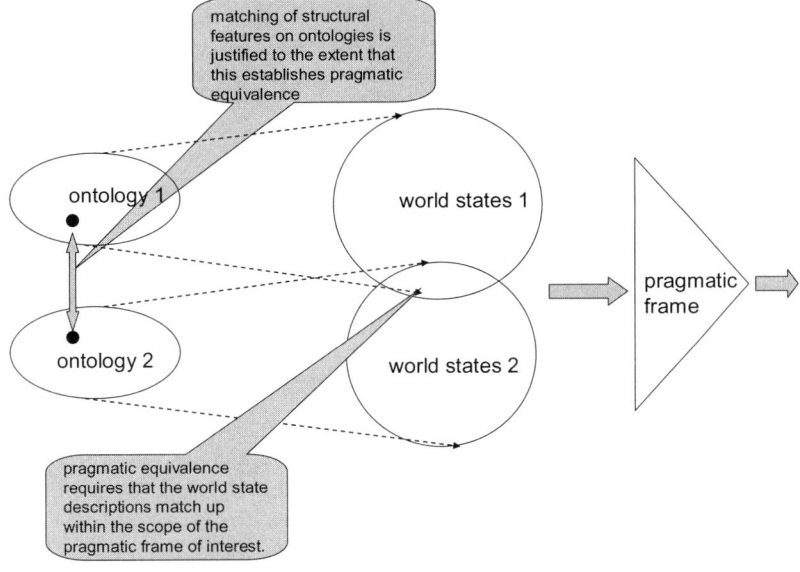

Figure 13.3

Pragmatic Equivalence of Ontologies as Viewed within the Information Exchange Framework

---

PRAGMATIC EQUIVALENCE
Two ontologies are pragmatically equivalent *in a pragmatic frame* if there is a one-to-one correspondence of their world state descriptions such that corresponding descriptions are used in the same manner within the pragmatic frame of interest.

In this conception of pragmatic equivalence, even though world state descriptions generated by the ontologies may differ, the manner in which they are processed downstream leads to the same results. For example, messages sent within one ontology might not differ from those of a second except in numerical precision. Consider corresponding number strings that are the same only up to a given number of significant digits. Pragmatic equivalence holds if both strings are treated equally by downstream processors. We say that this difference is absorbed within, or modulo, the pragmatic frame. Of course, another frame may treat these strings differently, leading to pragmatic unequivalence in this frame.

In a much stronger relationship, it might be that the sets of world state descriptions generated by the two ontologies are identical. In this case, we say that the ontologies are semantically equivalent. In other words, we define:

SEMANTIC EQUIVALENCE
Two ontologies are semantically equivalent if their sets of generated world state descriptions are equal.

It follows immediately that semantic equivalence implies pragmatic equivalence in the same frame. This is true because underlying the equality of sets is a one-to-one correspondence, viz., the identity mapping. Since corresponding state descriptions are in fact the same, their processing results are the same. Therefore, the ontologies are equivalent in any pragmatic frame.

Exercise

Consider the case where a commander's orders to his troops are intercepted by an adversary. Suppose also that both troops and adversary perfectly understand the orders. Of course, these parties will take different actions based on their understandings. Formally, there are two ontologies, one used by the troops and one used by the adversary. Argue that these ontologies are semantically equivalent but the pragmatic frames in which they are used are different.

A weaker requirement than set equality is set inclusion. This leads to the following definition:

SEMANTIC INCLUSION
An ontology is semantically included in a second ontology if the latter's set of generated world states is included within that of the former.

Consider the example of Figure 1.2, where the scope of a generic ontology includes both registration and inventory update frames, while that of a specialized one includes just registration. Indeed, suppose that the second generates a set of world states (registration notices) that is included in the first (registration notices together with inventory updates). In this case, the second ontology is semantically included in the first.

In practice, testing for semantic equality proceeds by testing for semantic inclusion in both directions. That is, we test whether ontology1 is semantically included in ontology2 and also whether ontology2 is semantically included in ontology1.

The concept of semantic equivalence underlies methods that seek to match and align features of ontologies[8-12] without regard to pragmatic considerations. In contrast,[6] Pohl stresses that ontology features that originate from perspectives other than the one of current interest should be filtered out in the harmonization process. By bringing the pragmatic frame into consideration we have formalized the concept of perspective and have arrived at the enhanced concept of pragmatic equivalence. In this approach, semantic matching and alignment should be informed by the pragmatic frames under which the source ontologies were developed and by the current frame of interest.

Harmonization of Data Formats

In the following discussion, we show how to introduce the role of pragmatic frame more centrally into matching and alignment procedures. As discussed in Chapter 3, many legacy formats for sending sensor data only specify flat files with lists of attributes. These formats tacitly reflect an ontology and pragmatic frame but these semantic and pragmatic features must be teased out as the process of harmonization proceeds. Let's examine a matching process for attributes in this context. This involves analyzing attributes in a pair of legacy formats with respect to the following dimensions:

1. *Names*: If the names are identical this is a promising indication that these attributes may match in all respects of interest to the pragmatic frame.
2. *Descriptions*: The metadata textual descriptions of the attributes might give clues as to whether the authors were trying to capture the same concept with the attribute.
3. *Indices*: The number of indices gives the number of dimensions in the space spanned by this attribute (e.g., the attribute might be energy-intensive with two indices indicating it is distributed in a two-dimensional space). It should have the same dimensions in both formats. If the attributes represent spatial coordinates then the reference frame should be the same.

4. *Range Sets*: To match up, the attribute should have the same range sets in both formats, including measurement units and precision.
5. *Usage*: This is the pragmatic frame test — the attributes should be used in the same way to support downstream processing. Examining the formats is not sufficient to make this judgment; it requires understanding the downstream process chain or getting a subject matter expert to make this determination.

Suppose that there is a one-to-one correspondence of attributes in the two formats and corresponding attributes match up exactly in each of the above dimensions. Then there is a good expectation that they are pragmatically equivalent. That is, the formats contain equal numbers of attributes that carry the same information and that play the same roles in the frame of interest. Although we offer no proof under these conditions, the formats are likely to generate the same world state descriptions as used within the scope of the pragmatic frame of interest.

The harmonization problem typically arises when such a one-to-one correspondence does not hold. Assume that two ontologies are targeted at the same domain — otherwise, there is little expectation for equivalence, whether pragmatic or semantic. Nevertheless, the ontologies may have differing scopes (i.e., they represent different subsets of elements in the domain), and within the intersection of scopes, represent elements somewhat differently. A major source of divergence is that the ontologies may have been oriented toward different pragmatic frames.

To proceed in this case, instead of requiring exact match-ups in the above five dimensions, we define a metric that measures how closely two attributes match up. The importance of employing appropriate metrics is discussed by Rosenthal et al.[13] The metric is a fuzzy function that yields 1 when exact matching occurs, 0 when there is no match, and values in the unit interval that reflect the extent of matching otherwise. We then define thresholds that help make decisions on treatment of the attributes. For example, if the metric exceeds 0.9, then place the attributes into correspondence and annotate the differences with additional metadata; if the metric lies below 0.5, then treat the attributes as distinct; otherwise, what to do depends on the approach to harmonization we are taking — many-to-one, or many-to-many (Figure 13.2). In the many-to-one case, we seek a common format that both formats can be mapped to. In the many-to-many case, we seek a mapping that translates between the two formats. Many of the same issues will come up, so let's consider the common format approach.

Table 13.1. Aligning Attributes That Differ Only in One Respect

Variance	Alignment in a common format
Attributes differ only in names	Create a representative in the common format and map each attribute to it. Define the name of the representative to be "representative" of each. (Unfortunately, this may not be an easy problem, since names may carry numerous associations for different community users of the legacy formats.)
Attributes differ only in descriptions	Create a representative in the common format and map each attribute to it. Write the description of the representative to imply both descriptions. This is possible since by assumption all the other dimensions were found equal, and therefore, it is likely that the attributes do play the same roles in the same way in the pragmatic frame of interest.
Attributes differ only in indices	Create a representative in the common format and map each attribute to it. Define the number of indices of the representative as equal to the maximum of the given numbers. Require that the unused indices be set to zero. If different coordinate systems are used, then choose a standard (that might be one of them) and require a transformation of original coordinate systems to the standard.
Attributes differ only in the range sets	Create a representative in the common format and map each attribute to it. Define the range set of the representative so that it includes both range sets.
Attributes differ only in usage	Retain only the attribute (if any) that is matched to the current pragmatic frame of interest.

Starting with a relatively simple situation in which attributes are assumed to differ in, at most, one of the dimensions mentioned above, we outlined the procedure in Table 13.1.

Need for More Structured Representation

We see that even in this simplified case, complexities arise that suggest that the resulting common artifact cannot be expressed as a list of attributes but must have characteristics requiring a more general ontology representation. For example, if range sets are disjoint, it would be necessary to include additional mechanisms for choosing which range set to use and for providing guidance as to how to make the choice. Moreover, consider going to the more general case where attributes might differ in more than one dimension. Here, a single attribute in one format might correspond to a group of attributes in the second, and further that the range set of the single attribute is at a lower level of resolution than the group, i.e., it can express only an aggregated representation of the information in the group. As suggested this will require procedures to enhance and/or refine the representations.[2,3] So again we see that this will require the use of structures, as offered by ontologies, to express the resulting representations.

Current Tools for Ontology Integration

Eklöf and Mårtenson[12] provides a review of ontology integration methods and tools in the context of the use of a common information exchange model to be discussed in Chapter 18. Two tools that are relevant to harmonization are based on the frame-based languages discussed in Chapter 2. Based on Ontolingua, Chimaera[18] supports merging multiple ontologies and diagnosing individual or multiple ontologies. It supports users in such tasks as loading knowledge bases in differing formats, reorganizing taxonomies, resolving name conflicts, browsing ontologies, and editing terms. PROMPT,[16] a plug-in for the Protégé environment, is an interactive ontology merging tool. It guides the user through the merging process, making suggestions, determining conflicts, and proposing conflict-resolution strategies. The initial suggestions are based on the linguistic similarity of the frame names. Iterating with the user, PROMPT determines the conflicts in the merged ontology that a user selected operation has caused and proposes possible solutions to the conflict. These and several other projects reviewed by Eklöf and Mårtenson[12] are in the research stage and tool support is still fairly weak compared to the labor that humans must put in.

Harmonization Supported by the SES

We have seen that in general, harmonization is a difficult problem currently requiring intense manual labor and in need of much research and tool development. In this light, let's consider some of the ways that the SES can be of help. First, because of its well-defined structure, the SES and the information exchange framework can allow more rigorous modeling and foster deeper understanding of the issues. Thereby, it can facilitate developing more advanced concepts and tools to address the problems that are identified. Of course, given the limitations of the SES as an ontology framework, one should expect such solutions to be limited to domains where the SES provides adequate representations.

Because SESs are finite recursively defined structures, they have useful decidable properties, that is, they allow development of algorithms that can do analysis and comparisons yielding definite results. For example, we can extract all the entities in an SES and compare its set of entities with that of another SES. Such comparison may involve set operations such as inclusion, complement, intersection, difference, and so on. The general problem of ontology comparisons requires advanced, computationally intensive, graph-theory isomorphism-checking techniques.[11]

EXAMPLE: *Profile Conformance to a Standard*

An example, illustrated in Figure 13.4, is the test for inclusion of mandatory standards classes in a profile. The ISO geospatial standards committees (Chapter 3) require that all application profiles include certain mandatory classes of the standard. Expressed in SES terms, this requires a tool that extracts the mandatory entities from the standard's SES and checks whether this set of entities is included in the set of entities extracted from the profile's SES. However, more than this, ISO requires that if an optional class is employed in the profile that itself has a mandatory class "below" it, then the latter must also be included in the profile. Let's call these classes the *inferred* mandatory classes.

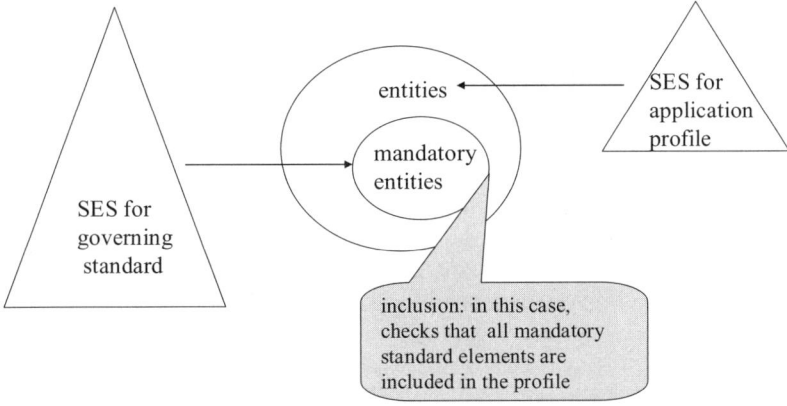

Figure 13.4

Checking for Inclusion of Mandatory Entities in a Profile

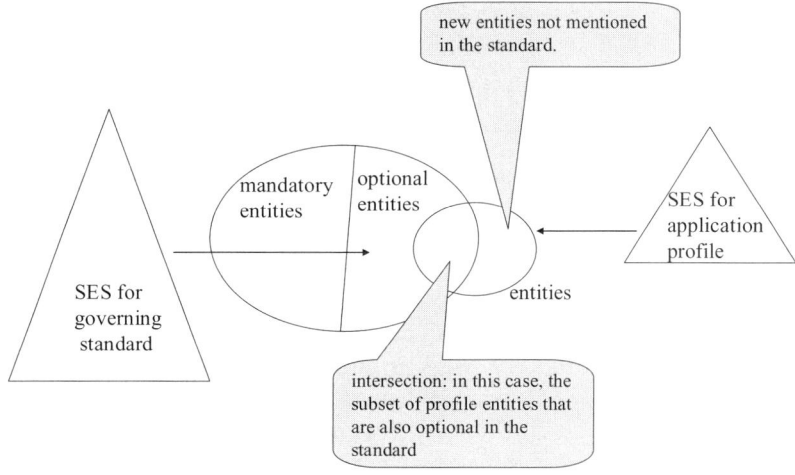

Figure 13.5

Intersection of Profile Entities and Optional Standard Entities

In SES terms, the relationship "below" is easily seen to be expressed as "in the substructure of."

To do such checking, we must be able to get the intersection of two sets of entities as illustrated in Figure 13.5. The first of these sets is the set of optional entities in the standard — which is the complement of the mandatory entity set. The second constitutes the profile's entities. The intersection of the two represents the entities in the profile that are optional in the standard. Note that the profile development rules allow adding new classes to a profile that are not within the standard (so these may well be non-standard profile entities outside the intersection).

As illustrated in Figure 13.6, we now need the substructures of each of the common optional entities — this is the SES hanging below it — in both the standard and the profile. The inferred mandatory entities can be now identified as the mandatory entities in the standard's sub-SES. These must be contained within the sub-SES in the profile's SES. Therefore, we develop a method that determines whether the profile conforms to the standard in relation to the inclusion of inferred mandatory entities. This method returns true if, for each of the common optional entities, the inferred mandatory entity set is included in the profile's sub-SES of the common entity.

Measuring Commonality

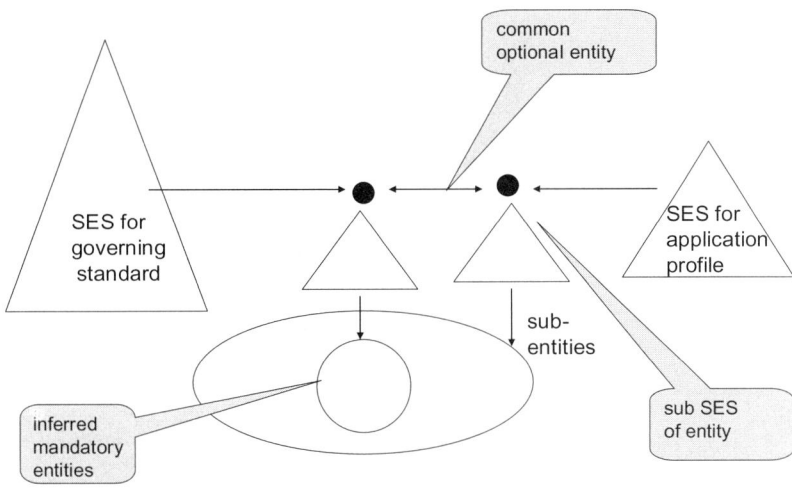

Figure 13.6

Comparing Substructures of SESs

Measuring Commonality

We have seen that measuring how well ontologies match up is an integral part of working toward their harmonization. In the framework of the SES, there are straightforward measures that can be defined. The intersection of the sets of entity names of two SESs is a measure of the degree of commonality between them. However, this is only a surface measure, since it does not require that the entities having the same names have the same substructure. For this comparison, we need to extract the sub-SESs of the common entities in each SES and compare them for equality. This is similar to the equality comparison illustrated in Figure 10.5.

Exercise

Write an algorithm for testing whether two SESs are equal. *Hint:* Use the hierarchical recursive pattern in Chapter 6 to check whether the root entities have the same names, variables, aspects, multiAspects, and specializations. For each of the latter, apply the same recursive check for equality of their child entities.

Chapter 13 Harmonizing Data Representations and Ontologies

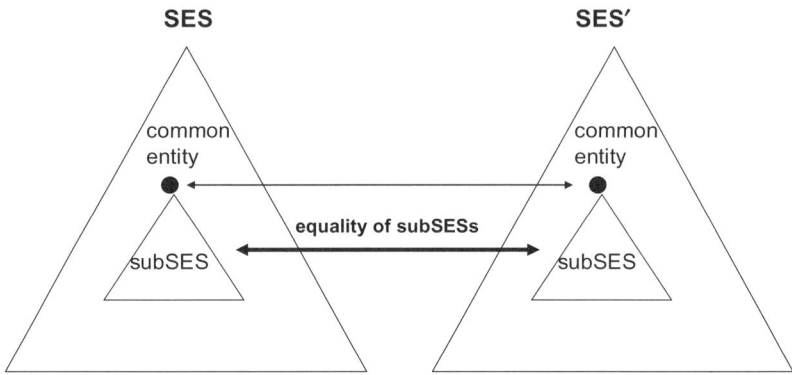

Figure 13.7

Commonality Measure Based on Shared Substructures

Having this algorithm, we define commonality as the fraction of entities with shared substructures relative to the total number of entities in both SESs, where the latter is measured as the cardinality of the union of the entity names of the two SESs (Figure 13.7). This fraction will equal 1 when the SESs have exactly the same entity names (so their union is just one of the sets) and when all entities with the same names have equal sub-SESs.

Harmonization: Increasing Commonality Via Restructuring

With the measure of commonality in hand, we may view harmonization as a process by which an SES (the slave) is restructured in order to increase its commonality with a master SES (such as an existing common reference model in the many-to-one approach). In the process, a variety of renamings and more radical restructurings may be applied individually and in combinations. We now discuss some of the possibilities.

Pragmatic Frame-based Filtering

An obvious place to start is to filter out entities and items in general that do not relate to the pragmatic frame of interest.[6] A

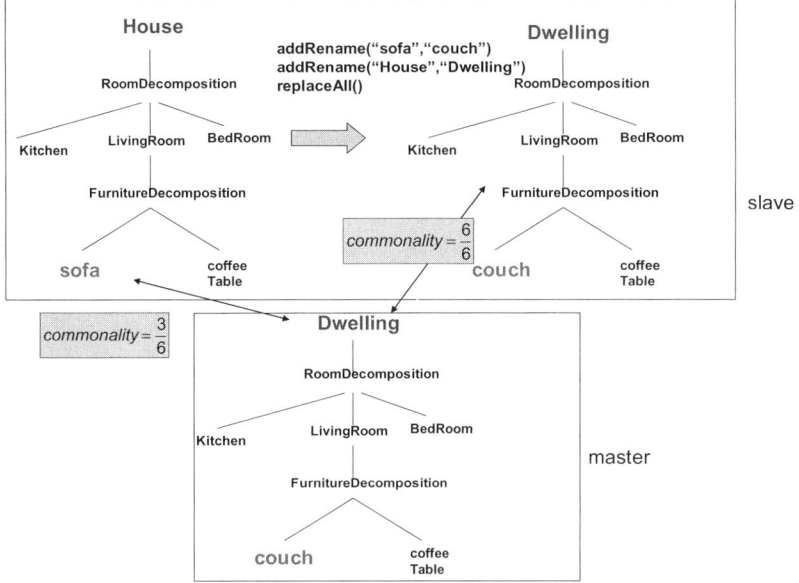

Figure 13.8

Renaming to Increase Commonality

method to remove an entity and its sub-SES from an SES can be used in this process.

Renaming

A second step is to look for entities that appear to have the same substructure in the two SESs but have different names. Commonality might then be increased by renaming entities in the slave SES to match those with which they appear to correspond in the master SES. Figure 13.8 illustrates a renaming method that makes substitutions according to pairs in a specified list. The example shows a case where simultaneous renaming of sofa and House in the slave achieves full harmonization with the master.

Exercise

Compute the commonality increase for each of the individual renamings. Why does the individual renaming of House to Dwelling not increase commonality?

Substructuring

Renaming works well when master and slave structures are isomorphic — they have the same structure except for labels. Typically this will not be the case. One reason may be that some substructures match up not in the sense of strong equivalence but in the weaker inclusion equivalence discussed earlier. In the SES, we can test whether a sub-SES of the slave is included in a sub-SES of the master under an entity with the same name in each. If such a test reveals such a sub-SES, we can check with subject matter experts whether the larger sub-SES can replace the smaller one in the slave.

Restructuring

If neither strong nor weak structural equivalence can be established, more radical restructuring will have to be brought to bear. Any of the restructurings discussed in Chapter 7 might be applicable. For example, it might be that two developers conceived of the same basic structure for their SESs but one went on to apply a restructuring that resulted in non-isomorphic structures. In this case, the problem becomes one of finding the inverse restructuring to restore the original. In Table 13.2, we formulate how the traces of a restructuring might be identified in a pair of SESs. The idea is that if the "marks" of a restructuring are detected, then applying the inverse restructuring would bring the slave back into correspondence with the master. Of course, the restructuring might have been applied to either the master or slave so the table does not distinguish between the two.

The last row in Table 13.2 mentions aggregation as a source of restructuring. This is our next topic.

Aggregation Restructuring

Aggregation is a form of abstraction commonly employed in modeling in disciplines ranging from physics to economics. In such applications a group of variables having the same range sets is replaced by a single variable that represents a summary of their values. Such summary variables include the sum, average, maximum, and so on. A more disaggregated form of summary is a distribution of values, e.g., a histogram, in which the range set is broken into intervals and counts are given for each interval. This is a higher resolution representation since the overall summaries can be derived from it. In the context of the SES, such restructuring takes the form illustrated in Figure 13.9(a). Here the substructure of an entity is removed and is replaced by one

Table 13.2. Effects of Restructurings

Source restructuring	In one SES	In the other SES
Expanding multiAspects via their entity specializations	Look for a single multiAspect with a specialization attached to its generating entity	Look for a number of peer multiAspects with different entities
A variable is replaced by a specialization whose entities represent values in the range set	Look for a variable with a discrete range set	Look for a specialization with the entities having the same names as the range set
A variable is replaced by a specialization whose entities have variables representing subintervals of the range set	Look for a variable with a continuous range set	Look for a specialization that appears to cover the range of the variable
A group of variables is replaced by an aspect whose entities correspond to the variables	Look for a group of variables that represent a single concept such as a coordinate system	Look for an aspect that is reused under many entities (using uniformity)
Specializations are replaced by their product	Look for specializations under the same entity	Look for a single specialization that has values that match the product of the group under the same entity
Specializations are increased in specificity	Look for a specialization whose entities differ in only a few respects	Look for a specialization that is more specific with fewer entities
Aggregation	Look for a set of variables with a common range set	Look for a single variable whose name suggests a distribution or a summary of such a distribution

or more aggregation variables that are attached to the entity. A common example occurs where a multiAspect has been expanded based on a specialization of its entity (Table 13.2) as shown on the left of Figure 13.9(b). Since the specialization entities all have the same variables, it is possible to create distributions and summaries of them. For example, on the right of Figure 13.9(b), we have a distribution of calories for the different toppings, for which a summary of interest might be the total caloric content of the toppings. Successive iterations of such aggregation restructurings are possible. For example, the calories of the toppings can be added to those of the crust to obtain the total caloric content of the pizza.

We note that aggregation is an information destroying process. For example, knowing the caloric content of the pizza does not

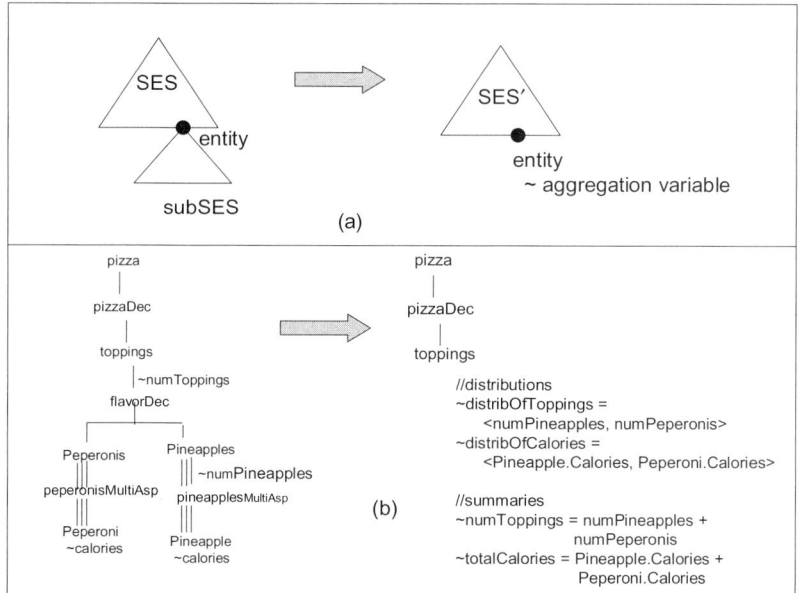

Figure 13.9

Aggregation Restructurings

tell us the number of calories contributed by each component. Mathematically, an information preserving function is a one-to-one mapping that has a functional inverse. In contrast, a function that destroys information is many-to-one mapping with only a relational inverse. Depending on the pragmatic frame of interest, loss of information may, or may not, be acceptable. In the inventory update frame of Chapter 1, aggregation to the level of vehicle model types is acceptable, but further aggregation to that of overall vehicles sold is not. However, the developers of a universal standard for sensor systems in Chapter 14 imposed a requirement that all raw data processing functions be information lossless. For a general discussion of aggregation in the context of dynamic model behavior preservation, see Ref. 17.

Tracking Restructurings

In the course of the harmonization process, a collection of restructurings will have been defined that together constitute a mapping

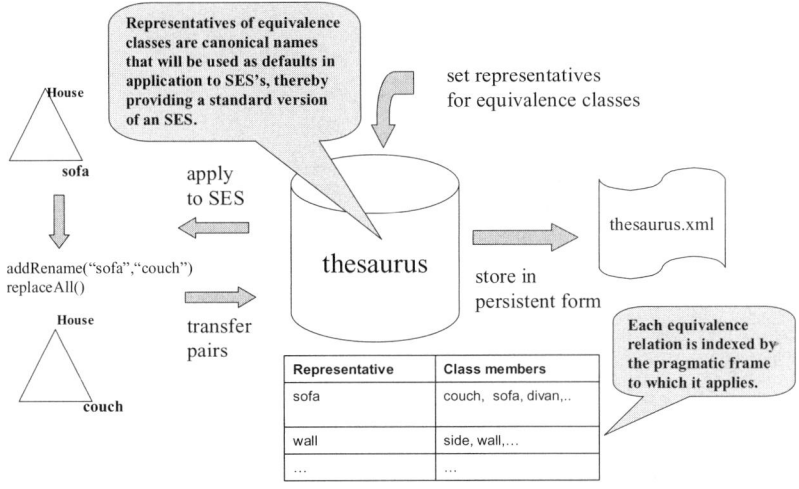

Figure 13.10

Keeping Track of Equivalences in a Thesaurus as Harmonization Proceeds

of the slave into the master SES. Indeed, there may be numerous restructurings for each master/slave pair under focus — with many more pairs for the many-to-many approach than for its many-to-one counterpart. At this writing, the restructurings, together with their assumptions and compositions, have to be described informally in accompanying metadata. Future research should aim to develop a more formal means to deal with this situation that would allow better quality control (verification and validation) and the capability to maintain integrity as standards and ontologies evolve over time, an aspect of ontology management.[12,15]

In the case of renamings, Figure 13.10 illustrates a concept of a thesaurus that records name equivalences and evolves as more are discovered in the harmonization process.

Synchronizing Prunings from Harmonized SESs

One of the tasks required after harmonizing a pair of SESs is retaining consistency between instances of different schema representing the same domain. We'll show how relation-based

238   Chapter 13   Harmonizing Data Representations and Ontologies

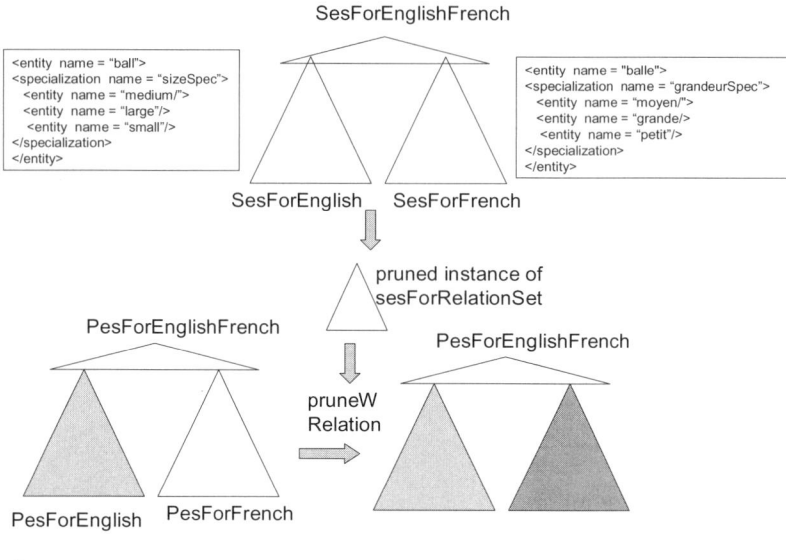

Figure 13.11

Synchronizing PESs of Harmonized SESs

pruning (Chapter 9) can provide a ready-made solution. As a metaphor, consider synchronizing pruned entity structures between representations of the same SES in two different languages, such as English and French. In Figure 13.11, we sketch an SES that has two substructures, one for English, the other for French (it can be composed from the two stand-alone SESs using the composition operation in Chapter 12). For example, for English there might be an SES for ball, with a specialization for size. On the French side, there would be an equivalent SES whose items are translations of those used in English.

Now consider pruning SesForEnglishFrench under the restrictions of the relation represented by Table 13.3 (namely, the translation from English to French of the size terms). Then, any pruning of the English SES will automatically prune the French SES in a corresponding manner. Imagine, for example, that a multinational company sends out production orders to its different branches in the languages prevalent in the countries in which they are located. Then, sending out equivalent orders to all branches is straightforward with relation-based pruning that uses relations derived from applicable dictionaries.

Table 13.3. Translation Relation

sizeSpec/grandeurSpec	petit	moyen	grande
small	x		
medium		x	
large			x

Exercise

Show that to force a translation from French to English in Figure 13.11, all that has to be done is to toggle the converse slots of all restrictive relations.

Summary

A recent comprehensive survey of semantic integration[11] states that current ontology alignment systems still tend to use syntactical features of the targeted ontologies and must move toward increased capture and exploitation of semantic "richness" to achieve quality results. The approach discussed in this chapter considers that beyond semantic richness, methods for ontology integration must include pragmatic richness, i.e., capture and exploitation of pragmatic information. We showed how the concept of pragmatic frame leads to pragmatic equivalence as the way to formalize the goal of ontology harmonization. The concept of pragmatic equivalence supports the harmonization process by filtering out features that are of interest and matching up such features in master and slave ontologies. This approach was employed to develop the standard discussed in Chapter 14. The use of the SES helps to formalize metrics of commonality and to suggest restructurings that might explain apparent mismatches of ontologies that are intended for the same aspect of reality. Such restructurings are readily formulated in a framework, such as the SES, in which decomposition and specialization are well-specified concepts that can be directly manipulated. They are not so apparent, nor directly supportable, in more general frame-based representations. In particular, the SES supports dealing with aggregation restructurings that are ubiquitous in real-world applications, where varying levels of resolution are often present.

References

1. IEEE Technical Committee on Data Engineering, http://www.ipsi.fraunhofer.de/tcde/ (accessed Nov. 2006)
2. Tolk, A., and S. Y. Diallo, Model Based Data Engineering for Web Services. *IEEE Internet Computing*, Volume 9, Issue 4, pp. 65–70, 2005
3. Tolk, A. XML Mediation Services Utilizing Model Based Data Management. *Proceedings IEEE Winter Simulation Conference*, pp. 1476–1484, IEEE CS Press, 2004
4. Turnitsa, C., and A. Tolk, "Evaluation of the C2IEDM as an Interoperability-Enabling Ontology," Proceedings of Fall Simulation Interoperability Workshop, 2005
5. Turnitsa, C., S. Kovurri, A. Tolk, L. DeMasi, V. Dobbs, and W. P. Sudnikovich, "Lessons Learned from C2IEDM Mappings within XBML," Proceedings of Fall Simulation Interoperability Workshop, 2004
6. Pohl, K. J. "Perspective Filters as a Means for Interoperability among Information-Centric Decision-Support Systems," Office of Naval Research (ONR) Workshop hosted by CADRC (Cal Poly) in Quantico, VA, June 5–7, 2001
7. Pohl, K. J. "A Translational Solution To Semantic-Interoperability Among Expressive Systems," 10th World Multi-Conference on Systemics, Cybernetics and Informatics, WMSCI 2006, July 16–19, Orlando, Florida, USA, 2006
8. Alexiev, V., M. Breu, J. de Bruijn, D. Fensel, R. Lara, and H. Lausen, "Information Integration With Ontologies: Ontology Based Information Integration In an Industrial Setting," Hoboken, NJ: John Wiley, 2005
9. Cruz, I., H. Xiao, and F. Hsu, "An Ontology-Based Framework for XML Semantic Integration," Proceedings of IDEAS2004, 2004
10. An, Y., A. Borgida, and J. Mylopoulos, "Constructing Complex Semantic Mappings Between XML Data and Ontologies," Proceedings of ISWC2005, 2005
11. Kalfoglou, Y., B. Hu, D. Reynolds, and N. Shadbolt, Capturing, Representing and Operationalising Semantic Integration (CROSI) project — final report. Technical Report, ECS, University of Southampton, 2005, http://eprints.ecs.soton.ac.uk/11717/ (accessed Nov. 2006)
12. Eklöf, Martin, and C. Mårtenson, "Ontological Interoperability Methodology Report," Command and Control Systems, FOI-R — 1943 — SE, 2006
13. Rosenthal A., L. Seligman, and S. Renner, "From Semantic Integration To Semantics Management: Case Studies and a Way Forward," SIGMOD record 33(4), 44–50, 2004
14. Department of Defense Chief Information Office (CIO). Department of Defense Net-Centric Data Strategy, 2003, www.afei.org/pdf/ncow/DoD_data_strategy.pdf (accessed Nov. 2006)

15. Noy, N., and M. Musen, "Ontology Versioning in an Ontology Management Framework," IEEE Intelligent Systems, 2004
16. The PROMPT tab for Protégé, http://protege.stanford.edu/plugins/prompt/prompt.html (accessed Nov. 2006)
17. Zeigler, B. P., T. G. Kim, and H. Praehofer, *Theory of Modeling and Simulation*, 2d ed., New York: Academic Press, 2000
18. Ontolingua, http://www.ksl.stanford.edu/software/ontolingua/ (accessed Nov. 2006)

14

Geospatial Sensor Data: The Universal Phase History Data (UPHD) Standard

This chapter discusses an example of the application of SES-based methodology to a significant real-world problem domain. Continuing the exposition of geospatial sensing first introduced in Chapter 3, we discuss a standard that has been developed for Synthetic Aperture Radar (SAR) systems, an important subclass of geospatial sensors. The problem is that many kinds of SAR sensors have been developed, each representing substantial investment and with its own idiosyncratic means of encoding its data for downstream interpretation. Clearly the development of a common format that allows data from such sensors to be shared and to be fused into much more perceptive observers of terrestrial properties and activities would be beneficial. Over a period of many months a standard was developed to provide such a format employing the methodology described in this book. Before discussing the standard in some depth, we provide some background in the problem domain to suggest the complexities that had to be handled in its development. This background is aimed at the non-expert in SAR and is intended to provide only enough terminology and concepts to render the discussion comprehensible. At the end of the chapter, we'll review the challenges posed by the problem domain and the ways in which the SES-based methodology helped to deal with them.

Background: Synthetic Aperture Radar Sensing

Synthetic aperture is the term used to describe the effective magnification of the radar antenna opening due to the coordinated processing of the returns from multiple pulses (see Wikipedia for a brief introduction). As depicted in Figure 14.1, advanced processing of remotely sensed SAR data fundamentally exploits the information in the phase histories of Fourier transformed frequency components as a pair of collecting antennas move through space and time. For example, interference fringes are generated from the near simultaneous radar pulse returns collected by a pair of satellites in orbit over a time interval. Further processing yields a time series of phase differences that

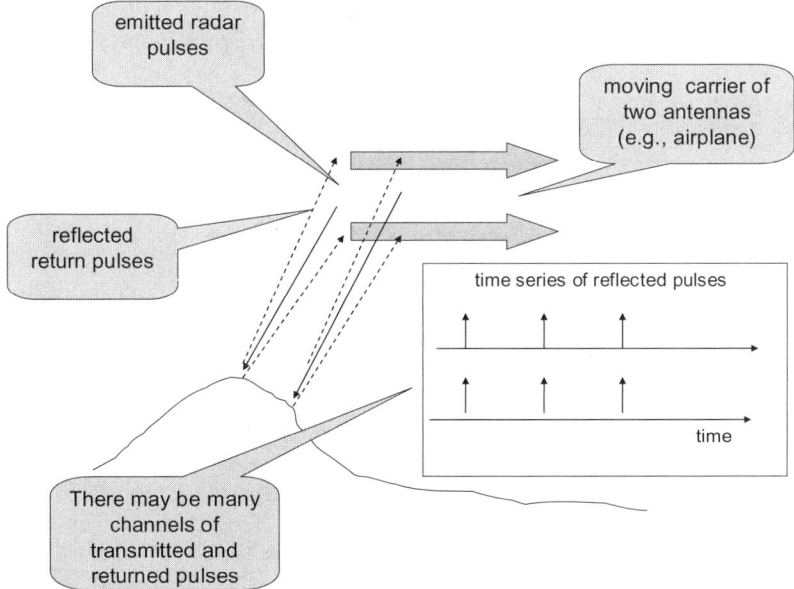

Figure 14.1

Principle of SAR Phase History Generation

contains information about the elevation of the terrain below. Consequently, this phase history can be transformed to produce accurate digital elevation models of the Earth's surface.[2]

Due to its power to detect small changes in spatial scenes from one location to another or from one event to another, SAR sensors can produce high discrimination/high resolution images. For this reason, SAR has become a key source of commercial and non-commercial geospatial data and is rapidly evolving a host of valuable scene-characterizing products and tools. However, the synthetic aperture principle requires that the SAR sensor be on a highly mobile platform and, therefore, many sources of SAR data exist. Generally, these systems were individually designed and built for specific and limited user communities without regard to interoperability. However, it has become increasingly evident that to produce timely, relevant, and accurate geospatial information, such stove-piped assets must be integrated by adopting a universally accepted set of standards. In particular, for SAR sensors, the *Universal Phase History Data* (UPHD) standard[1] is intended to coordinate their data production architecturally to reduce the costs of processing and exploitation and to allow sensors, services, and products to interoperate at the data level. Moreover, due to SAR's unique capabilities and all-weather, 24/7/365 operation, it can be anticipated that there will be significant growth in the technical designs for these sensors.

As illustrated in Figure 14.2, SAR phase history collection interval is called a Coherent Data Period. The acronym CDP is used to refer to this period and all times are referenced to the CDP, not the time of the start of collection. A SAR CDP is composed of a (large) number of Inter Pulse Periods (IPPs). An IPP is the time period between a successive pair of pulses. The actual duration of an IPP may vary from IPP to IPP within a CDP. The IPPs include a single transmitted pulse time period within each IPP interval so the "pulse index" and "IPP index" have the same values and meaning.

The Impulse Response (IPR) is the image domain representation of the return in a (time domain) phase history from a point scatterer. When a finite Fourier transform image formation technique is used, the IPR of a uniformly weighted return of a properly compensated point source phase history is a finite representation of the SINC function $\frac{\sin x}{x}$. More generally, the ambiguity function of the radar system may have multiple peaks in response to a point source. This also represents the IPR and

Goals for UPHD Standard Development

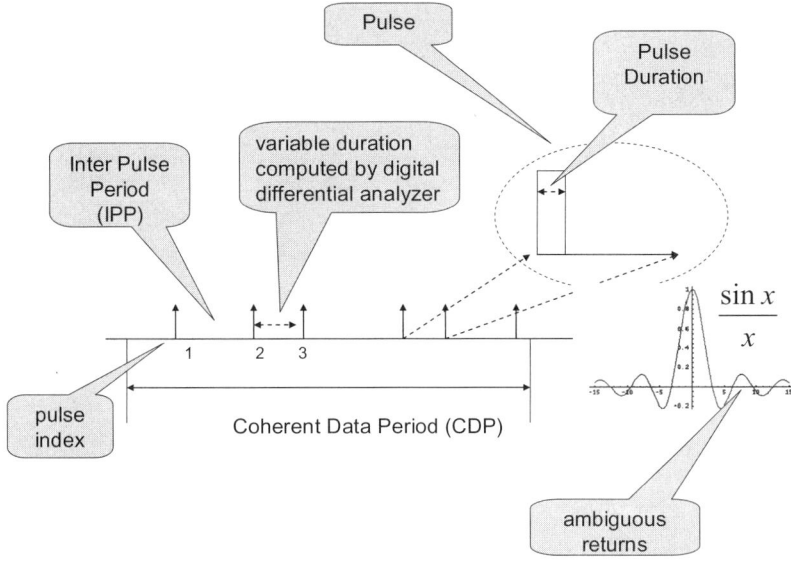

Figure 14.2

Illustrating the Timing Considerations in SAR

it is customary that quality requirements limit the relative power associated with the principal return and the ambiguous returns as well as the spacing of these peaks in the image domain.

The pulse stream can be expressed as a discrete event segment (Chapter 11). The successive pulse generation times are computed using polynomial expressions. In iterative form, the next event time is determined from the current event time by adding a time interval. The latter is determined with a polynomial expression derived as a difference of the original polynomials. An algorithm called the differential digital analyzer (DDA) is used to perform this iterative computation. A significant portion of the UPHD standard is devoted to capturing the coefficients of the polynomials employed to enable downstream processing to properly interpret the data collected in response to the pulses.

Goals for UPHD Standard Development

Recall the pragmatic framework introduced in Chapter 1 in which a producer sends world state information to a consumer. As in Chapter 3, where we first applied this framework to sensor systems, the framework is useful to keep in mind as we discuss

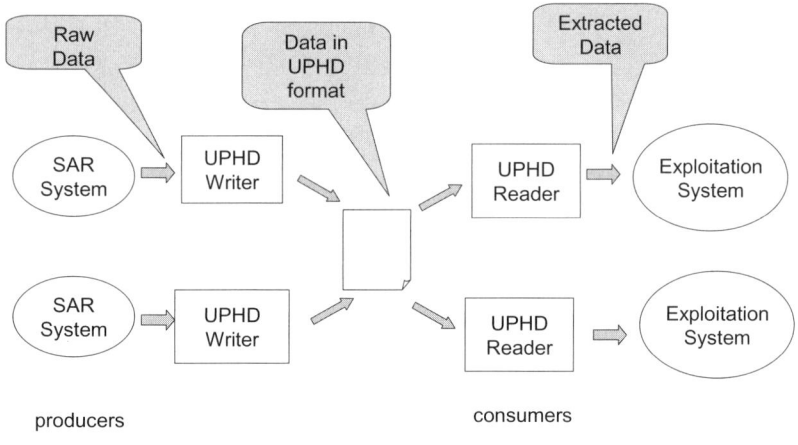

Figure 14.3

Many-to-Many Producers to Consumers Via UPDH Standard

the UPHD. As illustrated in Figure 14.3, the UPHD standard specifies how to covert metadata, support data, and wideband radar data from sources for any SAR sensor system to a single volume of data to be known as Universal Phase History Data (UPHD). UPHD is generated, for a given SAR sensor system, by a writer according to the conventions and definitions within the standard. A UPHD reader imbedded within a UPHD-compliant SAR processor then reads UPHD from any UPHD writer and interprets the data according to this document.

Stated goals of UPHD include

- Supporting interoperability of SAR processing and exploitation networks by producing a set of harmonized metadata elements necessary to fully characterize the technical parameters of phase history data generation and capture.
- Supporting the definition of downstream metadata profiles for processed SAR data, at the complex (level-2) and image product (level-3) stages (see Chapter 3). This would provide a sound foundation from which to inherit metadata to engineer downstream product standards.
- Affecting the designs of future SAR collection systems and the metadata from those systems and to facilitate meeting downstream processing and exploitation requirements in a cost-effective manner.

The UPHD standard effort is destined to become the front end of a data engineering, end-to-end, life cycle approach for SAR and extends to other sensors as well. This effort is in line with the adoption of common standards for exchange of data between producers (Chapter 3). In this chapter, we present some details of an application of this data-centric view that is intended to shift the system development paradigm from "change the architecture for every sensor" to "populate needed data in a common standards-based architecture."

Universality and Scope of the UPHD Standard

As its name suggests, the UPHD standard is intended to be *universal* for SAR systems. This means that it should represent a SAR data set as a generic set of sensor state parameters and sensor measurements. Moreover, the standard should act as an interface presented by the producer to the consumer in the sense of hiding details about the producer's sensor characteristics that are not relevant to downstream processing. Indeed, the guiding requirement is that *no specific knowledge of the sensor design other than that revealed through the UPHD data parameters and their values can be required to process the UPHD to any possible derived product.*

Recall the discussion in Chapter 1 where we asserted that an ontology developer has to decide whether to take a specialist or a generalist approach to determining the *scope* of the ontology. The pragmatic frames characterizing the intended downstream use of the data is critical in making this choice. Given the goals of development just outlined, the UPHD developers naturally opted for the generalist approach since it would accommodate the large variety of legacy and anticipated SAR systems. Their approach takes the form illustrated in Figure 14.4 for the hypothetical case of a standard that can accommodate both color and black and white (BW) TV. Here the data model must contain elements to support color TV reception that are not needed for less sophisticated BW reception. This means that the price of including both forms of TV is that some efficiency due to unused parameters is lost when BW data are transmitted to a BW receiver. Conversely, this would also be true in the case of the color and BW technologies being totally distinct. However, if the more capable color technology is downward-compatible with a less capable BW system, then it can usefully employ the BW data elements with no overhead.

In any event, a well-designed data model can provide the flexibility needed to tailor the data description to the least expensive

248   Chapter 14   Geospatial Sensor Data: The UPHD Standard

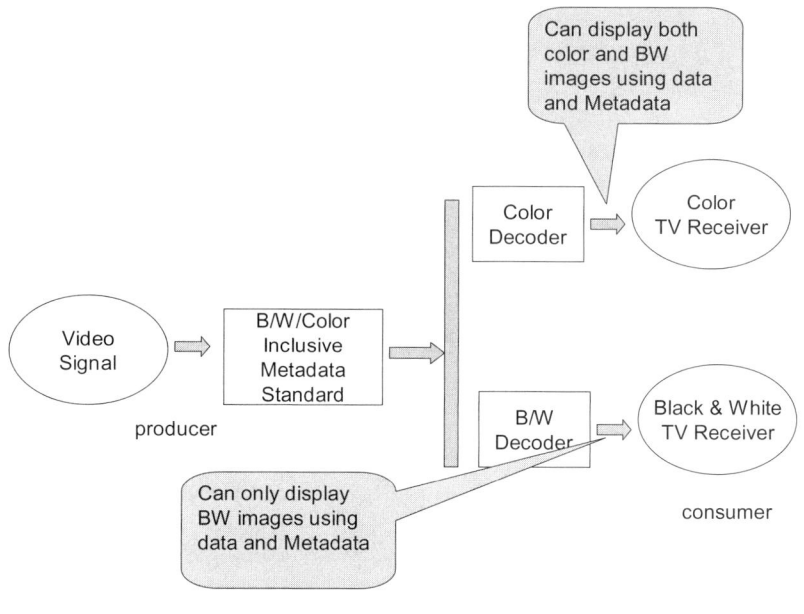

Figure 14.4

Generic Formulation of TV Set Standard

form required by discarding elements that are not needed in any particular case. Indeed, the SES mechanism of *pruning* provides a key concept and toolset for implementing this flexibility.

Indeed, the UPHD standard includes fields that may not apply to a specific SAR sensor, but are needed to describe SAR systems in general. Thus, any particular data set may use only a subset of the possible parameters in UPHD. This subset is determined by the capabilities of the source sensor and the state that sensor was placed into in order to collect the phase history component of the UPHD dataset. For example, an along-track interferometric SAR requires multiple channels of phase history while less capable SAR systems process only a single channel SAR phase history. To be generic, the UPHD must represent multiple channels while at the same time allowing pruning to just one.

To implement a generic approach, the UPHD developers had to make specific decisions on the scope of the standard, namely, on identifying just those elements that are truly generic, and therefore need to be included, versus those that are not. Specific cases had to be examined on a case-by-case basis but the identification of *pragmatic frames* and the principles outlined in Chapter 1 helped to make such decisions.

The UPHD was designed with two basic types of pragmatic frames in mind: *exploitation* and *engineering-related activities* such as assessing vehicle health and revising operating parameters.

- If the UPHD is configured for downstream analysis, some data, such as those related to temperature at the sensor instrument, should not be incorporated because they cannot be used any further in downstream processing.
- If the UPHD is configured for engineering purposes, elements such as the payload temperature telemetry readouts may be retained for long term reuse and the resulting file may be retained by the operators of the producer.

The UPHD standard must include sufficient elements to accommodate both types of frames. Therefore, in any particular case, the most challenging of the frames demands satisfaction. However, as discussed in Chapter 1, the standard should include no more than the minimum needed to satisfy the most challenging requirements of the chosen pragmatic frames.

The conclusion here is that the UPHD writer must exercise all of the sensor and platform-specific models and post the results into the UPHD dataset so that only generic processing is required downstream. However, it is not responsible for compensating for lack of generality in a less capable processor which is able to read UPHD.

The generic formulation may "push back" additional responsibilities on the designer of a new sensor or the UPHD writer for an existing one. The standard tries to capture existing systems as alternatives that can be spanned by the configurations it sets up. However, sometimes it may encounter cases where the variation in current practice is too great to capture in a workable abstraction. In this case, it is the sensor designer's responsibility to reconcile its algorithms with those of the standard. For example, if the sensor control is not designed to compute IPPs in the manner designated by the payload specification, then the UPHD writer must perform an appropriate translation.

Information Losslessness

We discussed the distinction between information-preserving and information-destroying operations in Chapter 13. The SAR data as represented in the UPHD must be lossless from a wideband information point of view. No irreversible transformation or filtering of the data is permitted for UPHD. This property, taken together with completeness, means that all possible SAR products that can be obtained from the original SAR data can be created to the same level of quality based on the UPHD representation of that data.

Approach: Syntactic, Semantic, and Pragmatic

SAR sensors are large, complex, dynamic systems. The core of the UPHD standard is the specification of a syntactic crucible to capture and describe the data product of a SAR sensor for further downstream processing. The eventual form of this crucible is an XML schema. The approach to developing such a schema is to employ the SES to represent appropriate structures and data elements and to obtain the final schema using the SES-to-XML transformation tools. However, as we have stressed throughout this book, the syntactic aspect of characterizing a real-world system such as an SAR sensor is not sufficient in itself to support use in practice. To fill in a schema instance with sensor-taken data, a UPHD writer, or its human designer, must understand the meaning of the schema slots (tags and attributes) — semantics — and the procedures and constraints governing their treatment and intended use — pragmatics. Accordingly in the UPHD document, each data field is described by four facets:

1. *Brief Definition*: a summary description of the slot and its role — typically, this is an element representing a leaf entity
2. *Indices*: the parameters that index the entity's occurrences, e.g., as a multidimensional array
3. *Units/Representation*: the measurement units or symbolic values that the slot can/must assume, e.g., watts/floating point
4. *Treatment/Usage*: a detailed discussion of the manner in which the slot(s) is/are to be filled in or its/their values selected, the constraints on such selections, other rules and procedures, informally presented. Importantly, the UPHD designer's intent is spelled out. For example, in the discussion of payload decode table, we see, "An intent of the UPHD standard is to remove the responsibility for general processors to process real sampled data, having the responsibility for conversion to complex data pushed upstream to the UPHD writer."

In addition, links to related data elements are provided, giving an indication of the holistic nature of the underlying domain decomposition.

Included within the pragmatics of the UPHD are its requirements for product quality. Given that UPHD will be not only a standard interface for emerging SAR systems, but an expected transition for a number of existing systems that have previously established downstream data processing chains, as described in Chapter 10. A significant portion of the standard outlines the quality comparison criteria for UPHD-based operation for systems with existing data chains. Any specific acquisition of a UPHD writer and the UPHD-compliant processors intended to ingest UPHD will need separate specifications for quality and a profile of support within this UPHD standard. The writer controls the ultimate product quality through choices it can exercise within the UPHD standard. These generally involve accuracy, granularity, and data sequences versus polynomial representations (see the UPHD documentation[1] for more detail).

Relation to Systems Modeling and Simulation

As suggested earlier, and as perusal of the UPHD document readily reveals, a SAR sensor is a hybrid of classical systems signal processing and discrete event systems. Indeed, it can be usefully placed within the conceptual framework of modeling and simulation, similar to the tactical data standard to be discussed in Chapter 17. From this perspective, the syntactical characterization provides only the *static* description, setting the stage for the specification of *dynamics*, the temporal behavior that governs how the sensor interacts with its environment to collect the data. This suggests that future extensions of the UPHD might attempt to conceptualize more of the system domain (Chapter 11) with an extended SES that exploits its more in-depth, simulation-based ontology.

Generic Formulation of an SAR System

An SAR is an active system in that it generates energy waves toward a target and then examines the effects that the latter has made on the transmitted waves. This suggests that any SAR sensor will have two major components: 1) pulse train formation/generation and 2) return pulse reception, correction, and processing (Figure 14.5). These components, as well as the physical platform that carries the sensor, serve as a way to organize

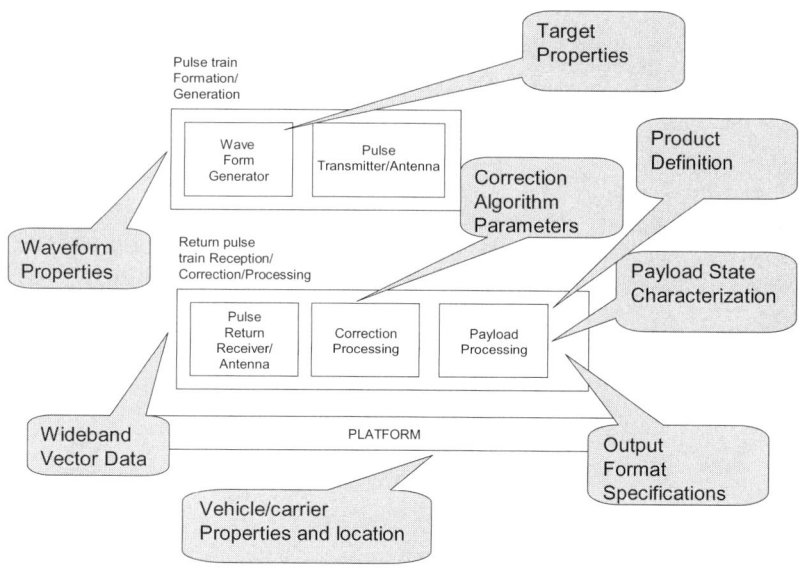

Figure 14.5

Generic SAR System Structure and Descriptors

the kinds of data and metadata that the UPHD standard must consider.

Components of the SES

The data/metadata elements associated with the major components in Figure 14.5 form the top level entities in the SES developed for the UPHD. As shown in Figure 14.6, the eleven major components for the SES are:

1. MD — metadata, principally for query and correlation purposes, not processable
2. PL — payload state parameter
3. SV — spatial data about the platform/vehicle
4. PD — product definition data including ground reference polynomial (GRP), planned lines, and samples, etc.
5. WF — wave form parameters
6. DC — deterministic corrections
7. TG — target geometric parameters
8. VH — wideband vector header data such as channel, pulse index, or time of collect

Wave Form Parameters (WF)

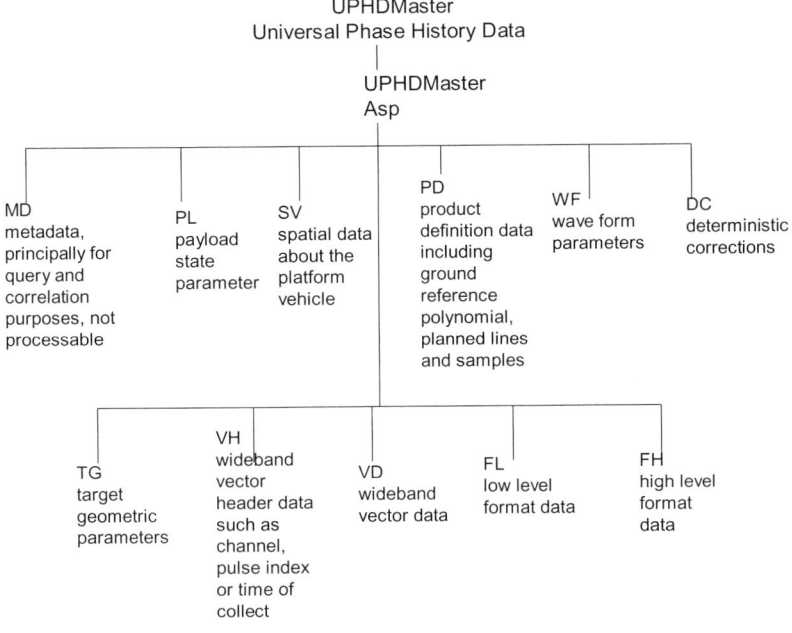

Figure 14.6

Top Level Components of UPHD Standard

9. VD — wideband vector data
10. FL — low level format data
11. FH — high level format data

Wave Form Parameters (WF)

Our previously stated objective is to give a sense of the complexities that are involved in capturing the data elements and structures in the SAR domain and how the developers of the UPHD dealt with them using the methodology in this book. It is not practical for us to review each of the eleven SES components and we refer the interested reader to the standard and its documentation for a detailed exposition.[1] However, we will discuss the WF, VH, and SV components in some depth to illustrate the domain and the methodology employed to capture it.

The WF component must deal with the generation of the pulses that constitute the wave form, transmission of the pulses, reception, and recording of the returns. In addition, accounts for

Chapter 14 Geospatial Sensor Data: The UPHD Standard

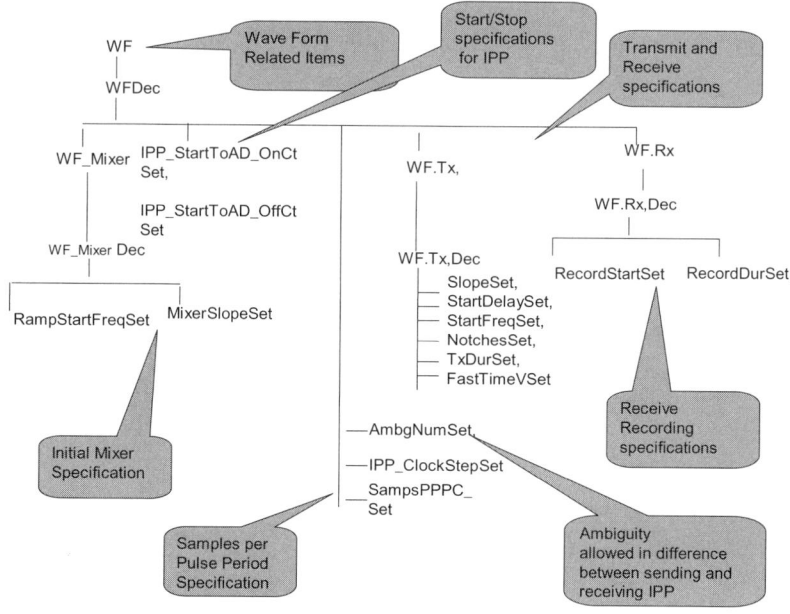

Figure 14.7

SES for Wave Form Parameters

"second order" considerations including ramping up of the transmitter power and the ambiguity (mentioned above) in pulse return to be allowed.

Figure 14.7 depicts the SES developed for the wave form component.

A natural language rendering of the SES is given in the following:

SES Natural Language Description of WF:

WF consists of,
IPP_StartToAD_OnCtSet, IPP_StartToAD_OffCtSet,
SampsPPPC_Set,
IPP_ClockStepSet,
AmbgNumSet,
WF.Tx,
WF.Rx, and
WF_Mixer

> WF.Tx consists of SlopeSet, StartDelaySet, StartFreqSet, NotchesSet, TxDurSet, and FastTimeVSet
>
> WF.Rx consists of RecordStartSet and RecordDurSet
>
> WF_Mixer consists of RampStartFreqSet, and MixerSlopeSet

We note that entities ending in "Set" or "set" are multi-entities, i.e., are decomposed by multiAspects. For readability we have omitted the elaboration of these entities as well as the variables attached to them. The abbreviations Tx and Rx refer to transmission and reception, respectively.

The natural language description is obtained using the sesRelation method, backToNatLang, after creating an instance for the SES to be translated:

```
sesRelation WF = new sesRelation(folder+"sesForWF.xml");
  WF.backToNatLang( folder);
```

Recall that the SES axioms imply that a uniquely labeled path is associated with every entity. It is helpful to prefix its path to an entity when discussing its meaning and role since the path provides an indication of its location and neighboring entities in the structure. You can locate the entity by tracing down from the root taking the branch indicated by the path information at each node. To keep such identifiers short and informative, we omit all aspects, specializations, multiAspects and multi-entities from the paths.

After restoring an SES into memory, we use the method:

```
getPathsHavingType( "entity", "Set");
```

to generate all paths from the root to entities in the SES, skipping over nonentity items and non-multi-entities (ending with "Set" or "set").

WF Leaf Entities

We draw upon the UPHD document to provide descriptions of the leaf entities in the WF component SES as identified by their shortened path identifiers. These descriptions are provided in Tables 14.1–14.5.

Wideband Vector Header Data (VH)

The VH component deals with the partitioning of pulses into blocks according to channels that organize received data. The term "channel" is used to distinguish a time series of phase data measurements. A single UPHD must contain the data for all channels that must be co-processed to properly form the azimuth IPR. Thus, channels associated with multiple polarizations get their own individual UPHD files while ambiguous channels associated with separate along track aperture centers all need to be included in a single UPHD file. Identification of the pulse blocks and channels is critical to determining the time of transmission of the originating wave form and provides the key information needed to determine what the geometry was for a reception.

The SES for the VH component is shown in Figure 14.8 (on page 260) and its natural language descriptions, simplified as discussed earlier, is given in the following:

> *SES Natural Language Description*
>
> VH consists of DDA, ChanNumSet, SpeclPulseBlkNumSet, and SpeclBlkIPP_Num
>
> DDA consists of IterCtSet, and LoadNumSet

VH Leaf Entities

Once more, we provide brief descriptions of the leaf entities in the VH component SES as identified by their shortened path identifiers. These descriptions, extracted from the UPHD document, appear in Table 14.6.

Spatial Data about the Platform/Vehicle (SV)

The final component to be discussed, SV, organizes the information concerning the location of the sensor platform in space and the effective aperture location and orientation with respect to the

Table 14.1. Interpulse Periods

WF.IPP_ClockStep	WF.IPP_ClockStep is the time step for each advance of the payload clock for marking intervals in the IPP. If the payload design does not use the basic clock rate for marking IPP level counts, a multiplier is presumed so that the process is transparent and WF.IPP–ClockStep is directly interpretable as an intra-IPP time step and is the unit of "clocks" or "counts" at this level. This is the reciprocal of the effective IPP driver clock frequency. This may be the output of a temperature-dependent CDP level clock calibration process for some designs and a simple database value for others.
WF.IPP_StartToAD_OffCt	WF.IPP_StartToAD_OffCt(n) is the payload clock count that marks the end of A/D conversion of signal in the output from the digital fast time data stream for the receive window that occurs in each IPP of a specialized pulseblock.
WF.IPP_StartToAD_OnCt	WF.IPP_StartToAD_OnCt(n) is the payload clock count that marks the start of A/D conversion of signal in the output from the digital fast time data stream for the receive window that occurs in each IPP of a specialized pulseblock.
WF.PRF_Fac	WF.PRF_Fac(n) allows special pulse blocks to have an integer multiple of the base DDA PRF. The processor computes pulse start times and related parameters for pulses in a specialized block according to a time interval given by the DDA load for the block, the iteration count of the pulse, other fine tuning timing parameters in the UPHD, and WF.PRF_Fac(n). In the calculation, the number of uniformly timed pulses per iteration count is WF.PRF_Fac(n). When WF.PRF_Fac(n) = 1, computation proceeds as if the parameter did not exist. The inter-pulse period (IPP) is IPPbasic/WF.PRF_Fac(n) where IPPbasic is computed directly from the DDA and the iteration count and is the difference between two consecutive pulse start times.
WF.SampsPPPC	WF.SampsPPPC(n) is the number of (complex, if applicable) samples per pulse per receiver channel and gives the length of the wideband vectors representing the output of the receiver(s) during the listen interval within an IPP.

platform. This parameter is necessary for calculation of the illumination pattern on the ground for all channels. Each receive channel has an effective earth-centered orientation of the normal to the effective mechanical aperture plane. This orientation takes on different definitions for different alternative designs. The

Table 14.2. Ambiguity

WF.AmbgNum	WF.AmbgNum(n) is the ambiguity number for specialized pulse block, n. The ambiguity number is the difference between the IPP number that a pulse is transmitted in and the one that the target range swath is received in. Thus, when a listen window is capturing the pulse transmitted in the same IPP, AmbgNum = 0 and it is not possible to receive returns from closer ranges at the same time as the return from the intended range. The pulses will be organized into specific sequential groups of sequential pulses. It is possible for the ambiguity number to be varied among these specialized pulse block groupings.

Table 14.3. Transmission

WF.WF.Tx.Dur	WF.Tx.Dur(n) for any specific pulse block, n, gives the Transmit pulse duration. WF.Tx.Dur(n) is the number of clock intervals for which the transmitter is on. When multiplied by the clock step size, WF.Tx.Dur(n) provides the duration in seconds for the transmit pulse.
WF.WF.Tx.FastTimeV	WF.TxFastTimeV(m,n) is the non-parameteric transmitted wave form represented as a time domain complex fast time vector. UPHD treats the wideband as Vector Quantized codes for complex voltage measurements. Thus these are bit strings whose interpretation is specified in the VQ decode parameters. WF.TxFastTimeV(n) describes the pulse as if it were directly received in the receiver and recorded from the same A/D conversion process used for the scene return (as if reflected from a single point scatterer). It is represented via the VQ decode table just as received phase history is. It is, thus, a sequence of words of length given by PL.VQ.DecodeTableHdr(m,1). That is, the case PL.VQ.DecodeTableHdr(m,1) = 8 could indicate 8 bit I,Q pairs, 8 bit reals, or 8 bit Vector codes. FH.VQ.length(4) = 64 implies I,Q pairs (32 + 32) will be output from the decoding process when VD.RgData is read and used in the decode lookup table. The successive words in a WF.TxFastTimeV string for a single pulse represent the fast time ordered output of the transmitter channel. Data is of type indicated by VD.RgData_type. Regardless of type, words are of length PL.VQ.DecodeTableHdr(m,1). The indices m and n are described as optional to denote that there may be only one WF.TxFastTimeV(m,n) due to using a single channel (m = 1 or absent) and the same transmit fast time pulse pattern (n = 1 or absent), i.e., voltage sequence, for each and every IPP. If WF.TxFastTimeV(n) is listed in the pulse indexed

Spatial Data about the Platform/Vehicle (SV)

Table 14.3. Transmission—cont'd

	parameter list, then there will be a WF.TxFastTimeV(n) string of VQ samples in each channel IPP in the wide band file so that every pulse is allowed to change transmit wave form. If the writer has not put the parameter name, WF.TxFastTimeV(m,n) in the pulse indexed list, then only one pulse of data is needed per transmit channel and this will be placed in the narrow band file rather than the wide band file. When there is a slow time variation, this parameter shows up in a slot below the VH data for each pulse following all of the other pulse indexed parameter data items and before the received range data, VD.RgData.
	This is only a pulse of data and therefore has a shorter fast time duration than the receive data which lasts for a pulse time plus a swath time. WF.Tx.Dur(n) should be used to represent this duration in receiver clock counts, just as for parametric wave form representations.
	They are arranged in sequences of digitized receiver measurements such that there is a sequential list of fast time samples for every interpulse period.
WF.WF.Tx.Notches	WF.Tx.Notches(nIndx, nParm) supplies the parameters describing any Tx notches used to avoid transmitting in discrete frequency intervals for whatever reason.
WF.WF.Tx.Slope	In the cases when a payload uses a linear FM transmit wave form signal, WF.Tx.Slope(n,m) gives the slope of the transmitted signal.
WF.WF.Tx.StartDelay	WF.SampsPPPC(n) is the number of (complex, if applicable) samples per pulse per receiver channel and gives the length of the wideband vectors representing the output of the receiver(s) during the listen interval within an IPP.
WF.WF.Tx.StartFreq	WF.Tx.StartFreq(n) is the start frequency for the transmit pulse.

Table 14.4. Reception

WF.WF.Rx.RecordDur	WF.RxRecordDur(n) is the number of payload clock counts during which the phase data is recorded to produce a range vector.
WF.WF.Rx.RecordStart	WF.Rx.RecordStart(n) is the interval of time after IPP start for the start of recording the mixer output in payload clock counts.

Table 14.5. Mixer

WF.WF_Mixer.MixerSlope	WF.Mixer.Slope(n,m) is the mixer frequency slope. The presence of this parameter allows flexible stretch mode P/L design (i.e., slightly differing Tx and mixer slope) as well as using mixer slope = 0 to act as a chirp mode (de-ramp in the processor) flag. This parameter will only be present if the receiver employs a mixer with a ramped frequency reference to process the signal before recording it. (U) De-ramping a received signal return from a linear FM transmitted wave form assigns power to filtered range bins but results in a fast time skewing of frequencies whose parameters can be calculated from the parameters of the IPP wave form, including WF.Mixer.Slope(n,m) and WF.Tx. Slope(n,m). Alternative pulse compression methods will use different wave form representations in the UPHD and may not involve fast time frequency skewing.
WF.WF_Mixer.RampStartFreq	WF.Mixer.RampStartFreq(n) is the RF starting frequency of the equivalent single stage local oscillator based mixing process that leads to de-ramped or basebanded IF signals.

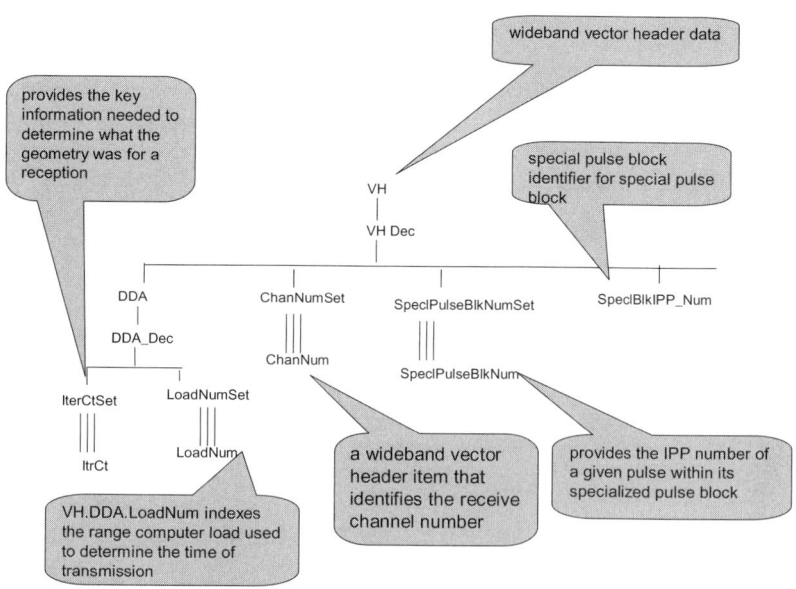

Figure 14.8

SES for the VH Component

Table 14.6. VH Entities

VH.ChanNum	VH.ChanNum is a wideband vector header item that identifies the receive channel number for the following range vector data string corresponding to a single channel's receive window data stream for an inter-pulse period.
VH.DDA.IterCt	VH.DDA.IterCt is a wideband vector header item that links a range vector to a slow time value and thus provides the key information needed to determine what the geometry was for the reception of the following data string.
VH.DDA.LoadNum	VH.DDA.LoadNum indexes the range computer load used to determine the time of transmit (or IPP interval start) for the IPP whose range vector data immediately follows.
VH.SpeclBlkIPP_Num	VH.SpeclBlkIPP_Num provides the IPP number of a given pulse within its specialized pulse block.
VH.SpeclPulseBlkNum	VH.SpeclPulseBlkNum is the special pulse block identifier for the special pulse block with which the wide band data vector belonging to this header is associated.

UPHD writer must work with the definition that is appropriate for its particular sensor.

The SES for the SV component is shown in Figure 14.9 and its natural language description, simplified as discussed earlier, is given in the following:

SES Natural Language Description

SV consists of SV_Rx and SV_Tx
SV_Rx consists of AperMechOrient,
   EffAperCentToCentMass, and Eph
AperMechOrient consists of more than one UnitVector
AperMechOrient has nSize, mSize, and jSize
The range of AperMechOrient's nSize is int (etc.)

Eph consists of

ECF_EUM_X_set,
ECF_EUM_Y_set,
ECF_EUM_Z_set,
ECF_LL_X_set,
ECF_LL_Z_set, and
ECF_LL_Y_set

> EffAperCentToCentMass consists of CentMass_Y_set, CentMass_X_set, and CentMass_Z_set
>
> SV_Tx consists of AperMechOrient, EffAperCentToCentMass, and Eph

SV Leaf Entities

The leaf entities in the SV component SES are identified by their shortened path identifiers. The entities for transmission provide slots for coordinates in the following list:

SV.SV_Rx.Eph.ECF_LL_X
SV.SV_Rx.Eph.ECF_LL_Y
SV.SV_Rx.Eph.ECF_LL_Z
SV.SV_Rx.Eph.ECF_EUM_Y
SV.SV_Rx.Eph.ECF_EUM_Z
SV.SV_Rx.AperMechOrient.UnitVector
SV.SV_Rx.EffAperCentToCentMass.CentMass_X
SV.SV_Rx.EffAperCentToCentMass.CentMass_Y
SV.SV_Rx.EffAperCentToCentMass.CentMass_Z

We note that the uniformity property of the SES is employed to use the same definitions of center of mass coordinates and aperture properties for both transmission and reception. This is evident in Figure 14.9, where the same aspect, SVx_Dec hangs under the Rx and Tx entities. This implies that each of the leaf entities below TX has the same name as a counterpart below Rx. For example, there are two occurrences of CentMass_X. Recall from Chapter 5, that unique path labeling will distinguish these occurrences and that the minimal context to do so can be obtained with the method, computeTopsSes. As expected, in this case, it finds that the top entities, Tx and Rx, are sufficient to uniquely identify the multiple occurrences for pruning purposes.

UPHD Global Range Definitions

The UPHD sets forth broad conventions for the treatment of numerical range sets as follows:

- All UPHD parameters that represent counts of some item are IEEE 32 bit integers.

UPHD Global Range Definitions

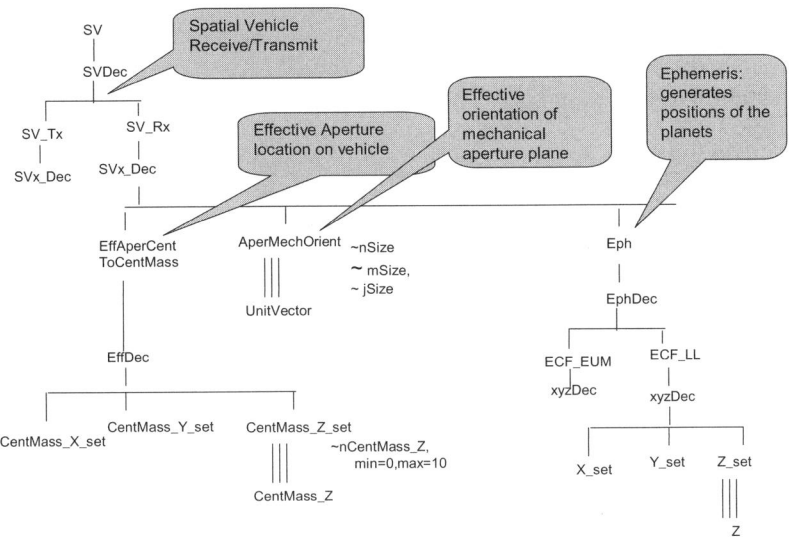

Figure 14.9

SES for Data Related to Vehicle Position in Space (some elements are omitted to simplify the diagram)

- All UPHD parameters that represent frequencies, gains, or other electronic parameters are IEEE 32 bit floating point numbers.
- All UPHD parameters related to computing positional parameters relative to the center of the Earth are IEEE double precision floating point numbers.

Within these conventions, there remains a large number of special cases where slot values have to be restricted to various forms. For example, a particular field requires five digits of precision distributed such that two digits appear before the decimal point, and three digits appear after it. To manage such detailed specifications, we use the approach discussed in Chapter 6, that exploits XML's ability to express such specifications combined with the "front-end" nature of the SES mapping to XML to ease such specification. As illustrated below, the UPHD contains an initial section that provides the global complex and simple type definitions for re-use as needed throughout the document. The following fragment illustrates the structure of the master file, UPHDTop.xml.

```
<top>
  <!-Global complex and simple type definitions->
  ...

  <complex name="complexFloat_m_n_k" entityType="float">
    <var name="m" rangeSpec="int" use="required"/>
    <var name="n" rangeSpec="int" use="required"/>
    <var name="k" rangeSpec="int" use="optional"/>
  </complex>

  <complex name="decimalGroup" entityType="decimal"/>
  ...

  <complexReference name="complexFixed_XX.XXX" restriction-
  ValuePairs="totalDigits, 5, fractionDigits, 3"
    restrictionBase="decimalGroup"/>
  <complexReference name="complexFixed_XXX.X"
    restriction ValuePairs="totalDigits, 4,
    fractionDigits, 1" restrictionBase="decimalGroup"/>
  ...

  <entity name="UPHDmaster">
    <aspect name="UPHDmasterAsp">
    ...
    </aspect>
  </entity>
</top>
```

Managing the UPHD Master and Its Components

As a prototypical large scale SES, we can use the tools described in Chapter 12 to help manage its complexity.

To decompose and recompose the UPHD to/from its top level components we use the merge methods discussed in Chapter 12. For example,

mergeOps.mergeMultipleSes("UPHDTop",
        new String[] {
        "MD",
        "PL",
        "SV",

Complexity Measures of the UPHD and Its Components

"PD",
"WF",
"DC",
"TG",
"VH",
"VD",
"FH" },
folder);

To decompose the UPHD into its top level components, we use the method:

SESOps.extractSesFromFirstEntities(folder)

to write the first level component SESs into the specified folder.

To generate the natural language description of UPHD top level components, we use the method:

SESOps.printSesTextsOfFirstEntities(folder);

to write the natural language descriptions of the first level component SESs into the specified folder.

Complexity Measures of the UPHD and Its Components

As discussed in Chapter 6, alternate schemas can be generated from the same SES. In the following discussion, we use the label-erasing method in Chapter 7 to produce the most compact representation.

To get a sense of the size and complexity of the UPHD we use the method:

XMLToDom.printDocStats(document, folder)

to generate statistics of a document. Applying this method to the SES as represented in XML as well as to the Schema, we get the measurements in Tables 14.7–14.10.

Here we see that the SES document contains 1250 elements while its schema representation expands this number to 1529. The respective file sizes, measured in kilobytes reflect this expansion. The expansion factor of approximately 20 percent is consistent with the manner in which XML schemas are constructed

Table 14.7. Overall Measures

Representation	SES (*.χml)	Schema (*.χsd)
Number of elements	1250	1529
Depth of root node	18	24
Maximum breadth	56	83
File size	73K	89K

Table 14.8. SES in XML

SES (*.χml) elements	Number of occurrences
complex	26
addvar	40
aspect	87
complexReference	16
complexVar	121
entity	438
multiAspect	121
numberComponentsVar	121
simpleVar	29
top	1
var	210

from their primitives. More detail is provided by the following tables in which the numbers of occurrences of each tag are displayed. We see that the use of complexType as a container for most constructions in Schemata accounts for most of the expansion.

To break down the complexity measures for the individual components in the SES, we use the method:

SESOps.printDocStatsOfFirstEntities(folder)

The results are given in the Table 14.10.

We see that the file size of each component document correlates well with the number of its elements. The largest document is that of DC (deterministic corrections) which has 340 elements, broken down into 114 entities, 20 aspects, 38 multiAspects, and 70 variables, as well as other supporting elements. The components range in size down to the smallest, VD, with 7 elements, with an average size of 100 elements. The wide range in size

Table 14.9. Schema Generated from SES

Schema (*.χsd) elements	Number of occurrences
all	1
attribute	254
complexType	288
element	465
enumeration	51
extension	65
fractionDigits	3
length	1
maxExclusive	1
maxInclusive	2
maxLength	9
minExclusive	1
minInclusive	12
pattern	4
restriction	49
schema	1
sequence	205
simpleContent	81
simpleType	33
totalDigits	3

reflects the diversity of the system aspects covered and the technical sophistication needed to describe them. Indeed, the largest component deals with the data and metadata associated with the error correction processing of the radar returns, an elaborate set of procedures that are necessary to obtain the most reliable and accurate images and downstream products possible.

Although the UPHD SES contains over 100 multiAspects, they do not fall into deeply nested patterns. We have developed a method to obtain the recursive depth of a multiAspect and its volume as illustrated in Figure 14.10. Large recursive depth and large volume indicate a potential for exponential growth of pruned entity structures.

Applied to the UPHD, we found that the average depth and volume is found to be small. This reflects the fact that multiAspects are mainly used for indexing of data at leaves of the SES. The implication for pruning is that the growth in schema instances is not likely to be a significant issue.

Table 14.10. Breakdown of Measures to Individual Components

Component	MD	PL	SV	PD	WF	DC	TG	VH	VD	FH
Number of elements	295	197	84	44	107	340	30	26	7	27
Depth of root node (longest path from root to a leaf)	12	11	11	8	8	15	9	8	6	9
Maximum breadth (largest number of children)	26	17	6	6	9	11	3	4	2	4
File Size	25K	17K	8K	4K	10K	32K	3K	3K	1K	3K
Element occurrences										
complexVar	52	27	4	9	15	35	0	5	1	1
entity	130	69	28	18	36	114	12	11	3	16
aspect	27	13	5	4	4	20	3	2	1	7
var	31	31	15	3	20	70	3	0	0	1
numberComponentsVar	19	24	10	5	16	38	3	4	1	1
simpleVar	7	4	6	0	0	9	3	0	0	0
addvar	10	5	6	0	0	16	3	0	0	0
multiAspect	19	24	10	5	16	38	3	4	1	1
Multiple Occurrences	yes	yes	yes	no	no	yes	yes	no	no	yes

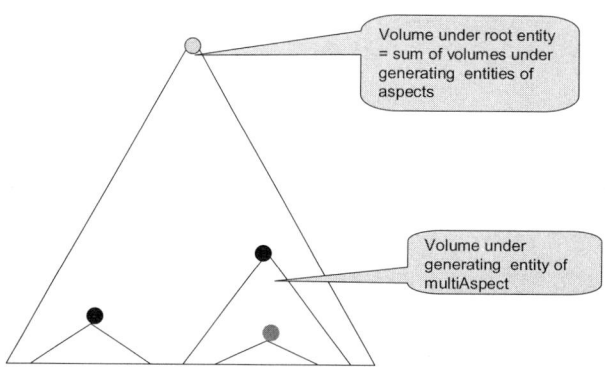

○ = root entity
● = first level multiAspect
● = second level multiAspect

Figure 14.10

Recursive Depth of a multiAspect and Its Volume

Summary

We was employed the SES-based methodology to model a standard for capturing, producing, and interpreting data sensed by Synthetic Aperture Radar (SAR) systems. Extensive contributions to data modeling, metadata harmonization, and XML naming conventions were needed to take the data model that had been developed in textual form into a generic schema and data dictionary applicable across the wide spectrum of SAR sensor systems. The SES-based methodology was supported by an intimate interaction between the knowledge representation developers and the subject matter experts. Indeed, in such knowledge engineering endeavors, the desired results can only be obtained from experts who enable the depth of understanding needed to come up with the abstractions that express the truly common elements shared by a wide variety of similar systems.

References

1. Universal Phase History Data (UPHD) Standard Document #STDI-0007 Version 0.90,14 SEP 2006, NGA/IID
2. Einar-Arne Herland, "Operational use of SAR interferometry for DEM generation and land use mapping," VTT Automation, Remote Sensing Group, P.O. Box 13031, 02044 VTT Finland, http://www.geo.unizh.ch/rsl/fringe96/papers/herland/ (accessed Nov. 2006)

Appendix A: Abbreviations

- Num or num: number
- Targ: target
- Nom: nominal
- Loc: location
- Orient: orientation
- Expl: exploitation
- Az: azimuth
- Rg: range
- Sl: slant
- Ref: reference
- Dir: direction
- flg: flag

- blk: block
- pwr: power

Appendix B: UPHD Table of All Paths

.UPHDmaster
.UPHDmaster.DC
.UPHDmaster.DC.Bm_ISL
.UPHDmaster.DC.Bm_ISL.DeltaTheta
.UPHDmaster.DC.DC_AGC
.UPHDmaster.DC.DC_AGC.Flg
.UPHDmaster.DC.DC_AGC.Flg.Domain
.UPHDmaster.DC.DC_AGC.T_Delay
.UPHDmaster.DC.DC_AGC.T_DomainComp
.UPHDmaster.DC.DC_AGC.XferFunc
.UPHDmaster.DC.DC_AGC.XferFunc.AGC_FreqStep
.UPHDmaster.DC.DC_AGC.XferFunc.CompAmp
.UPHDmaster.DC.DC_AGC.XferFunc.CompPh
.UPHDmaster.DC.DC_AGC.XferFuncStartFreq
.UPHDmaster.DC.DC_Rx
.UPHDmaster.DC.DC_Rx.BmFrmr
.UPHDmaster.DC.DC_Rx.BmFrmr.AmpComp
.UPHDmaster.DC.DC_Rx.BmFrmr.PhComp
.UPHDmaster.DC.DC_Rx.RxBm
.UPHDmaster.DC.DC_Rx.RxBm.Angles
.UPHDmaster.DC.DC_Rx.RxBm.Angles.ElemPattCent_Az
.UPHDmaster.DC.DC_Rx.RxBm.Angles.ElemPattCent_El
.UPHDmaster.DC.DC_Rx.RxBm.Angles.RelSteer
.UPHDmaster.DC.DC_Rx.RxBm.Angles.RelSteer.Az
.UPHDmaster.DC.DC_Rx.RxBm.Angles.RelSteer.AzAxis
.UPHDmaster.DC.DC_Rx.RxBm.Angles.RelSteer.El
.UPHDmaster.DC.DC_Rx.RxBm.Angles.RelSteer.ElAxis
.UPHDmaster.DC.DC_Rx.RxBm.Angles.RelSteer.RelSteerRef
.UPHDmaster.DC.DC_Rx.RxBm.Angles.RowsToEvector
.UPHDmaster.DC.DC_Rx.RxBm.BmAntennaType
.UPHDmaster.DC.DC_Rx.RxBm.ElemGain
.UPHDmaster.DC.DC_Rx.RxBm.ElemPattern
.UPHDmaster.DC.DC_Rx.RxBm.Gain
.UPHDmaster.DC.DC_Rx.RxBm.Gain.OnAxis
.UPHDmaster.DC.DC_Rx.RxBm.Gain.SteerFactor

Appendix B: UPHD Table of All Paths

.UPHDmaster.DC.DC_Rx.RxBm.GridRowDir
.UPHDmaster.DC.DC_Rx.RxBm.IncAng
.UPHDmaster.DC.DC_Rx.RxBm.IncAng.ElemPattAz
.UPHDmaster.DC.DC_Rx.RxBm.IncAng.ElemPattEl
.UPHDmaster.DC.DC_Rx.RxBm.IncAng.PatternAz
.UPHDmaster.DC.DC_Rx.RxBm.IncAng.PatternEl
.UPHDmaster.DC.DC_Rx.RxBm.ISLR
.UPHDmaster.DC.DC_Rx.RxBm.Pattern
.UPHDmaster.DC.DC_Rx.RxBm.PatternSpacingFactorAz
.UPHDmaster.DC.DC_Rx.RxBm.PatternSpacingFactorEl
.UPHDmaster.DC.DC_Tx
.UPHDmaster.DC.DC_Tx.TxBm
.UPHDmaster.DC.DC_Tx.TxBm.Angles
.UPHDmaster.DC.DC_Tx.TxBm.BmAntennaType
.UPHDmaster.DC.DC_Tx.TxBm.ElemGain
.UPHDmaster.DC.DC_Tx.TxBm.Gain
.UPHDmaster.DC.DC_Tx.TxBm.IncAng
.UPHDmaster.DC.DC_Tx.TxFt
.UPHDmaster.DC.DC_Tx.TxFt.PwrProfile
.UPHDmaster.DC.DC_Tx.TxFt.PwrProfileInc
.UPHDmaster.DC.DC_Tx.TxPulsePwr
.UPHDmaster.DC.IFFilter
.UPHDmaster.DC.IFFilter.IFFilterXferFunc
.UPHDmaster.DC.IFFilter.IFFilterXferFunc.IFF_AmpComp
.UPHDmaster.DC.IFFilter.IFFilterXferFunc.IFF_FreqStep
.UPHDmaster.DC.IFFilter.IFFilterXferFunc.IFF_PhComp
.UPHDmaster.DC.IFFilter.IFFilterXferFunc.IFF_StartFreq
.UPHDmaster.DC.IonoTECV
.UPHDmaster.DC.Mixer
.UPHDmaster.DC.Mixer.MixerXferFunc
.UPHDmaster.DC.Mixer.MixerXferFunc.FreqStep
.UPHDmaster.DC.Mixer.MixerXferFunc.XferFunc_AmpComp
.UPHDmaster.DC.Mixer.MixerXferFunc.XferFunc_PhComp
.UPHDmaster.DC.Mixer.MixerXferFunc.XferFunc_StartFreq
.UPHDmaster.DC.TropoN0
.UPHDmaster.FH
.UPHDmaster.FH.Endian
.UPHDmaster.FH.FH_Bm
.UPHDmaster.FH.FH_Bm.BmRx
.UPHDmaster.FH.FH_Bm.BmRx.NumValues
.UPHDmaster.FH.FH_Bm.BmTx
.UPHDmaster.FH.FH_Bm.BmTx.NumValues

.UPHDmaster.FH.FH_Rx
.UPHDmaster.FH.FH_Rx.ChanMap
.UPHDmaster.FH.FH_Tx
.UPHDmaster.MD
.UPHDmaster.MD.BE_Num
.UPHDmaster.MD.CntryCd
.UPHDmaster.MD.CollectorName
.UPHDmaster.MD.Cov
.UPHDmaster.MD.Cov.PosECI
.UPHDmaster.MD.Cov.VelECI
.UPHDmaster.MD.Cov.VelPosCrossECI
.UPHDmaster.MD.CR_Num
.UPHDmaster.MD.CSD
.UPHDmaster.MD.CSD.CDR
.UPHDmaster.MD.CSD.CDR.EphCorrCoef
.UPHDmaster.MD.CSD.CDR.EphErrDecorrRate
.UPHDmaster.MD.CSD.CDR.IonoErrCorr
.UPHDmaster.MD.CSD.CDR.IonoErrDecorrRate
.UPHDmaster.MD.CSD.CDR.RgBiasDecorrRate
.UPHDmaster.MD.CSD.CDR.RgBiasErrCorr
.UPHDmaster.MD.CSD.CDR.TropoErrCorr
.UPHDmaster.MD.CSD.CDR.TropoErrDecorrRate
.UPHDmaster.MD.CSD.IonoRg
.UPHDmaster.MD.CSD.IonoRg.ErrStdDev
.UPHDmaster.MD.CSD.IonoRg.RateErrStdDev
.UPHDmaster.MD.CSD.IonoRg.RgRateCorrCoef
.UPHDmaster.MD.CSD.PMM
.UPHDmaster.MD.CSD.PMM.AzStdDev
.UPHDmaster.MD.CSD.PMM.RgAzCorrCoef
.UPHDmaster.MD.CSD.PMM.RgStdDev
.UPHDmaster.MD.CSD.RgBiasStdDev
.UPHDmaster.MD.CSD.STE
.UPHDmaster.MD.CSD.VDF
.UPHDmaster.MD.GMTI
.UPHDmaster.MD.GMTI.CDP_SubDwells
.UPHDmaster.MD.GMTI.FAR
.UPHDmaster.MD.GMTI.GMTIPD
.UPHDmaster.MD.GMTI.GMTIPD.GMTI_PD
.UPHDmaster.MD.GMTI.GMTIPD.PD_AzWidth

15
Processing Networks and Pragmatic Frames

Recall Figure 8.2 that recast the information exchange framework in the context of the SES ontology framework. In a web-service environment, the producer sends SOAP messages containing XML documents generated by pruning an SES or its schema implementation. The consumer receives and interprets these messages using the same schema in which they were sent. Such a message encodes a world state description (or changes in it) that is a member of a set delineated by an ontology. The ontology must support the pragmatic frame, i.e., a description of how the information will be used in downstream processing.

In this chapter, we'll formulate the network of nodes that carry out such processing levels within the SES/DEVS framework of Chapter 11. This formulation will allow us to analyze the aspects of pragmatics that relate to data exchanges among producers and consumers. This analysis will then provide a framework for modeling real processing collaborations that can be set up on a Service Oriented Architecture (SOA[1], see Wikipedia definition) and for answering design questions about such collaborations.

---

[1] Pronounced soh-ah.

Processing Network Pragmatics

To better understand the pragmatics of processing we will consider a collection of producers and consumers that are collaborating with some common purpose. We'll call such collaboration a *processing network* and will depict it by a data flow diagram such as Figure 15.1. We have seen that this kind of network can be represented by an SES and transformed to a DEVS-coupled model (Chapter 11). The figure depicts a scenario in which a sensor observes a target and forwards its observations to a shooter with the objective of hitting the target. The same sensor subsequently observes the target and now sends its observations to an assessor that compares the state of the target before and after the shot has fired (i.e., we assume the sensor has also sent the original state to the assessor). A diagnoser records the action for further downstream use such as later review. We'll refer to this scenario as "simple" because it simplifies the interactions that might be characteristic of a real military exercise.

In such a processing network, data is generated, transmitted, modified, stored, and/or retransmitted as it moves from one

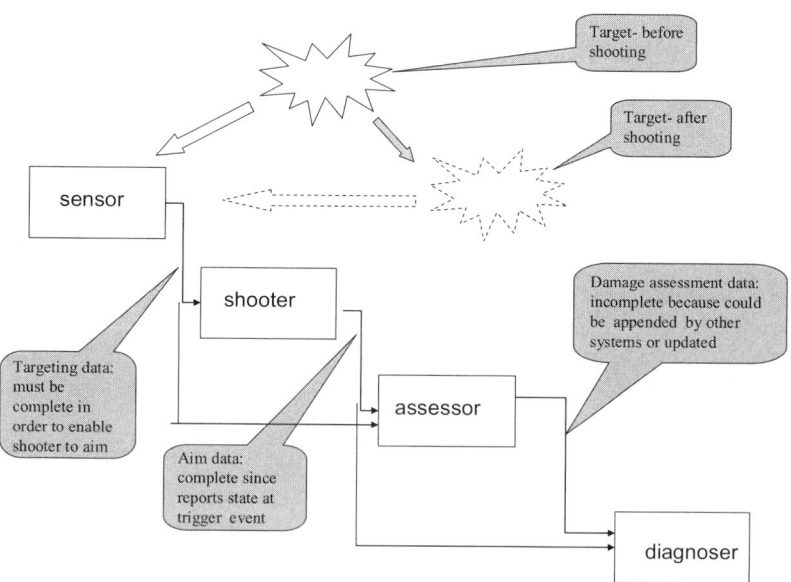

Figure 15.1

A Simple Processing Network

Table 15.1. Pragmatic Frame Attributes for Producer/Consumer Pairs

consumer → producer ↓	shooter	assessor	diagnoser
sensor	**Use** — to aim at the target **Intent** — change world state **Completeness** — complete **Urgency** — high **Tense** — future	**use** — compare the target information used by shooter with the target state after being shot **intent** — describe world state **completeness** — incomplete **urgency** — moderate **tense** — past	
shooter		**Use** — diagnose the assessment result based on the aim data generated by shooter **intent** — describe world state **completeness** — incomplete **urgency** — moderate **tense** — past	
assessor			**use** — to record damage assessment data generated by assessor **intent** — describe world state **completeness** — incomplete **urgency** — low **tense** — past

processor to the next. An account of the pragmatics must include not only the use to which the data is put as it moves along, but also other factors that characterize the information surrounding this use, such as the *intent* that a producer has in sending the data to a consumer, the state of *completeness* of the data, its *urgency*, and its *tense*, whether the data refers to events in the past, present, or future. Let us try to understand these factors more completely. Table 15.1 enumerates the producer/consumer pairs present in Figure 15.1. For example, sensor/shooter is such a pair since it targets data flow from the first to the second.

The table entry for each such pair includes the following attributes:

- **Use** — an indication of how the consumer will use the data.
- **Intent** — an indication of whether the data are intended to be used to change the state of the world or describe it. So when the sensor sends its data to the shooter, the intent is to change the world, but when it sends data to the assessor, the intent is to describe it.
- **Completeness** — an indication of whether the data have reached the final stage of processing or are subject to further processing. So, in our simple scenario, targeting and aim data are complete because at some point you have to shoot based on the best information available. However, the target damage assessment report may be complete or incomplete depending on whether there is to be further action based upon it. In the simple chain of Figure 15.1, we have not included a feedback loop to continue until some level of target damage has been realized. However, in general, assessments continue to be updated or appended for some time after a mission execution.
- **Urgency** — an indication of the degree to which the data must be attended to. In the simple scenario underlying Figure 15.1 we assume that it is more urgent to get data from the sensor to the shooter than from the sensor to the assessor, and the logging has lowest priority of all.
- **Tense** — an indication of the timing of the use of the data relative to its transmission. For example, the sensor sends its observations of the target for use in the future by the shooter. In contrast, the assessor and diagnoser deal with the firing event that occurred in the past, i.e., before the data transmission.

Let's focus for now on the use requirements of the producer/consumer pairs that appear in a processing chain such as those described in Table 15.1. Consider the accumulated requirements of all downstream users of a producer's data. This is illustrated in Figure 15.2 for the sensor where we see that shooter, assessor, and diagnoser are downstream consumers of the sensor's data, whether firsthand, such as the shooter, or secondhand, such as the diagnoser. The accumulated use requirements of the downstream processors consist of:

- the use requirements of the shooter that needs a sufficient description of the target state to enable an effective aim,

Layered Processing Networks

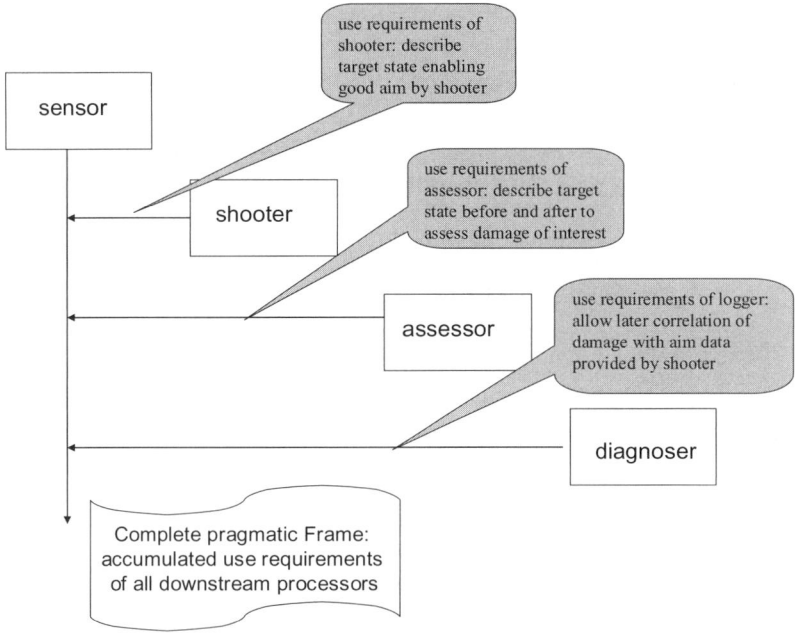

Figure 15.2

Accumulated Use Requirements for Sensor

- the use requirements of the assessor that needs a sufficient description of the target state before and after to allow it to assess damage of interest, and
- the use requirements of the diagnoser that will allow later correlation of damage estimated by the assessor with aim data provided by shooter. Such correlation might provide a diagnosis; for example, damage was light because the target was missed because the aim was off.

These accumulated use requirements constitute the complete pragmatic frame that an ontology for the sensor would have to support. However, in a SOA environment, there can be more tailoring of producer data to consumer needs. Therefore, we need a more formal way of characterizing individual and collective requirements, or pragmatic frames.

Layered Processing Networks

We now consider processing networks where distinct levels will be identified. We have seen examples of such multilevel chains as in the area of geospatial information processing (Chapter 3), where identified levels include Pre-Processed, Geo-Referenced (or Geo-Rectified), and Ortho-Rectified. We first assume that all pragmatic information as outlined in Table 15.1 is included within the metadata elements to support and augment the actual data transmitted from producer to consumer. In a conventional broadcasting environment, all metadata (as well as the data they describe) have to be sent to all consumers. However, a SOA will allow much more flexibility to meet the individual needs of consumers. For example, since different exploitation algorithms require different data collection parameters, a particular data set may be collected to support a particular exploitation algorithm. Here the metadata can provide the information that characterized the data collection constraints. This information allows a downstream user to know the parameters that governed the collection of a particular data set and therefore to assess whether the data set is applicable to the problem at hand. Yet more advanced is the situation where a generic ontology is developed that allows raw data to be captured in a very inclusive format that will support a variety of consumer processing requirements (see Chapter 14). In this case, users can "pull" data from a repository by specifying the metadata needed for their particular pragmatic frames.

We model a processing node by a function

$$f : I \to O$$

where $I, O \subseteq M$, a set of elements representing a universal set of metadata.

Figure 15.3 depicts this concept together with shooter, assessor, and diagnoser examples. We see that the input, or domain, of the function represents the metadata elements that a processor of the function needs in order to perform its task. Likewise, the output, or range, is a set of metadata elements that represents the data produced by the processor and which downstream processors can use as input.

A processing network is an abstraction of a DEVS coupled model in which the atomic models are represented by processing nodes. Formally, we have:

$PN = \langle N, F, IC, EIC, EOC \rangle$

where

PN is a processing network
N is a set of processing nodes

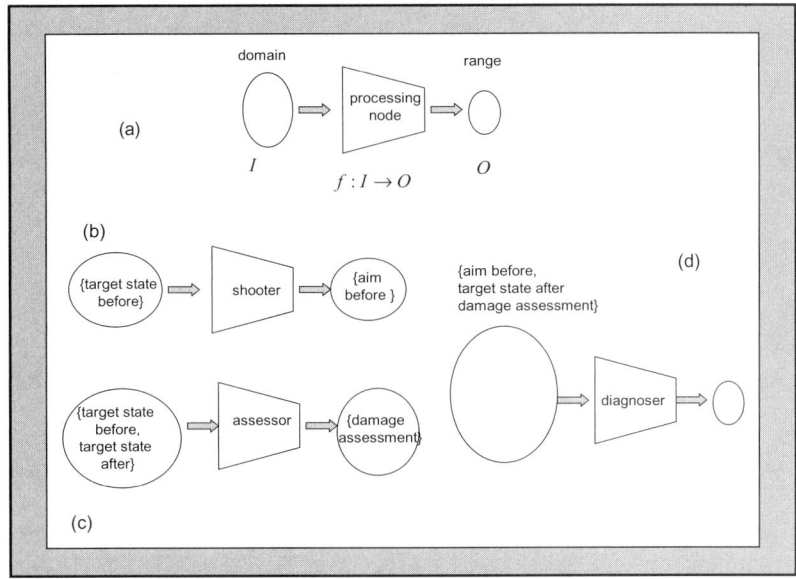

Figure 15.3

Models of Processing Nodes

$F = \{f_n / n \in N\}$, the set of processing functions, one for each node.
$IC = \{Inf_n / Inf_n \subseteq N, n \in N\}$, the internal coupling, given by a set of influences or nodes for each function,
$EIC = \{Inf_n / Inf_n = N, n \in N\}$, the external input coupling, allowing all external inputs to be injected into each node
$EOC = \{Inf_n / Inf_n = \{n\}, n \in N\}$, the external output function, allowing each node's output to be available externally

We require that the directed graph set up by the internal coupling is *acyclic*, i.e., there are no feedback cycles within it. (This is very similar to the strict hierarchy axiom in the SES.)

EXAMPLE: The target tracking processing chain in Figure 15.1 is represented by the following model:

$PN = \langle N, F, IC, EIC, EOC \rangle$

where

$N = \{\text{shooter, assessor, logger}\}$
$F = \{f_{\text{shooter}}, f_{\text{assessor}}, f_{\text{logger}}\}$

where $f: I_n \to O_n$ and

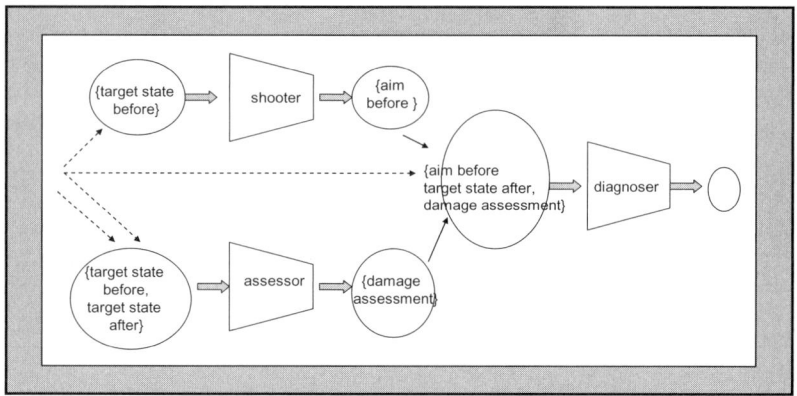

Figure 15.4

Processing Net Model – Dotted Lines Indicate Node-Specific Frames

$I_{shooter}$ = {target state before}, $O_{shooter}$ = {aim before}
$I_{assessor}$ = {target state before, target state after}, $O_{assessor}$ = {damage assessment}
$I_{diagnoser}$ = {aim before, target state before, target state after},
    $O_{diagnosis}$ = { }
IC = {$Inf_{shooter}$, $\varnothing$, $Inf_{assessor}$ = $\varnothing$, Inf_{logger} = {shooter, assessor}}

This model is illustrated in Figure 15.4.

Given a processing network, $PN = \langle N, F, IC, EIC, EOC \rangle$, we can compute the metadata needed by each node that are not supplied by its influencers.

$$\hat{I}_n = I_n - \bigcup_{i \in Inf_n} O_i$$

this is the difference between the set of inputs and the union of all outputs of its influencers.

In general, $\hat{I}_n$ is the metadata deficit that must be supplied to node n from sources external to the processing network. In the example of Figure 15.4. we have:

$\hat{I}_{diagnoser}$ = {aim before, target state after, damage assessment}
    − {aim before, damage assessment}
    = {target state after}
$\hat{I}_{shooter}$ = {target state before} − $\varnothing$ = {target state before}
$\hat{I}_{assessor}$ = {target state before, target state after} − $\varnothing$
    = {target state before, target state after}

Table 15.2. Metadata Deficit for Each Processing Node

Processing node	Input metadata requirements	Influencers	Output metadata supplied by influencers	Metadata deficit = needed sensor metadata
diagnoser	aim before target state after, damage assessment	shooter, assessor	aim before, damage assessment	target state after
shooter	target state before			target state before
assessor	target state before, target state after			target state before, target state after

In this example, the sensor is the only relevant external source. Therefore, it must make up the metadata deficit for each of the nodes. The results are summarized in Table 15.2.

The combined processing functions of a node and all its precursors must be considered in computing the external metadata needed to adequately support the node's processing. The combined deficit of node n is the union of the deficits of all precursors of the given node, i.e.,

$$Deficit(n) = \bigcup_{i \in Precursors(n)} \hat{I}_i$$

where

$$Precursors(n) = \bigcup_{i \in Inf(n)} Precursors(i)$$

The definition of precursors sets up a recursion in which the influencers of a node are its precursors together with their precursors. This recursion terminates at nodes that are leaves, i.e., for which the influencer set is empty.

Exercise

Show that termination is guaranteed by finiteness of the node set and absence of cycles.

A subnet for a node, n, $Subnet(n)$, is defined as the restriction of the overall network to the precursors of n. Figure 15.5 depicts a subnet (shaded) of a network.

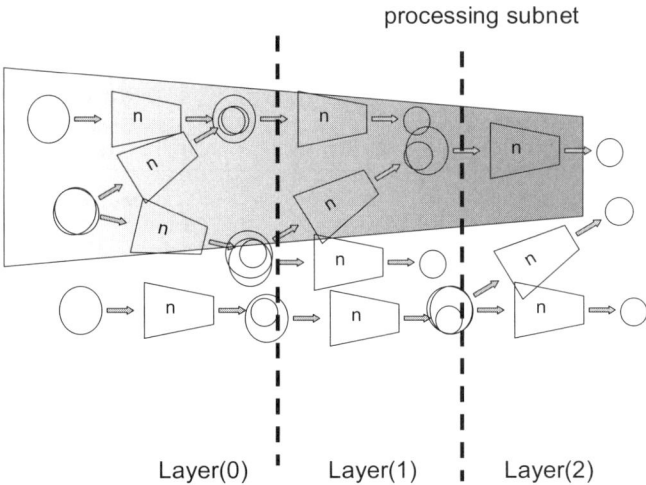

Figure 15.5

A Processing Subnet (Shaded) and Layered Decomposition

Exercise

Delineate the subnets of the target tracking example in Figure 15.4.

Since the combined deficit is the accumulation of these of its precursors, a node's deficit can be expressed as its own immediate deficit together with the deficits of its precursors:

$$Deficit(n) = \hat{I}_n \bigcup_{i \in Precursors(n)} Deficit(i)$$

Therefore for all $i \in Precursors(n)$,

$Deficit(i) \subseteq Deficit(n)$

Another way to look at the combined metadata deficit for each node is that it constitutes the pragmatic frame that external sources must satisfy to support the processing at the given node. Thus, we define the *pragmatic frame associated with node n*:

$$Frame(n) = Deficit(n)$$

and the family of frames associated with PN is:

$$Frames(PN) = \{Frame(n)/n \in N\}$$

Exercise

Use Table 15.2 to define the pragmatic frames for each of the nodes in Figure 15.4.

A way to organize the processing net is by decomposition into a family of layers.

Layer (0), ..., *Layer* (i), *Layer* ($i + 1$), ...

where *Layer* (0) consists of the leaf nodes:

Layer (0) = $\{n/Inf_n = \emptyset\}$

and where *Layer* ($i + 1$) consists of the nodes that have all their influencers in *Layer*(i):

Layer($i + 1$) = $\{m/Inf_m \subseteq Layer(i)\}$

We require that all nodes be included in the decomposition:

$$N = \bigcup_{i=0}^{i=\max} Layer(i)$$

where there must be a maximum layer index since the node set is finite.

The requirement that all influencers of a node lie within the layer below places a strong stratification on the processing net. It requires that all internally processed metadata needed to support a processing function be made available from the layer below.

$$Layer\ Deficit(l) = \bigcup_{i \in Layer(l)} Deficit(i)$$

Exercise

Derive the layered decomposition for Figure 15.4 and show the sensor metadata needed at each layer.

The processing structured by a layered decomposition is visualized in Figure 15.6. At each layer, some metadata are required — these are the LayerDeficit or accumulated deficits for all nodes in that layer. These metadata must be supplied by external sources to adequately support processing at this layer. Some metadata may also be copied directly from one layer to the next to preserve the strong layered structure. Finally, some metadata may be discarded because they are not needed for processing at higher layers.

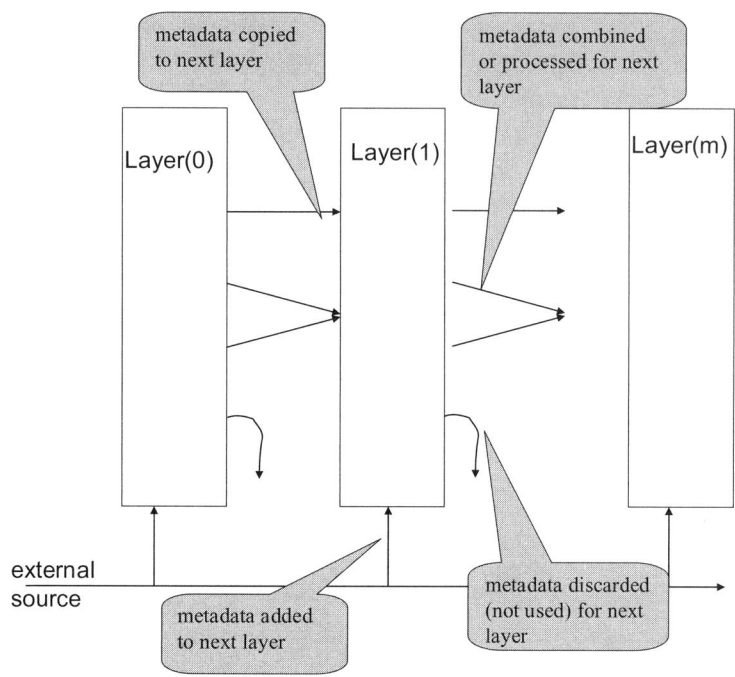

Figure 15.6

Metadata Processing and Transmission from Layer to Layer

Repository of Discoverable Data

Suppose now that all external data is available from one or more repositories. Note that the pragmatic frame associated with a node provides the metadata specification of the external data that the node needs for its processing function. Therefore data in the repositories should be tagged with this metadata to enable nodes to discover all, and only, the subset of data needed for their functions. For example, a single repository holding all raw sensor data should employ an organization of all external metadata elements needed in the network. Individual repositories holding particular types of data should employ an organization of all external metadata elements specific to the type of data they hold. In our information exchange framework, an SES, implemented as an XML schema, is used for such an organization. We then associate a pruned entity structure, in the form of an XML instance document, with each specific data set that is stored. A matching process is needed to support a discovery process in which a user's pragmatic frame is matched against frames that are represented in the repository for the closest match.[1,2]

Dynamically Evolving Processing Networks

A network processing model should not be expected to remain static. It should be able to accommodate new data collection technologies that emerge and new processing functions are demanded as time goes by. Developers of new processing algorithms should be able to compare the algorithm's metadata requirements with the metadata currently offered in the network. When an algorithm is integrated into the processing network the model should be restructured to reflect the function's presence. Moreover, having such an explicit representation at an enterprise level should facilitate the development of novel and important applications that in turn stimulate further enhancement of the network.

Exercise

Consider the evolution of a book from initial concept through several revisions and reprints as in Figure 15.7. Consider each stage in the evolution to be a processing node in a network as modeled above. This network represents some aspects of the work flow of a book publisher. Let's track a book as it moves through this network by associating with each node, the metadata, in the form of an SES, needed to process the book at that stage. For example, the metadata needed for the book in manuscript form is an elaboration of the SES for content in Figure 8.5. Likewise, the SES with physicalDec in Figure 8.5 can be elaborated to describe physical layout. Since printing requires both content and physical layout, both SESs must be merged to support the printing stage. Now consider a second printing as the next layer of processing. For example, if the book first came out in hard cover, then a second printing might be in soft cover. Subsequently, a revision followed by a third printing constitutes the layer of processing after that. Consider the revised edition to be given in the form of the change in content from the first edition, not the new content itself.

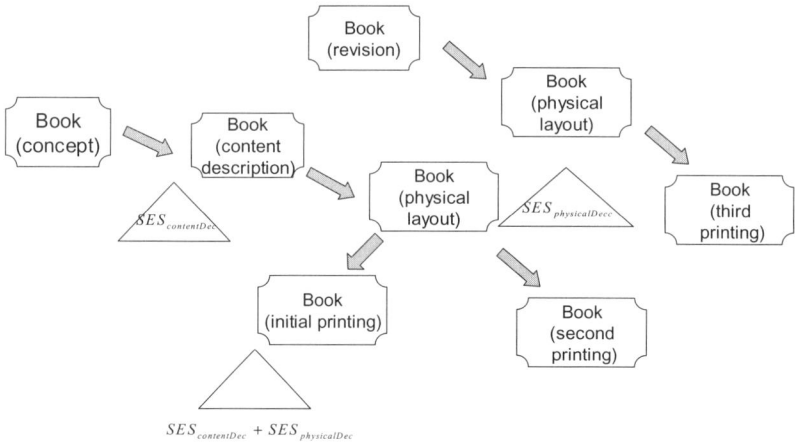

Figure 15.7

Evolution of a Book

1. Draw a layered network showing the SES for metadata required as input at each node in each layer.
2. Describe the pragmatic frame associated with each node — consider the author and layout designer as external sources of data and metadata.
3. Consider editing of content and printing to be separate services available in a SOA. If the book data is held in a repository, how should the data be organized to supply only that part needed by an editor? by a printer? *Hint:* Which metadata characterizes the needs of each service?
4. Describe how the processing network might change as the book evolves through further reprints and editions.

Summary

In a SOA environment, data dissemination may be dominated by "user pull of data," discovery using metadata, and automated retrieval of data to meet user pragmatic frame specifications. This chapter has developed a framework for modeling the processing networks that emerge in SOA environments. We discussed the framework's implications for supporting repository-based data dissemination. We also consider the dynamic nature of processing network evolution. Chapter 16 will use this background to consider in greater depth the nature of pragmatic frames in such dynamic environments.

References

1. Yilmaz, Levent, On the Need for Contextualized Introspective Models to Improve Reuse and Composability of Defense Simulations, http://www.scs.org/pubs/jdms/vol1num3/Yilmaz.pdf (accessed Nov. 2006)
2. Yilmaz, Levent, and Swetha Paspuleti, Toward a Meta-Level Framework for Agent-Supported Interoperation of Defense Simulations, http://www.scs.org/pubs/idms/vol2num3/Yilmaz-Paspuleti-pp161-175.pdf (accessed Nov. 2006)

16

Dynamic Pragmatics: Issues and Methodology

Definitions that distinguish metadata and data are often ambiguous. We understand metadata to be data that describes other data. Raw data are data that resist further decomposition for the context in question. In other words, metadata is a hierarchical concept in which metadata are a descriptive abstraction above the data it describes. From here on, we will use the term metadata in this general sense.

How good or how appropriate is a metadata field or a collection of metadata fields in existing data exchange standards for a range of possible applications or missions? And, how do we manage change of goodness or appropriateness? These are complex, but important, pragmatic level questions that must be asked when operating in the data-centric environment of Service Oriented Architecture (SOA) and Web Services. Before the widespread use of XML to carry information, data engineers worked at the code level for specific systems in specific contexts. This information was unlikely to be shared by other systems or services in its raw "untranslated" format. For example, in a traditional sequential format, one program receiving data from another program might be instructed to interpret the second block in a message as an urgency code in the range from 0 to 5, with 5 being the most urgent. Any number outside that range would be interpreted as error. This illustrates the fact that in traditional data engineering, syntactical conventions are defined very rigidly and are hard-coded into the programs of both sender and receiver. As a consequence, the need for translation from the sender's format to the receiver's format is eliminated. However, this hard coding greatly restricts the flexibility of information exchange. "Stovepiped" solutions typically employ new or revised proprietary syntax formats or abstruse ontologies with difficult to implement languages (Chapter 13). They may continually demand a faster processor and more memory to keep execution time under control. However, systems, especially in a SOA environment, are becoming much more complex. The dialogs and "polylogs" between systems are expected to convey meaningful, timely (sometimes critical), and precise information that enables quick and appropriate action in many different contexts.

The need to translate metadata from one application context or *pragmatic frame* to another is an important requirement that has heretofore received relatively little attention in the data engineering or modeling and simulation literature. Let us acknowledge that there are many possible translations that may be correct and some will be indeterminate.[1] Nevertheless, it is possible to establish the elements that must be present in any correct and usable translation. Indeed, this book presents an approach that attempts to harmonize existing standards that are intended for the same area (Chapter 13). This harmonization then becomes a basis for a new standard to be applied broadly and consistently. This approach is practical and useful for the engineering community to adopt. A successful product of this methodology, presented in Chapter 15, shows it can reduce cost and increase interoperability. The methodology is based on a mathematically rigorous, set theoretical process which goes well beyond symbols and syntax to examine semantics and pragmatics of metadata. It is important to remember, as the level of abstraction becomes higher, that the total information complexity exceeds the sum of the individual information fields. The semantics and pragmatics of a group of metadata fields is a function of their use in some specific context or set of contexts.

This chapter reviews and applies concepts from information and linguistic theory to enrich the formulation of pragmatic

---

[1] Arguments about the possibility of attaining truly equivalent translations of statements (formal or informal) have raged for centuries and are no closer to resolution in the worlds of philosophy/mathematics/linguistics.[1-4]

frame and pragmatic equivalence first presented in Chapter 13. On this foundation, we present a process for harmonizing data exchange standards that builds upon the earlier discussion in Chapter 3. The process begins with the identification of pragmatically comparable candidate metadata and continues through the data engineering process to include analysis which determines whether the metadata are interchangeable and to what degree. Although we do not go into detail, the approach provides a basis for applying recognized quantitative techniques from set theory, multivalued logic, linear algebra, and vector analysis. In particular, multidimensional scaling analysis and principal component analysis may provide effective means to measure functional similarity/dissimilarity.

Information Exchange between Humans, Systems, and Services

In the early days of digital information exchange, it was quickly recognized that error detection and correction were critical to effective exchange of information between systems,[5,6] even though the overhead of the additional data could grow to be larger than the data payload itself. Methods for detection and correction were developed at the syntactic or technical level. However, the complex information that systems will exchange in the evolving net-centric environment requires error detection and correction metadata at higher semantic and pragmatic levels. Symbolic-level error detection and correction are analogous to the way that students acquire new languages. They spend considerable time developing pronunciation, orthography, grammar, reading, and listening comprehension. These are the fundamental information exchange elements that allow errors to be detected and corrected by native speakers and teachers. The student must have this technical-level detection and correction capability before moving on to meaningful conversation, scientific literature, negotiation, or diplomatic exchanges.

In order to detect and correct errors, and ultimately to translate from one system of communication to another, it is necessary to add sufficient information, i.e., more metadata to existing symbolic metadata. This information allows systems and or humans to detect and correct errors in semantic and pragmatic level exchanges as the information traverses a set of services. It also supports the capability to provide valid and relevant translations where necessary. Unrelated groups of words or tags or fields are not quite adequate for effective services, just as they are not quite adequate for human communication. Human translators rely on dictionaries, thesauri, translation guides, and/or years of experience to provide good natural language translations that capture both the technical facts and the cultural nuances of translation from one language and perspective to another. The fundamental problem is that the meaning of a word, field, or combinations of words and fields may not be the same for all systems in all contexts. For example, consider trying to use dictionaries and thesauri alone to translate the English version of Tom Sawyer into Arabic. Without experience, training, or adequate understanding of the two cultures, you will produce a text that is unreadable and probably bizarre to the native speaker/reader of Arabic. Good translations require not only the technical error detection and correction of the right letters, word forms, and grammar, but also a more subtle understanding of "what means what" in both languages.[7-11]

The question, then, is how do we provide such semantic and pragmatic level correction mechanisms in a Service Oriented Architecture environment? Humans do it fairly well in static contexts, but even a small change in context will result in miscommunication. Such miscommunication will persist until all the words and combinations of words and their meanings become (implicitly or explicitly) mutually understood between communicating parties in a given context at a given time.[2] These concerns for uncontrolled miscommunication have existed for some time in machine-to-machine communication systems, but until recently, they have been considered as non-critical or "un-addressable." However, the developing operational expectations for net-centric, SOA-based environments have made addressing this problem at the pragmatic level critically important.

Let us continue with our language translation example to help bring the general problem more into focus. This will lead to an examination of sensors and a data engineering problem in the SOA. Imagine that a native speaker of French, a native speaker of German, and a native speaker of English all observe a car accident on the corner of Maple and Vine somewhere in the United States and have been called to be witnesses in a court proceeding on a specific date in 2006. It should be a simple matter

---

[2] This process of mutual adjustment is very evident in the language change found in groups trying to distance, or distinguish, themselves from a conventional or older group.

of collecting statements from the witnesses in their respective native languages and translating them so that they can be used in an English-speaking court. There are many communicative assumptions,[12] that must be made if these translations are to be used in court objectively and for the sake of justice. First, were all of the witnesses in a physical position to provide the court useful, factual information? Were all of the witnesses technically able to understand what they saw and convert what they saw into statements of fact? Finally, can the authority in charge of translation convert the perspectives, constraints, and peculiarities of each person and language system into a common expression of information and perspective that will be useful to the court? The answer is yes, to some degree. It is certainly not guaranteed. Some authority must ultimately be the interpreter of meaning, which has been shaped and formed by witnesses, translators, technical experts, police officers, and attorneys. Once provided to the court, could the data still be misinterpreted or manipulated? Yes, but the careful examination of the metadata and precise coding will support a sincere outcome. Table 16.1 depicts the analogy and an abstraction that will be developed for the net-centric, SOA environment.

The analogy in Table 16.1 is basically as follows. Witnesses at a trial are comparable to sensors of geospatial information. Compounding and reconciling witness testimony to obtain a verdict is like fusing data from such sensors to obtain an accurate world state. Court translators brought in to take foreign language testimony are like the data harmonization engineering team. The activities of the court officials are like those of downstream consumers of sensor information. *The point is that all these aspects have to be taken into account to achieve useful harmonization.*

General Pragmatics

Pragmatics is defined as the use of metadata in relation to metadata structure and context of application. It is based on Speech Act Theory,[8-11] and focuses on elucidating the intent of the semantics constrained by a given context. For example, suppose that I say: "I see the plane." There is not enough context here to determine whether the word "plane" refers to a flat space defined by at least three points, an airplane, or a wood working tool, which are all valid semantic values for the word "plane." It does not make sense to examine, in detail, the low-level semantics of attributes of the word "plane" when an examination of the use of the plane will obviate further examination of the details. The addition of either metadata or contextual information will help

Information Exchange between Humans, Systems, and Services

Table 16.1. Pragmatic Frame Analogy

Language analogy example	Military example	General example
Witnesses and qualities	**Sensors and qualities**	**Abstraction**
French-speaking witness — Engineer — Good vision — Bad hearing — Far from event	UAV, satellite, etc.	Sensor 1 providing data in a specific format from a given perspective with various defined qualities
German-speaking witness — School teacher — Bad vision — Good hearing — Near to event	Seismic sensor	Sensor 2 providing data in a specific format from a given perspective with various defined qualities
English-speaking witness — Construction worker — Fair vision — Fair hearing — In between French and German witnesses	Ground-based multisensor system	Sensor 3 providing data in a specific format from a given perspective with various defined qualities
Court translators	**Data engineering/ harmonization team**	**Syntax, semantics, and pragmatic frame interpretation**
French to English (very good — experienced)	Fixed Record Length to XML	Proprietary to common
German to English (adequate — newly hired)	HDF4/5 to XML	Proprietary to common syntactic system
Court officials	**Analysis and execution**	**Analysis and execution**
Prosecutor	Operator/Analyst	Analyst, technical interpretation
Defense	Operator/Analyst	Analyst, technical interpretation
Judge	Airborne warning and control system	Analysis/Command authority
Jury	Airborne warning and control system	Analysis/Command authority
Jailor	Weapons systems	Executive/Action authority

with the determination. I could write in English with the help of my word processor: "I see the plane$_{(woodworking\ implement)}$," or I could add information in the sentence: "I see the plane, and it needs to be sharpened." Either additional source of information will serve to isolate the context from the general and ambiguous, but valid, individual semantic possibilities. In English, the structure should also be faithful to conventional English grammar (syntax) rules of *SVO*: subject in the nominative singular case, verb in the indicative mood, present tense singular, noun, accusative case singular.[6] For example, consider "Me the plane see." This sentence does not follow widely held rules of English grammar. We would probably be able to figure it out though, just as we do if Yoda were to say it. However, we humans are particularly good at solving such problems. Net-centric systems and services are not (yet) good at such problem solving. This example illustrates the need for additional metadata as a mechanism to describe intent in a given context to disambiguate the contents of data fields within a conventional structure or rule set.

Data Engineering and Metadata Harmonization

Metadata harmonization is the detailed analysis and comparison of metadata that is similar in structure and function from disparate standards and sources in order to create metadata that is at once more expressive and interoperable. Harmonization efforts can take place to support new design, import of new sources, or to make existing systems function similarly to share metadata. A harmonization effort would include an examination of symbolic (spelling choice, capitalization), syntactic, semantic, and pragmatic level information from several standards, documents, or actual systems output files (where no standard exists). According to The Technical Committee on Data Engineering (TCDE) of the IEEE Computer Society: "Data Engineering is concerned with the role of data in the design, development, management and utilization of information systems. Issues of interest include database design; knowledge of data and its processing; languages to describe data, define access and manipulate databases; strategies and mechanisms for data access; security and integrity control; and engineering services and distributed systems." Model and Simulation-based data engineering relies on rigorous principles to ensure that metadata structures can be designed, represented, constructed, and implemented in such a way that facilitates automated comparison and analysis, translation, and implementation.

Static Pragmatic Frames

Before we can examine dynamic metadata we must first understand a concept called the "static pragmatic frame." As discussed in Chapter 13, the comparison of two metadata entries from distinct standards begins with a general examination of how the standard is conventionally used and then individual metadata entries are examined in context of use. In linguistics, languages are often examined in terms of how they express three concepts: *Mood* (indicative, imperative, etc.), *Tense* (past, present, and future), and *Aspect* (action which is either completed or not). Based on these concepts from linguistics, we suggest defining three parts to the pragmatic frame: *Intent, Tense,* and *Completeness*. We will describe Completeness and Tense first. Intent is the least familiar and most complex of the three so we will focus the remainder of our discussion on it. Completeness refers to whether the data described in the field are complete in terms of its processing. Are the data complete, or can they be processed further downstream by other processors and be changed, deleted, or appended? Tense refers to whether the metadata or data is relevant in the past, present, or future. Once the pragmatic frame is analyzed, the semantic and syntactic levels can be examined in more detail. The pragmatic frame may be attached orthogonally to any metadata element. The pragmatic frame may be thought of as another dimension of a metadata hierarchy.

Our formulation of the pragmatic frame relies on the System Entity Structure (SES) as an "implementable" semantic and pragmatic ontology. The axiomatic formulation of the SES (Chapter 5) allows for both hierarchical and logical expressions in a formal mathematical/logical format such as Gödel numbers. We will show that the use of Gödel numbers allows for translations that claim to be equivalent to be examined rigorously. An SES for a pragmatic frame is given in Figure 16.1.

The intent component in Figure 16.1 has the following sub-components:

Authority (A) refers to the quality of the information provided by the sensor or the sensor's capability.

Urgency/consequences (U/C) refers to the consequences of the proposition being sincere and something going wrong or having it executed improperly.

Relationship (R) refers to the relationship between the sensor and the consumer.

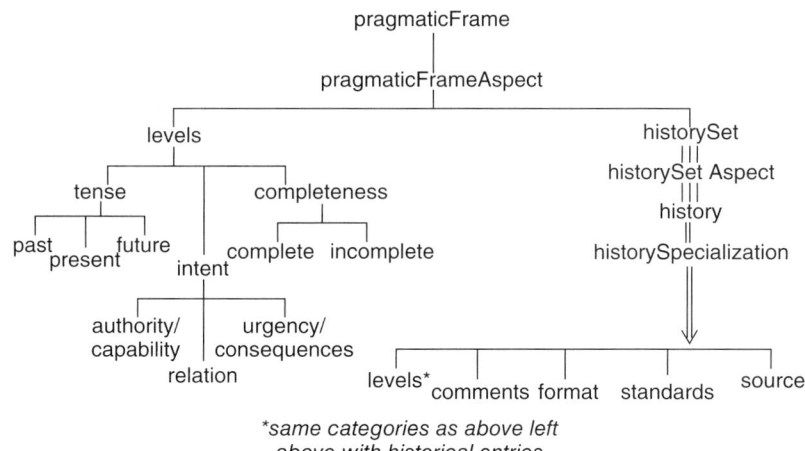

Figure 16.1

General Pragmatic Frame SES Graphical Representation

Table 16.2. Pragmatic Frame Elements

For a given sensor and message	Authority	Urgency/ Consequences	Relationship	Tense	Completeness
Consumer$_1$					
Consumer$_2$					
Consumer$_3$					

In addition we have the entities mentioned earlier:

Tense (T) refers to the time of the process: does it take place in the past, present, or future?

Completeness (C) refers to whether the data/information in the proposition is complete or will be appended before it is consumed.

Therefore in a given application context, a Pragmatic frame (PF) = {A, U/C, R, T, C} as illustrated in Table 16.2.

EXAMPLE: *Evaluating Service Candidates*

The rationale for the examination at the pragmatic level first is similar to looking for the right candidate for a job or a service. Let's say that a general practitioner has told you that you need a neurosurgeon for a neurosurgical procedure. You might look in the phone book under physicians and then for the specialist you need. At this point you must rely on certain certification agencies to assure you that this person is qualified at the lower levels of requisite detail. The neurosurgeon has made it through many technical and administrative hurdles before he or she can be called a certified neurosurgeon. Note that it is not necessary to examine what grade he may have received in a specific course in medical school. At this point, it is enough to say that he or she is qualified, certified, and has a certain number of years of board-certified experience (demonstrated performance).

In the SES in Figure 16.2, we present an abbreviated hierarchical arrangement of neurosurgeons and have assigned a pragmatic frame value based on three variables. The following terms are chosen based on similarities to counterparts in general use, but the meaning is specific to this context. The first variable A is the authority level of the doctor. In this case authority is equivalent to some documented capability. The second variable U refers to the urgency level which refers to the doctor's

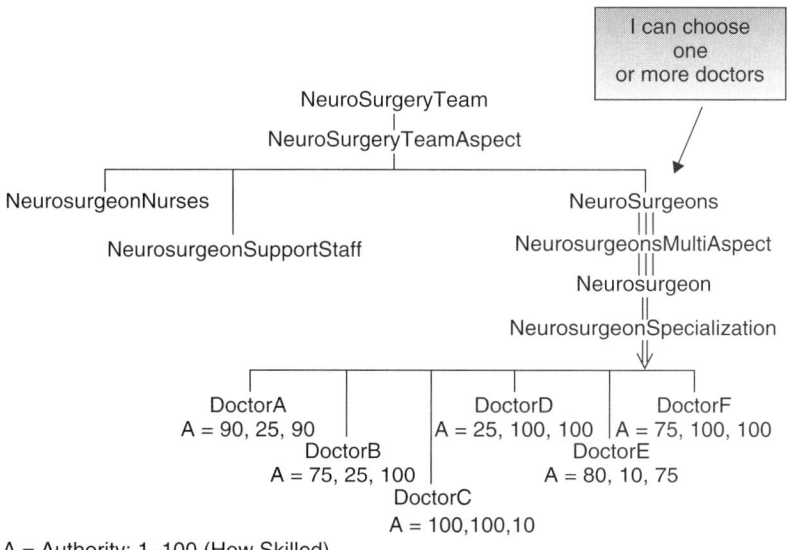

Figure 16.2

NeuroSurgeryTeam SES with Pragmatic Frame Intent Values

availability to fulfill the task. This can also be thought of as the consequences that arise if a doctor is not available. The third variable **R** refers to the relationship level between the doctor who provides the service and the consumer (the patient). Normally this relation occurs through the patient's insurance company. A large number indicates that the doctor is part of the patient's HMO. A small number means the patient will be paying more, if not all, out-of-pocket — *not the normal or ideal relationship*.

This capability scale allows several types of automated analyses to take place. First, I can simply choose the best doctor based on my needs. Second, I can choose back-up doctors, or a team of doctors. As an administrator, I could determine who may need more certification courses, what insurance companies need to be represented, or how closely certain doctors should live from the hospital (measuring their ability to show up for emergencies or meet a certain schedule). Also, if the information were available, I could drill down to the next level to determine what factors give a doctor a certain rating and how I translate or substitute one capability to another or indeed, whether it is possible to translate or substitute capabilities. In a given scenario, either hypothetical or practical, the relationship between the sensor/provider of information or service, the proposition, and the consumer of information or service are considered.

Doctor A is the second most capable, sometimes available and closely associated with my insurance.
Doctor B is exceptionally capable, sometimes available, and on my insurance list.
Doctor C is the most capable and most available, but I will have to pay most of the fee.
Doctor D is not very capable, very available, and on my insurance list.
Doctor E is exceptionally capable, rarely available, but available to my insurance company.
Doctor F is exceptionally capable, very available, and on my insurance list.

Remember, all of these doctors have met the minimum criteria to be certified. It is up to you now to make a decision, based on your priorities. If you want the very best (assuming that evaluation was fair and complete), then you will have to pay out-of-pocket. If you want maximum availability (minimal consequences for delay), you have to choose between D and C and F. F looks to be the best choice there. The thresholds of decision-making will be decided by you, the client.

This process allows quality analysis at a given level based on a given context. You can drill down further if you wish to find out what scores each of the doctors received on their medical school admissions test if you had the time or access to the information. The fact that it is possible does not mean that it should be done. You would drill to that level only if that criterion was somehow important, or if all things were equal to that point and a decision had to be made as a "tie breaker."

Let's now make the analogy to harmonization explicit — doctors represent sensors, treatments represent sensor services, and patients represent application requirements. The pragmatic frame concept allows us to formulate an approach to harmonizing among existing sensor standards in the same way that we can select among doctors or teams of doctors to meet specific patient requirements. Tagging each metadata field in a standard using the A, U, R, T, C parameters allows us to characterize the standard's applicability to the service associated with that field. Rolling up these parameter values over all metadata fields allows us to compare standards for pragmatic equivalence — their abilities to describe world states taking into account the context in which applications might occur (Chapter 13). The ability to find very similar pragmatic frames leads to better analysis at the semantic and syntax levels. Suppose that two standards are equivalent in various pragmatic frames but their syntactical structures are different. It would then be reasonable to build a processor to translate the syntax of one to that of the second so that multiple systems using the two standards could share the ubiquitous information more or less seamlessly. If the differences were only semantic, then another translation system could be built to address that level.

In conclusion, in performing harmonization of disparate standards, we must ask, "What are the functions of the different standards in the context of the systems sharing information?" The concept of pragmatic frame and the related notion of pragmatic equivalence allow us to rigorously answer this question. Characterizing such functionality in this manner enables us to drive the harmonization process (negotiate and change meanings of one or more metadata entries to meet important requirements), translate from one to the other, or abandon completely harmonization in favor of a new system of metadata.

EXAMPLE: *Harmonizing Standards for Targeting*

Let's consider the harmonizing standards with application to targeting. Say that I have a metadata entry called ***position*** in three different standards (Figure 16.3). Standard 1 is a targeting

300 Chapter 16 Dynamic Pragmatics: Issues and Methodology

Pragmatic Frame for Targeting

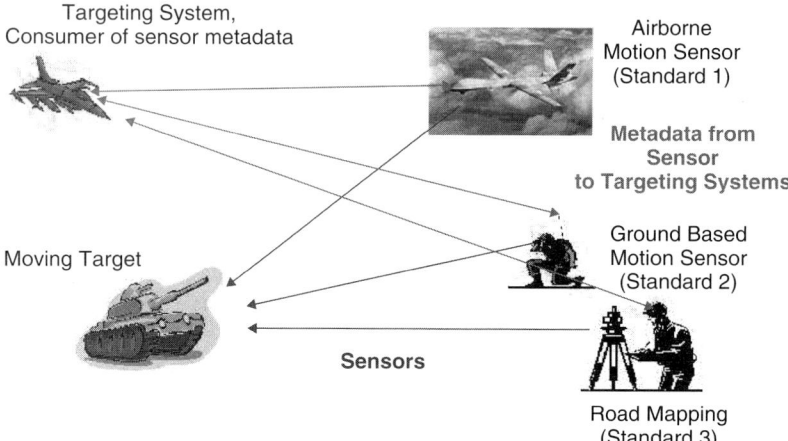

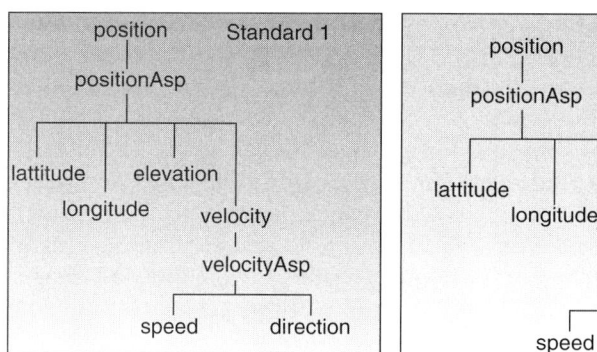

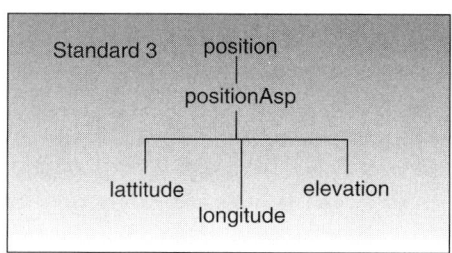

Figure 16.3

SESs for Targeting Mission Example

system related standard; Standard 2 is a two-dimensional motion analysis standard; Standard 3 is a road map design standard for position description. To share information between standards we must first examine the pragmatic frames of the standards to ensure that they could be used interchangeably for a given task by a consumer system. Here are the pragmatic frame values for intent.

Standard 1 (UAV): Authority = 90; Urgency = 90; Relation = 20
Standard 2 (Ground Sensor): Authority = 70; Urgency = 90; Relation = 70
Standard 3 (Road Map): Authority = 25; Urgency = 25; Relation = 10

The sensors and targeting system form a processing network as modeled in Chapter 15. Within such a network we can ask whether one standard for information exchange can replace another that already exists in the network. Let us say that we want to replace the feed to a targeting system which normally uses standard number 2. The relationship level is determined by the frequency of use by the consuming system with the producing sensor system. If they regularly work together or were intended to work together as defined by some specification where they are listed together, then the relationship is high. If they are never even mentioned together in documentation then this value will be low.

If you imagine that a dialogue is taking place between the sensor system and the targeting system, the information provided by standards 1 and 2 conveys a message more appropriate and more likely to be executed by the consuming system than standard 3 does.

Managing Pragmatic Equivalence within the SES

Figure 16.4 illustrates how pragmatic equivalence can be managed within the SES framework. Individual SESs for the standards to be harmonized are developed and enhanced with pragmatic metadata. For example, using merging capabilities (Chapter 12), the SESs for the sensor standards in Figure 16.4 are enhanced with the metadata concerning authority, urgency, and relation within applications such as targeting, analysis, and documentation. Then common entities can be established using the methods

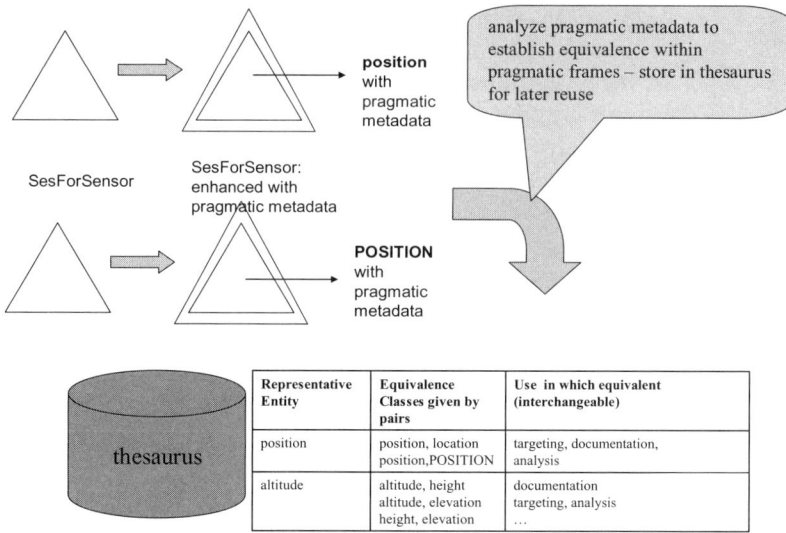

Figure 16.4

Enriching Semantic Equivalence with Pragmatic Equivalence

just discussed. This approach will allow entities in different SESs to be considered common despite the fact that they have different names (such as position and POSITION) and somewhat different substructures. In Figure 16.4, a thesaurus holds equivalence classes of common entities, with each class indexed by a representative. The equivalence classes are given by sets of pairs, i.e., by a mathematical equivalence relation. For example, position, location, and POSITION (from standards 1, 2, and 3, respectively) might be in an equivalence class that is given by the pairs (position, location) and (location, POSITION). The axioms governing equivalence relations (reflexivity, antisymmetry, and transitivity) are built into the thesaurus so that only some of the pairs have to be explicitly given in order for the equivalence class to be inferred. Based on the pragmatic metadata analysis discussed above, a pair of equivalent entities can be further indexed by the application context in which they are judged to be exchangeable. For example, the pair (position, location) might be interchangeable in the targeting application, while (position, POSITION) are interchangeable in analysis. *In this way, a thesaurus, which is a repository based on semantic equivalence, is enhanced to support pragmatic equivalence as well.*

Static Pragmatic Frame Analysis

The analysis of static pragmatic frames focuses on the comparison of one standard to another to determine how similar or dissimilar two frames, subtrees, or trees are from one another. Focusing again on the Intent component of a pragmatic frame, a given assessment of a field, subtree, or tree's pragmatic frame yields a value that can be graphed in three dimensions. Assume we have three standards. One is a reference $\mathbf{a}$ and the other two ($\mathbf{b},\mathbf{c}$) are to be compared to it using a dot product of the pragmatic frames. This provides a measurement of similarity of any two fields when they are to be compared to a third. For example, the (A, U, R) values for Standard $\mathbf{a}$ = (10, 25, 90), Standard $\mathbf{b}$ = (25, 50, 80) and Standard $\mathbf{c}$ = (19, 24, 88). In this case, $\mathbf{a} \cdot \mathbf{b} = a_1b_1 + a_2b_2 + a_3b_3 = 250 + 1250 + 7200 = 8700$. While $\mathbf{a} \cdot \mathbf{c} = a_1c_1 + a_2c_2 + a_3c_3 = 190 + 600 + 7920 = 8710$. This shows that $\mathbf{c}$ is slightly more similar to $\mathbf{a}$ than is $\mathbf{b}$.[3]

Dynamic Pragmatics

Dynamic pragmatics refers to the change in state of the pragmatic frame due to a change of context occurring in continuous time, discrete time change, or due to a discrete event. For example, the English words "cool stream" in 1900 would probably refer to flow of moderately cold water such as in a brook or a creek. In 2006, the meaning may be the same as 1900, or in a certain context, it could refer to music or video that is delivered in real time across the Internet. As in our previous example, the context is not strictly defined, but the words in isolation may have multiple meanings. Metadata must be appended or revised in order to capture the meaning change over time or due to events. There is a substantial overhead associated with this additional metadata, but like error detection/correction mechanisms associated with all modern data transfer protocols, the overhead is necessary and is mitigated by faster processors and more memory. As

---

[3] It is appropriate to mention that the Cauchy-Shwartz-Buniakovskij inequality $|<\mathbf{a}, \mathbf{b}>| \leq \|\mathbf{a}\| \|\mathbf{b}\|$ applies here and demonstrates the whole is potentially greater than the sum of the parts. A cross product $\mathbf{a} \times \mathbf{a}$ indicates a resultant vector of 0. $\mathbf{a} \times \mathbf{b} = (-2500, 1450, -125)$ and $\mathbf{a} \times \mathbf{c} = (40, 830, -235)$. This value can be interpreted as an indicator of the relative difficulty that must be applied to $\mathbf{a}$ or $\mathbf{b}$ in order to translate from one to the other.[14]

is the case with static pragmatic frame metadata, once the harmonization has been completed, this metadata can be discarded.

There are two primary uses of dynamic pragmatic metadata. The first is for harmonization from one static standard to another, and the second is to harmonize metadata over time within or across standards. The Pragmatic Frame construct described earlier allows for a pragmatics-based history of metadata use in relation to metadata structure and context of application to be captured over time. The frame can be analyzed in two different ways. The first converts the Intent variables to a scalar value so that it can be combined easily with the Completeness and Tense variables. Or, since all three are independent variables, concentration can be applied to the Intent variable. We will apply numerical techniques to examine clustering trends, or determine how to change a standard in order to reach a specific pragmatic frame value.

Dynamic Pragmatic Frame Comparison and Analysis

Metadata definition may not be static and the relationships may change over time depending on the context of use and intent of the producer/consumer. Consider a case in which a metadata field is required to change in order to reflect a new intended application.

In Figure 16.5, let *Metadata*(T_0) indicate that at time T_0 a specific data field is used to support targeting. We'll assume that it is a field that is based upon past requirements and, is disposable and complete.

Now suppose that at time T_1 the same Metadata field, *Metadata*(T_1) now indicates that the same data field is to be used to support damage assessment. *Metadata*(T_1) is now a field that is based upon current requirements, is not disposable, and may be appended or changed.

In terms of our framework, the new field, *Metadata*(T_1) is now part of a **different** pragmatic frame — the intended downstream processing is now concerned, not with targeting, but with assessing the damage caused on the target (see Figure 16.6).

Mathematically, when we harmonize dynamically, we try to determine what must change in terms of the values of the variables in one pragmatic frame in order to reach another existing frame or a hypothetically ideal frame. An analogy to solving a goal-seeking control problem is given in Appendix A.

In harmonization and legacy transition, decisions have to be made as to whether to change, append, or delete existing

Dynamic Pragmatics

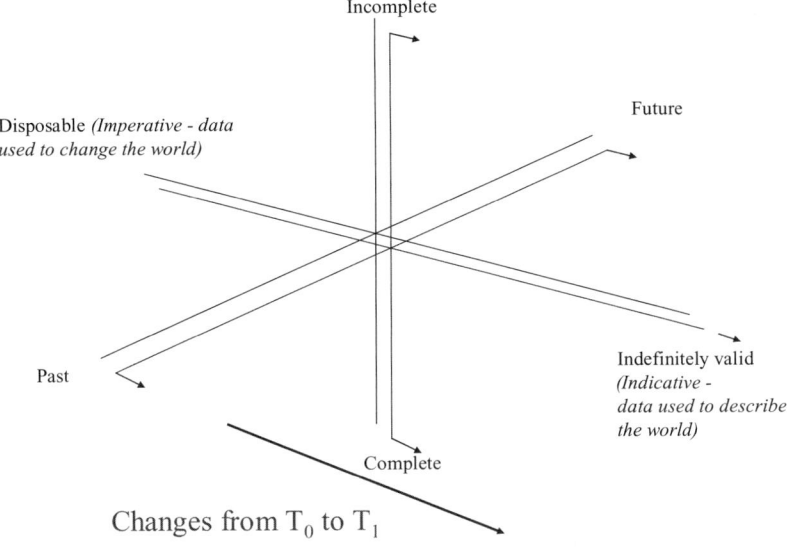

Figure 16.5

Shifts of the Axes over Time

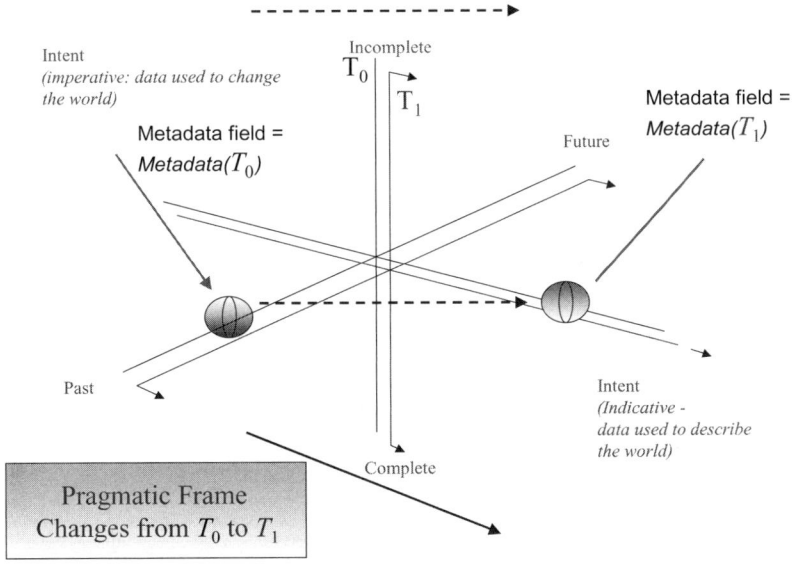

Figure 16.6

Pragmatic Frame Changes from T_0 to T_1

metadata schemas or individual fields and their content definitions. These decisions should be based on a solid principle. The "Intended Use," i.e., pragmatic *frame compatibility* should be the determining design principle. In other words, ask yourself, "To what quantitative extent and to what qualitative extent is the current pragmatic frame different from the target pragmatic frame?" For example, if you are required to move from one syntactic system, e.g., fixed length fields to XML without any other semantic or pragmatic concerns, then you could reasonably abandon fixed file syntax requirements and constraints in favor of recognized XML design principles. However, if there are requirements to retain consistency with legacy or other data and metadata standards or practices, then the data engineer is not free to abandon the past entirely.

How Pragmatics Enhances Semantics

Consider the context: I have a wife and sister that happen to be named Saffron. How would you interpret the following statement?

I love Saffron.

Here:

Love has 3 meanings:

$Love_1$ = passionately adore,
$Love_2$ = have a family bond with,
$Love_3$ = enjoy

and Saffron has 3 meanings:

$Saffron_1$ = wife,
$Saffron_2$ = sister,
$Saffron_3$ = spice

An SES for the sentence is given in Figure 16.7.

The SES indicates that there are $3*3 = 9$ prunings or interpretations of the sentence. However, there are constraints on the possible interpretations (see Chapter 9). For example, consider that the selection of Saffron is made from SaffronSpec. This would give rise to 3 sentences, e.g.,

I love Saffron, my wife.
I love Saffron, my sister.
I love Saffron in my chicken and rice.

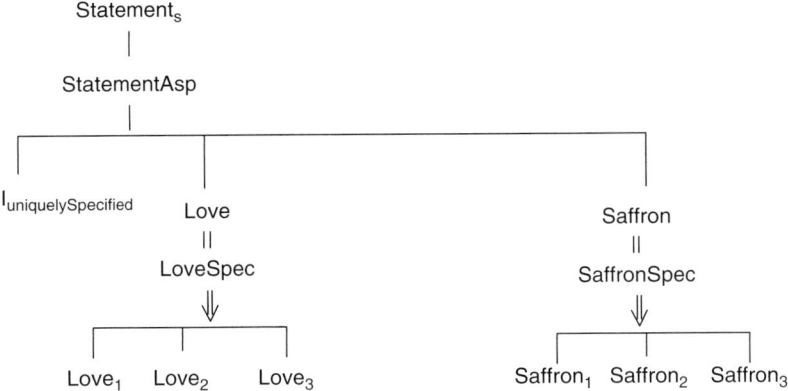

Each Love$_n$ and each Saffron$_n$ may be specified uniquely by a Godel number which is a combination of other Godel numbers. This sentence results in a new Godel number, and can be decomposed uniquely and the components compared to other statements and components.

Figure 16.7

Unique Specification of Expression in SES and Gödel Numbering

But in each sentence, the meaning of love is now unambiguous, as can be seen from:

I adore Saffron, my wife.
I have a family bond with Saffron, my sister.
I enjoy Saffron in my chicken and rice.

Similarly, we prune from the LoveSpec to generate:

I adore Saffron.
I have a family bond with Saffron.
I enjoy Saffron.

Once again, the meaning of Saffron in these sentences is unambiguous.

We see that due to pruning constraints, a partial pruning can provide enough information to complete the pruning in an unequivocal way:

I love Saffron + partial pruning = unique complete pruning

In harmonizing a data standard, we often come across a similar problem. Several standards use the same term very tersely in analogy to the short sentence *I love Saffron*. However, when we look more deeply, the intended meaning of the term may be different in each standard. We would like to find out the intended meaning so that standards with the same intended meaning can be regarded as equivalent. We will consider pragmatic frames as supplying the additional information needed to find the intended meaning in a particular use or application context. Then two standards that have the same meaning for a particular use under a pragmatic frame are pragmatically equivalent. Viewed in an SES perspective, the additional information concerns pruning selection. We would like to know the SES and its pruning constraints that will enable a partial pruning to imply a unique complete pruning.

One approach to solving this problem is to code SES items and relations in numeric form, called Gödel numbering. Appendix B discusses the possibility that mining through a large number of such instances over time may reveal patterns that can be used as partial prunings that can be unambiguously completed to full prunings, i.e., pragmatic frames. The consequence might be that harmonization becomes a more automated process relieving humans of much effort.

Summary

In the introduction we asked two related questions: How good or how appropriate is a metadata field or a collection of metadata fields in existing data exchange standards for a range of possible applications or missions? And, how do we manage change of goodness or appropriateness? This chapter gave an approach based on pragmatic frames to answering the first question and extended this to dynamic pragmatics to answer the second question. Metadata is good to the extent that it makes clear what are the data structures and values needed for each application context of interest. To manage changes in applications and uses over time requires more metadata and goal-seeking control as suggested by dynamic pragmatics. The dialogic nature of exchanges suggests that notions from philosophy, mathematics, and linguistics and, in particular, speech act theory are applicable. Supplemental semantic and pragmatic metadata is required to specify how the standard is actually used and should include information regarding intent (*mood*), tense, and completeness (*aspect*). Assignment of numerical values (Gödel numbers or similar) to pragmatic frame structures will enhance automation.

References

1. Wittgenstein, L. "The Blue and Brown Books: Preliminary Studies for the 'Philosophical Investigations.'" New York: Harper Torchbooks, 1960
2. Nagel, E., and J. R. Newman, *Gödel's Proof*. New York University Press, 2001
3. Quine, W. V. O., *Word and Object*. Cambridge, MA: MIT Press, 1960
4. Zadeh, Lotfi, "Fuzzy Sets as the Basis for a Theory of Possibility," *Fuzzy Sets and Systems* 1:3–28, 1978
5. Shannon, C. E., and W. Weaver, *The Mathematical Theory of Communication*. University of Illinois Press, 1949, 1998
6. Pierce, J. R., *An Introduction to Information Theory: Symbols, Signals, and Noise*. New York: Dover Publications, Inc., 1961, 1980
7. Austin, J., *How to do things with words*. Oxford: Clarendon Press, 1962
8. Searle, J., *Speech Acts*. Cambridge: Cambridge University Press, 1969
9. Searle, J., "Indirect Speech Acts." In P. Cole and J. L. Morgan (Eds.). *Syntax and Semantics*. Vol. 3: *Speech Acts* (pp. 59–82), New York: Academic Press, 1975
10. Searle, J., "The Classification of Illocutionary Acts." *Language in Society*, 8: 137–151, 1979
11. Harnish, R. M., "Mood, Meaning and Speech Acts." In S. L. Tsohatzidis (Ed.), *Foundations of Speech Act Theory* (pp. 409–449), London & New York: Routledge, 1994
12. Grice, H. P., *Logic and Conversation*. In P. Cole and J. L. Morgan (Eds.), *Syntax and Semantics*, Vol. 3: *Speech Acts* (pp. 59–82), New York: Academic Press, 1975
13. Hammonds, P. E., *Directive Speech Acts in Conflict Situations Among Advanced Non-native Speakers of English*. PhD Dissertation, University of Arizona, 2001
14. Adams et al., http://www.mathresource.iitb.ac.in/linear%20algebra/example7.4/index.html (accessed Nov. 2006)
15. Quine, W. V. O., *Elementary Logic*, Revised Edition, Cambridge, MA, Harvard University Press, 1980
16. Zhang, M., B. P. Zeigler, and P. E. Hammonds, "DEVS/RMI — A Dynamic and Flexible Distributed Simulation Environment," *ITEA Journal*, Vol. 26, No. 3, pp. 49–60, 2006
17. Kularini, A. D., *Computer Vision and Fuzzy-Neural Systems*, Upper Saddle River, NJ, Prentice Hall, 2001

Appendix A: Dynamic Harmonization Goal-seeking Control Theory Analogy

Let us say that we have a standard with initial (arbitrary for this example) pragmatic frame with values that we know are changing at an understood rate of $\mathbf{a}_{(0)}$ = (Authority, Urgency, Relation) = (8, 6, 4) and we have an existing standard or an ideally defined standard that gives us pragmatic frame value of $\mathbf{b}$ = (37, 6, 44). In order to reach that standard we must provide some factor that indicates something like the force acting on one component (in this case a requirement for a new relationship within a time period change) acting on the standard $\mathbf{a}$ to reach standard $\mathbf{b}$. This value, a second derivative Factor = (0, 0, –30), by integrating this we get a first derivative vector (0, 0, –30t) + C. Combining this with $\mathbf{a}$ with the initial Velocity vector$_{(t)}$ = (8, 6, 4 –30t). Integration of this gives us a Position vector$_{(t)}$ = (8t, 6t, 4t – 15t^2) + D. D = the initial position, which indicates a starting position for Relation (0, 0, 50), r(t) = (8t, 6t, 50 + 4t – 15t^2).

The coordinate (Relation) we want to match is 6 so we solve for t; 50 + 4t – 15t^2 = 6.

Which equals 15t^2 – 4t – 44 = 0 4t^2 – t – 1 = 0, t = 1.58 yields (37.4, 6.3, 44), which is very close to (37, 6, 44).

Appendix B: Pragmatic Frame Conjecture

Entries in the pragmatic frame fields indicate typical, conventional usage of a given field in a specified context or mission. History of use is also included.

In this example there are three possible "Love" contexts and three possible "Saffron" contexts. It is interesting that any one context uniquely determines the other (in polite, conventional society of the year 2006 anyway). Other aberrant possibilities exist, but they are eliminated by the initial constraint of "polite and conventional society."

Take the statement

I love Saffron.

Assume that the hearer is either acquainted with, or watching, the speaker at the time of utterance. **I** is not an ambiguous term, but **love** and **Saffron** could be, even in polite conventional conversation that takes place in 2006 in the United States. Love may have $\mathbf{n}$ = 3 meanings here and Saffron may have $\mathbf{m}$ = 3 meanings here, so it would seem that there are $\mathbf{n} \times \mathbf{m}$ = 9 possibilities for

Appendix B: Pragmatic Frame Conjecture

meaning, but the pragmatic frame in which the utterance occurs reduces this to a sum rather than a product or exponent problem.

1) I love Saffron$_1$. (She is my wife and I am deeply in love with her.)

{I | I is the speaker}
Saffron ∈ {Wives| Wives of the speaker}
{Saffron | Saffron is the wife of the speaker}
love is assumed to be a bijection f where $f: I \to$ Saffron

In this case Saffron is a person. The authority of the speaker would be high. He would know if he sincerely loves Saffron or not. In a given context the relationship of the speaker to the hearer is not very high, otherwise there would be no ambiguity. The urgency/consequences of making such a statement and then not meaning it or expressing it somehow would be high (legally it could be very expensive). The tense would be present and the completeness would be incomplete or ongoing. So for argument's sake we provide some arbitrary values on a scale from 0 to 100, where A = 100, U/C = 95, R = 20, Tense = 2 (Present), Completeness = 1 (incomplete meaning continuous).

2) I love Saffron$_2$. (She is my sister and I love her very much.)

{I | I is the speaker}
Saffron ∈ {Sisters| Sisters of the speaker}
{Saffron | Saffron is the sister of the speaker}
love is assumed to be a bijection f where $f: I \to$ Saffron

In this case Saffron is a person. The authority of the speaker would be high. He would know if he sincerely loves Saffron or not. In this context, also, the relationship of the speaker to the hearer is not very high, otherwise there would be no ambiguity. The urgency/consequences of making such a statement and then not meaning it or expressing it somehow would be high (it can be argued whether loving a sister or a wife is more important — legally, we have to say that the wife may be greater, but that may vary in other societies). For illustration's sake, we will say that love between this brother and sister is less urgent or consequential than love between a husband and wife. The tense would be present and the completeness would be incomplete or ongoing. A = 100, U/C = 80, R = 20, Tense = 2 (Present), Completeness = 1 (incomplete meaning continuous).

A = 100, U/C = 85, R = 20, Tense = 2, Completeness = 1.

3) **I love Saffron$_3$.** (Saffron brand Saffron thus the capitalization) (I like this spice very much with chicken and rice and olive oil.)

{I | I is the speaker}
Saffron ∈ {Spices| Spices available to the speaker}
{Saffron | Saffron is a spice}

love here is assumed to be an injection f where f: I → Saffron, but Saffron cannot love the speaker in a way we could understand.

In this case, Saffron is a spice and it would be difficult for one to argue sincerely that a person loved a spice more than a wife or a sister — jokes aside. A = 100, U/C = 25, R = 20, Tense = 2, Completeness = 1.

The three Saffrons have their name in common, but is there anything else in common?

Saffron$_1$ ∩ Saffron$_2$ = name and human and female
Saffron$_1$ ∩ Saffron$_2$ ∩ Saffron$_3$ = name only

Clearly, only **Saffron$_1$** and **Saffron$_2$** are arguably similar. This difference would be further distinguished by where the I and the Saffron would appear in some hierarchy of relationships and semantics. A sister is not equal to a wife (constraints imposed by society and legal status). So, even if someone claimed that the love was the same, the relative positions of the two Saffrons to the speaker are very different.

If it is known that Saffron is the wife of the speaker, then the Love context and semantics are uniquely understood.

If it is known that Saffron is the sister of the speaker, then the Love context and semantics are uniquely understood.

If it is known that Saffron is a spice, then the love context is of a completely different nature.

The same is true if it is known what kind of love is being expressed, but in this case the object of the verb is where we as native speakers usually focus to remove ambiguity.

Love$_x$ and Saffron$_x$

	Saffron$_1$	Saffron$_2$	Saffron$_3$
Love$_1$	Wife	Sister	Spice
Love$_2$	Wife	Sister	Spice
Love$_3$	Wife	Sister	Spice

Appendix B: Pragmatic Frame Conjecture

Conjecture/Theorem: In any exchange of information between systems (in this case a speaker and a hearer, but could be any systems that normally exchange information), operational meaning is determined not by the **n** × **m** possibilities, but by the sum of the maximum number of contextually ambiguous terms. Let us say in this case that the hearer understood that Saffron = **m** was a woman, but did not know whether the woman was the speaker's sister or wife. There would be only one possibility for conventional expressions of love once the relationship was clarified, but until that time there are two possible relationships, **m** + **n**. So it seems to be a simple sum of the ambiguous elements of a defined context. The addition of context through metadata, extra words in English, clarifying the relationship through amplification reduces the sum of the ambiguous terms. It is the convention and context (pragmatic frame) that serve to keep the ambiguity from being either a product or exponential relationship.

I love my wife, Saffron.
I love my sister, Saffron.
I love Saffron in my chicken and rice.

The ability to uniquely place a statement or assertion within a syntactic, semantic, and pragmatic frame is our goal.

The use of Gödel numbers is one example of how a sentence can be uniquely specified. It is important to remember that these numbers are only used for comparison. They are generated through a lookup table and then simply compared. No calculations are made with them except of the summation of parts. A detailed, yet accessible description of Gödel numbering may be found in Nagel[2] (see text references). A standard is a very large sentence consisting of the sum of many unique statements.

A Gödel number is a unique number for a given statement that can be formed as the product successive prime numbers raised to the power of the number corresponding to the individual assigned symbols that comprise the sentence. The standard example is the statement $(\exists x)(x = s\, y)$ that reads "there exists an x such that x is the immediate successor of y" can be coded based on the following table of usual interpretations.

314 Chapter 16 Dynamic Pragmatics: Issues and Methodology

Usual Interpretations of the First Twelve Gödel Numbers,[2]

Logical symbols	Numbers 1:12
¬	1 ("not")
V	2 ("or")
⊃	3 ("if..., then...")
	4 ("There is an...")
=	5 ("equals")
0	6 zero
s	7 the immediate successor of
(	8 Left/open parentheses
)	9 Right/close parentheses
,	10 comma
+	11 plus
X	12 times

$$2^8 \cdot 3^4 \cdot 5^{13} \cdot 7^9 \cdot 11^8 \cdot 13^{13} \cdot 17^5 \cdot 19^7 \cdot 23^{16} \cdot 29^9$$

The numbers in the set (8 = (, 4 = ., 13 = x, 9 =), 8 = (, 13 = x, 5 = =, 7 = s, 16 = y, 9 =)) correspond to (∃ x) (x = s y). The statement can be encoded into a value and the value can be uniquely decoded to the exact statement.

In our example here. Love is defined as some function, and is a combination of other more primitive functions and nouns, I is defined as some function which is a combination of functions and nouns, Saffron is defined as some noun which also consists of many functions and nouns. Each of these entities exist independent of the other, but when brought together in a specific context express meaning in that context. "I love$_1$ Saffron$_1$" has a uniquely specified numerical assignment based on its syntactic, semantic, and pragmatic composition. All statements so expressed can be compared to one another to see if they have the same or similar meaning. None of the three statements with the same English words could ever be considered synonymous with such a system. While statements that are equivalent (based on their entry in a thesaurus) would be immediately evident.

John S. Smith is of a specified address in a specified town in a specified country in the year 2006 with a specified identification number. This value is assigned to the variable "I." Saffron M.

Smith is of a specified address in a specified town in a specified country in the year 2006 with a specified identification number.

"Love" as verb or function would have uniquely defined characteristics that would ensure that it could only be applied to the statement$_s$ in one way given the identification of I and Saffron in one way for the given context. That is not to say that this context may not change, but then each element of the sentence would change correspondingly.

Statements of the type *A loves B* or *A loves B, but does not love C in the way that A loves B* can be evaluated in the abstract, or for actual instantiations of the variables A, B, C, and love. Constraints may be then placed on what kind of love (or any other function) can occur between any two or more variables. If the context changes and constraints change, then a unique value can be assigned to that function based on the new context. The pragmatic frame value is an expression of this context.

EXAMPLE: Our goal is to capture the syntax, semantics and pragmatics of some system or service in an SES format. This example is based on the well-known example from Quine[15] (see text references).

From the WVO Quine book
Elementary Logic:

Complex Statement = CS = If Jones is ill or Smith is away then neither will the Argus deal be concluded nor will The directors meet and declare a dividend unless Robinson comes to his senses and takes matters into his own hands.

1) J = Jones is ill.
2) S = Smith is away.
3) A = The Argus deal will be concluded
4) M = The directors will meet.
5) D = The directors will declare a dividend.
6) C = Robinson will come to his senses.
7) T = Robinson will take matters into his own hands.

316 Chapter 16 Dynamic Pragmatics: Issues and Methodology

If (J or S) then (neither A nor M and D unless C and T.)
p = (J or S)
q = (neither A nor M and D unless C and T.)
If p then q = ~(p. ~q)
neither A nor M and D unless C and T = ~(J or S. ~neither A nor M and D unless C and T)
(J or S) = ~(~J. ~S)
~(p. ~q) = ~(~(~J. ~S). ~neither A nor M and D unless C and T)
(neither A nor M and D) unless (C and T) = p unless q
p unless q = (p or q) = ~(p. ~q)
~(~(~J. ~S). ~~(~neither A nor M and D). ~ (C and T))
neither A nor M = ~p. ~q = ~A. ~M
~(~(~J. ~S). ~~(~ (~A. ~M and D). ~ (C and T)))
~(~(~J. ~S). ~~(~(~A. ~(M. D)). ~ (C. T)))
DONE

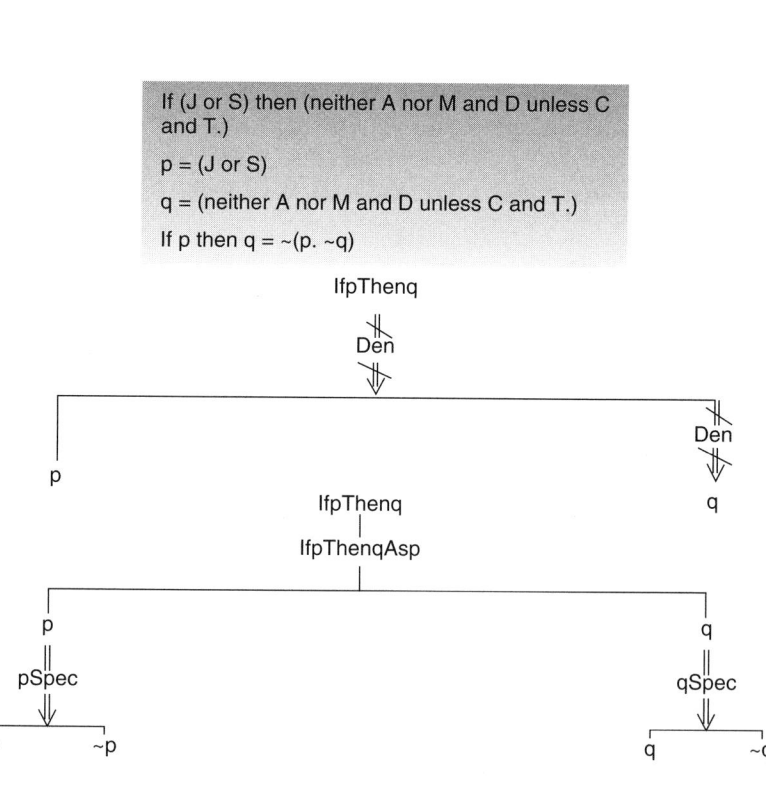

Appendix B: Pragmatic Frame Conjecture

Neither A nor M and D unless C and T = ~(J or S. ~neither A nor M and D unless C and T)

(J or S) = ~(~J. ~S)

~(p'. ~q') = ~(~(~J. ~S). ~neither A nor M and D unless C and T)

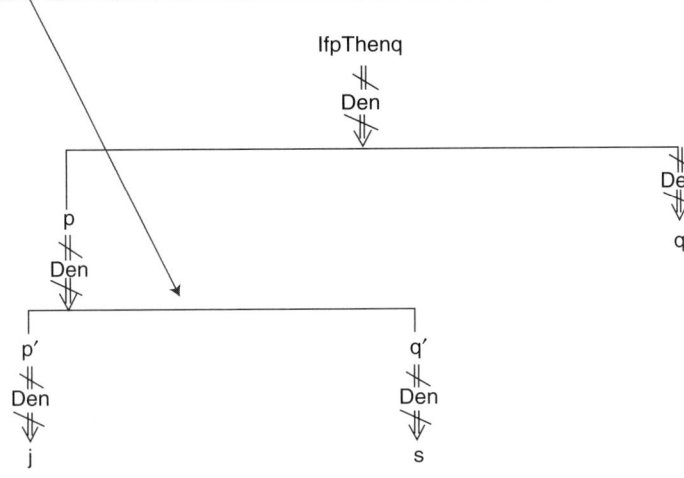

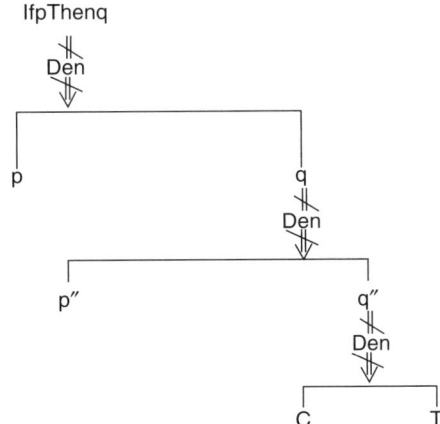

(neither A nor M and D) unless (C and T) = p" unless q"

P" unless q" = (p" or q") = ~(p". ~q")

~(~(~J. ~S). ~~(~neither A nor M and D). ~(C and T))

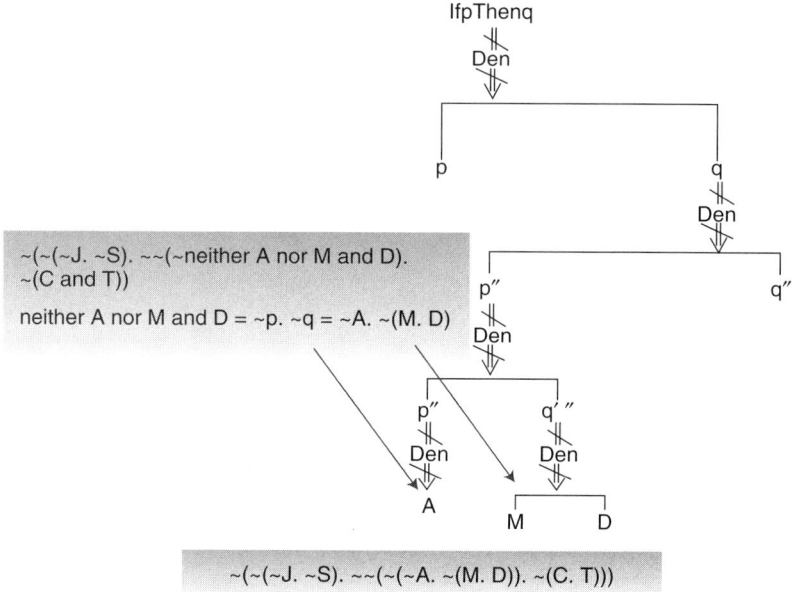

Using the alternative Gödel numbering[4] table below which includes logical AND. The Gödel number for this statement is

$2^1 \times 3^6 \times 5^1 \times 7^6 \times 11^1 \times 13^{12} \times 17^5 \times 19^1 \times 23^{15} \times 29^7 \times 31^5 \times 37^1$
$\times 41^1 \times 43^6 \times 47^5 \times 53^6 \times 59^1 \times 61^{18} \times 67^5 \times 71^1 \times 73^6 \times 79^{21} \times 83^5$
$\times 89^{24} \times 97^7 \times 101^7 \times 103^5 \times 107^1 \times 109^6 \times 113^{27} \times 127^5 \times 131^{30}$
$\times 137^{\ 7} \times 139^7 \times 149^7 = 2.6482455906120382979271758595314\mathrm{e}$
$+346.$

This is a large, but determinate integer represented here in scientific notation for convenience. The Gödel numbers can be auto-generated and attached to metadata and pragmatic frame data. This can be done on stand-alone computers for the most part, but the entire process can also be ported to a shared memory or cluster environment. Fuzzy adaptive logic can then be applied to search for a distributed manner on a cluster pattern in the domain of interest (Zhang et al.,[16] Kularini[17]). These quantifications serve only to assist in the establishment of use-based meaning (Wittgenstein[1] — in a given application and how that application may relate to other similar applications).

---

[4] Logical AND (P ∧ Q) is represented in the previous Gödel numbering system, via De Morgan's rule as ~(~P V~Q).

Appendix B: Pragmatic Frame Conjecture

Logical symbols	Numbers 1:12
¬	1 ("not")
∀	2 ("for all")
⊃	3 ("if, then")
∨	4 ("or")
∧	5 ("and")
(	6
)	7
S	8 ("is the successor of")
0	9
=	10
•	11
+	12

Propositional symbols	Numbers greater than 10 but divisible by 3
J	12
S	15
A	18
M	21
D	24
C	27
T	30

Individual variables	Numbers greater than 10 with remainder 1 when divided by 3
v	13
x	16
y	19

Predicate symbols	Numbers greater than 10 with remainder 2 when divided by 3
E	14
F	17
G	20

Part IV

Testing in Net-Centric Environments

Testing in a Net-Centric Environment: Technology Basis

This chapter exemplifies the concepts developed thus far in application to testing of complex systems. The increasing complexity and advanced decision capabilities of defense and civilian information technology-based systems requires that testing methodology has to become more rigorous, in-depth, and thorough. At the same time, to keep up with the rapid change and short development life cycles expected from the system builders, tests have to be ready to conduct in time scales compatible with the agile development strategies of new systems. We will show how formal data engineering, modeling, and simulation can address the polar requirements of increased rigor and faster test development in a net-centric environment.

Our development begins with a discussion of how the above concepts were critical in successfully introducing automation into traditional testing of distributed radar data sharing and coordination for conformance to military standards. This sets the background for the next stage of development: exploiting agent technology for web-services testing. In the final section, we place these applications in a broader methodology for integrated system development and testing. This chapter lays the technological basis for the next one where the discussion centers on future development that requires testing of multiple levels (syntactic, semantic, pragmatic) simultaneously in a net-centric environment.

Introducing Automation into Traditional Testing of Standards Conformance

In the early years, Standards Compliance and Interoperability tests were performed on pieces of equipment such as individual radios in a laboratory setting. However, as communication components grew in complexity and became integrated into larger command and control (C2) systems, the testing process likewise had to evolve to maintain its effectiveness. In addition to lab testing, it became necessary to test systems with many component systems (known as Systems of Systems) in their real operational environments. Such live testing especially applied to joint operations, where equipment and systems from the different military services (Army, Navy, etc.) had to interoperate together, a non-trivial requirement since each service has its own special requirements and associated ways of constructing, and contracting for, such equipment, even where there are standards that dictate the basic rules of operation.

Over the course of several decades, methods and software tools were developed that helped engineers and analysts to perform live tests of Systems of Systems for conformance to applicable communications command and control standards. This need led to the decision to employ formol modeling and simulation both to increase its capabilities for simulation-based testing and also as a basis to increase the automation of its testing processes (Figure 17.1). This would lead them to exchange messages carrying radar and other tactical data that could then be monitored by the testers and logged onto computers for later review. Such testing could detect certain failures of interoperability when, for example, messages failed to be properly relayed or never arrived at the intended destination. Furthermore, logged message streams could be intensively checked for conformance to tactical data message standards, particularly by noting whether their formats were correct and the values of their fields were within the prescribed limits. However, because the voluminous data could easily overwhelm human memory and perception capacities, the ability to check more dynamic aspects of information

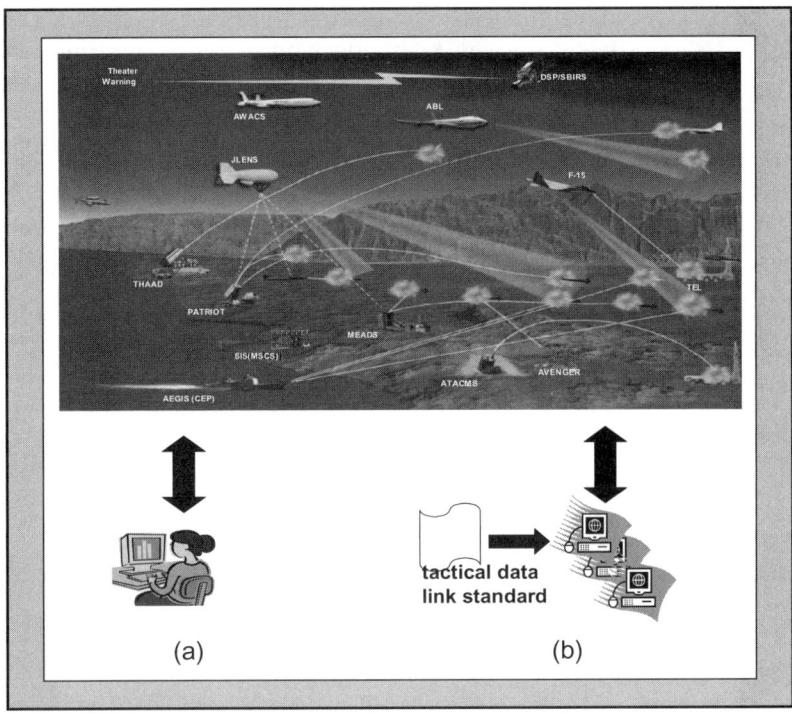

Figure 17.1

Testing Systems of Systems for Interoperability and Standards Conformance. (a) In the traditional approach voluminous message traffic exchanged among the players is monitored and analyzed by expert analysts. (b) In the new approach computerized models derived from standards documents participate in the exercise, driving it or watching for opportunities to test for conformance

exchange, such as whether messages were sent at the right time or in the right order was severely limited.

Challenge: Automated and Simulation-based Testing

The need to modernize traditional test methods became increasingly acute as simulation-based acquisition and the move toward net-centricity pose increasingly severe challenges to existing methods (see Appendix A: Infusion of Modeling and Simulation into Defense Acquisition). It was decided to take the initiative to employ formal modeling and simulation both to increase its capabilities for simulation-based testing and also as a basis to increase the automation of its testing processes. In this way the polar requirements of increased rigor and greater responsiveness could be addressed. It was decided to look for critical areas that lacked conventional solutions. The nationale was that if a new untested approach were successful in such a "Killer app," this would provide indisputable evidence that the approach was ready for adoption and integration into enterprise testing paradigms.

Two critical areas were identified; coincidently, they came together with a high degree of overlap and timeliness. One area was that of the Joint Single Link Implementation Requirements Specification (JSLIRS), an evolving military standard (MIL-STD-6016c) for tactical data link information exchange and networked command/control of radar systems.[1] For reasons discussed below, JSLIRS (or Link-16 as it is more popularly known) presented certain obstacles to traditional test approaches that had not been overcome. The second area was that the SIAP (Single Integrated Air Picture) project, whose objective was to improve the adequacy and fidelity of information, was to form a shared understanding of the tactical situation.[2] Although not immediately obvious, it was eventually realized that any system achieving SIAP requirements would have to be compliant with Link-16. Thus, the natural challenge presented itself — develop an M&S-based approach to automating the Link-16 testing process and apply it to SIAP system development.

Link-16: The Hard Nut-to-Crack

The Link-16 specification document states requirements in natural language, employing directives such as the "system shall . . . ," that are often open to ambiguous interpretations. As is apparent by examining an excerpt such as that in Figure 17.2, the document is broken into paragraphs and sub-paragraphs that are reminiscent of legal contractual language (that might be used in a rental lease for example), except that technical terms abound. In addition to such linguistic and technical complexities, the document is also voluminous, with many hyperlinked chapters and appendixes, thus rendering its interpretation labor intensive and prone to error. Because of its size and complexity, the specification as a whole is potentially incomplete and inconsistent. As a consequence, it is a major challenge to ensure that a certification test procedure developed from the specification document completely covers the requirements and can be consistently replicated across the numerous military service, national, and manufacturer contexts in which standard certification testing are executed. Traceability of the individual test cases of the procedure back to those specific parts of the specification from

> **4.11.13.12** <u>Execution of the Correlation</u>. The following rules apply to the disposition of the Dropped TN and the retention of data from the Dropped TN upon origination or receipt of a J7.0 Track Management message, ACT = 0 (Drop Track Report), for the Dropped TN. The correlation shall be deemed to have failed if no J7.0 Track Management message, ACT = 0 (Drop Track Report), is received for the dropped TN after a period of 60 seconds from the transmission of the correlation request and all associated processing for the correlation shall be cancelled.
>
> a. Disposition of the Dropped Track Number:
> (2) If own unit has R^2 for the Dropped TN, a J7.0 Track Management message, ACT = 0 (Drop Track Report), shall be transmitted for the Dropped TN. If the Dropped TN is reported by another IU after transmission of the J7.0 Track Management message, own unit shall retain the dropped TN as a remote track and shall not reattempt to correlate the Retained TN and the Dropped TN for a period of 60 seconds after transmission of the J7.0 Track Management message.

Figure 17.2

An Excerpt from the MIL-STD-6016c Document

which they derive is extremely important — when tested systems fail, and developers protest, it is crucial to be able to show in "chapter and verse" where the applicable requirements are stated.

Moreover, as mentioned before, while earlier predecessors focused on message formats, the heart of the Link-16 specification concerns more complex functionalities such as correlation of radar tracks to eliminate redundancies and shifting of reporting responsibilities to grant priority to information sources with the highest quality. Unlike the static framework which was sufficient to test formats, the new functionalities could only properly be understood as dynamic system behaviors. With this motivation, in the spring of 2003, it was decided to create a formal version of the document that could provide a complete and unambiguous account of its static and dynamic information processing complexities — and at the same time, enable more automation to be introduced into the development of tests of the standard.

Goals and Approach

The overarching goal of the Automated Test Case Generator (ATC-Gen) development effort was to increase the productivity and effectiveness of standards conformance testing. Recognizing the dynamic nature of the system behaviors specified by Link-16, it was natural to apply mathematical systems theory and modeling and simulation concepts, in addition to current software technology to seek to (semi-)automate portions of conformance testing. Systems theory provides well-formulated levels of system description (with formal equivalents for intuitive white and black box notions). The theory also underpins the Discrete Event Systems Specification (DEVS) formalism that provides a rigorous formal approach to building models and simulating them in various media — workstation, distributed simulation, real-time execution — with only a change in the simulation protocol. Thus the DEVS approach affords a computational framework base that enables the appropriate systems concepts to be implemented in software and hardware.[3]

An overview of the ATC-Gen methodology, as shown in Figure 17.3, shows that a standards document (in this case MIL-STD 6016C) is analyzed to uncover relevant elements of the DEVS specification that then become the basis for semi-automated test case generation. In outline:

- Trained analysts identify *if-then* rules from the document, casting them into XML, with associated condition and action variables. (XML, the eXtended Markup Language, has become a standard for structuring, exchanging, and storing data.)

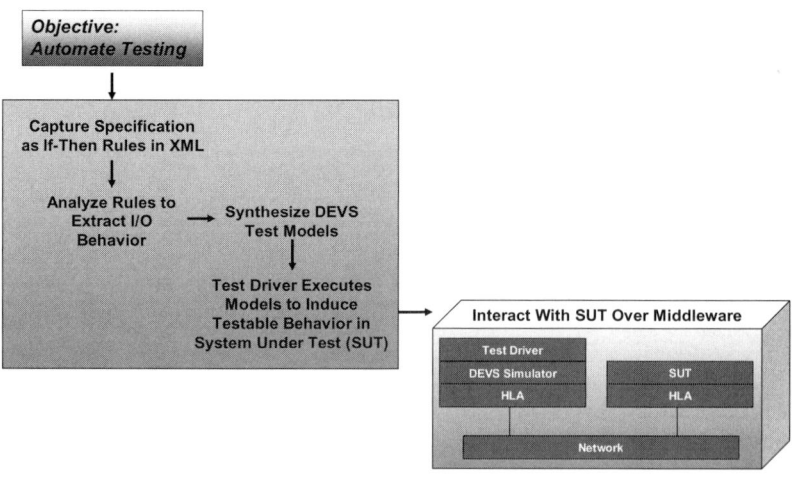

Figure 17.3

Overview of the ATC-Gen Methodology

- Automated analysis of the variable dependencies enables visualization of rule firings and selection of potential test sequences. The test sequences are expressed in XML and stored in a repository for combinational reuse.
- Test cases are represented as DEVS models that are semi-automatically generated from test sequences. DEVS hierarchical construction allows basic models from a small set of primitives to be coupled together in higher level models. An XML transformation automates the conversion of a set of test sequences into composite test packages that execute in the DEVSJAVA environment where they are tested against a stub representing the system under test (SUT).
- Finally, a Test Driver that is based on the DEVS simulation protocol executes the test models against the SUT in a distributed simulation infrastructure based on the HLA or other middleware.

Discrete Event Nature of Link-16 Specification

Since system theory and DEVS play central roles in the ATC-Gen, it is worth spending some time on how they apply particularly to Link-16. We begin with the recognition that the Link-16 specification is, in fact, a description of a dynamic system that tells how it should respond to stimuli in various situations. In contrast to the interpretation in the context of system design, the objective in the context of conformance testing is to transform this description into a large family of test procedures and to specify the required outcomes of tests and sequences of tests. In this context, sequences of tests can be regarded as the injection of stimuli by a test driver that will realize the test procedures. To execute a test plan for a message-based system requires building such a driver to inject the appropriate input messages to the SUT and observe the SUT's output messages. For testing conformance, a set of positive (to which the system should respond) and negative (to which it should not) test cases must be selected to provide the desired coverage. Further, the set of test cases should cover the normally expected interactions between the SUT and the rest of the world, as well as interactions that would not normally occur but cannot be excluded from consideration.

Figure 17.4 further illustrates the discrete event systems nature of the Link-16 specification. As a node in tactical data link network, the inputs to the system come at discrete instants from external sources, such as human operators, sensors, or messages

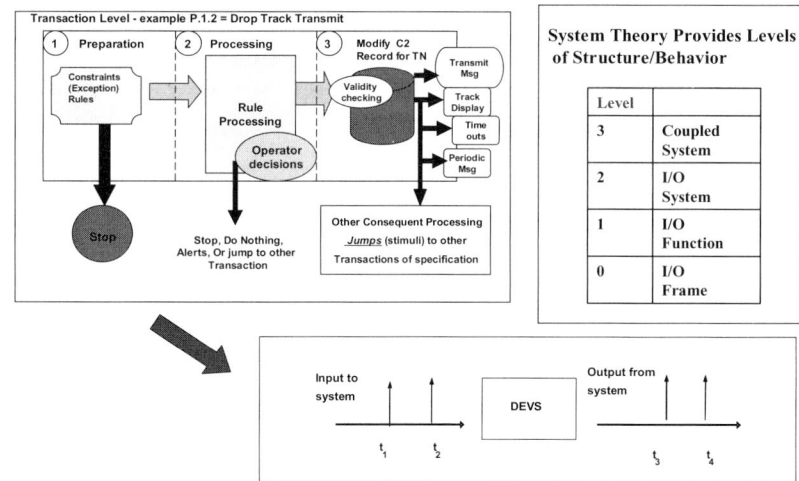

Figure 17.4

The Discrete Event Systems Nature of the Link-16 Specification

from other nodes on a network. The systems outputs also occur discretely in time as a result of scheduled events in the system, such as periodic updates of position. Other internal events such as timer expirations (time-outs) and detected anomalies, also result in outputs in the form of operator alerts. *A system that has discrete event input and output characteristics such as this can be characterized by a DEVS model.* Further, the message processing steps illustrated in the upper left of Figure 17.4 are formulated at the higher structural levels of the system description hierarchy (at the right of the figure), while the tests that need to be developed lie at the lower behavioral levels. Systems theory helped us develop a systematic process for deriving testable input/output behaviors from the information processing functions specified at the structural levels.

ATC-Gen Top-Level Architecture

The process flow in Figure 17.3 is realized by an architecture that is illustrated in Figure 17.5. Four major components are shown for capturing rules, formalizing rules, analyzing rule sets, and generating tests. In addition, the interfaces with outside actors (humans and software and hardware elements) are shown.

The ***Rule Capture Component*** supports the fundamental conversion of the MIL-STD into a large set of if-then rules that

Introducing Automation into Traditional Testing of Standards

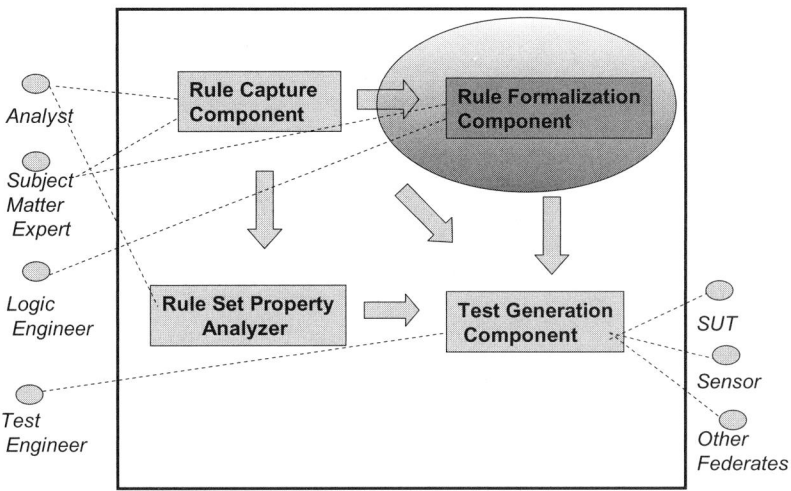

Figure 17.5

ATC-Gen Top-Level Architecture

forms the basis of automation for subsequent steps. The specification is parsed and interpreted by analysts with the help of subject matter experts into a collection of if-then rules expressed in XML with specific variable definitions that are required to completely define the behavior of the system. Rules are of the form:

If C is true now
Then do action A (and possible output O) within T
Unless E occurs in the interim

where C is a condition, A is an action, O an output, T is a time interval, and E is an exception. The interpretation of such a rule is that the condition dictates the circumstances under which the action should be scheduled to occur within a specified time interval; the action, including any output, should be carried out as prescribed unless some exceptional circumstances arise before it can be so executed. Figure 17.6 shows an example of a rule as expressed in XML that was interpreted by our analysts from the excerpt of the Link-16 document shown earlier.

Note that the condition is described in terms of variables whose values are checked for taking on certain values or falling within a range of values. Likewise the action is specified in terms of variables whose values are determined as a result of the execution of the action. The condition and action variables, when

```
<rule trans="4.11.13" stimulus="4.11.13.12" reference="4.11.13.12.a.2"
ruleName="R2 Unit transmits J7.0">
    <condition txt="Check for R2 own unit" expression="AutoCor==True and
    (CRair.TNcor.CORtest==3 and          J32.TNref.CORtest==3) and
    CRair. R2held==1 AND J72.MsgTx==True">
    </condition>
    <action txt="Prepare J7.0 Drop Air Track message" expression="J70.
    TNsrc=TNown; J70.TNref=TNdrop;           J70.INDex=0; J70.INDcu=0;
    J70.ACTVair=0; J70.SZ=0; J70.PLAT=0; J70.ASchg=0; J70.ACTtrk=0;
    J70.ENV=0; MsgTx(J70)">
    </action>
    <output txt="Transmit J7.0" outType="Message" outVal="J70"></output>
</rule>
<QA>
    <revisit name="DHF" date="10/16/04" status="Open">need to add timer
    for a period of 60 seconds in which correlation is not reattempted
    </revisit>
</QA>
```

Figure 17.6

The XML Translation of the MIL-STD-6016c Excerpt

collected together over all rules, form the state variables of the system. An essential responsibility of the analyst is to properly understand the document and correctly define (or reuse, if already defined) the right variables to capture the state of the underlying system. Moreover, as we will see in a moment, subsequent steps exploit the dependencies among the rules that are made evident through the condition and action variables.

Also note that XML representation includes fields pointing to the paragraphs from which the rule was gleaned. This information is carried through the rest of the process so that tests ultimately are traceable back to the source document. Tags for quality control are implemented as well. These help manage the numerous iterations that may be necessary to get analysts, experts, and controlling authorities to come to agreement about the rule's correct formulation.

The ***Rule Set Analyzer Component*** provides the capability to perform analyses of user-specified parts of the rule collection in the XML repository. This component provides user-friendly tools to compute various abstractions and aggregate properties, including input-output dependencies and global structure properties such as information flow through the various modules of the specification. Visual displays of the abstractions and aggregate properties are also provided to facilitate user comprehension of the elements of the specification and their interactions. A dependency graph, exemplified in Figure 17.7, is a key abstraction in, and starting point for, the test generation process just described. Rules are shown as boxes with colors (yellow for initialization, pink for generating output, and grey otherwise), input ports (for condition variables) and output ports (for action variables). A rule dependency is depicted as an arrow from an output port of one rule to an input port of a second rule where the two ports have the same name — signifying that a variable in the action set of the first is also found in the condition set of the second. Such a dependency portends, but does not necessarily guarantee, a possible influence that the firing of the first rule might have on the second. A path through the graph from an initial rule to an output rule suggests a possible sequence of rule firings that can be used as the basis for a conformance test.

By enabling the test engineer to visualize such possible sequences, this component supports generations of test paths and their integration into test sets. It also computes other information such as the efficiency and rule coverage attributes of such test sets. The goal of this component is to help narrow down quickly from a very large number of possible paths to a smaller, usable subset that can be used by a test engineer to construct the exact test cases needed for the conformance testing. This is the essence of the semi-automation expected from the ATC-Gen.

The ***Rule Formalization Component*** supports the development of test cases by assisting the test engineer in deriving data required for executing test models and verifying the correctness of test models. In Figure 17.5, the component is shown as shaded since its full capability is an ideal that may, or may not, come to fruition. In principle, it provides the methods and tools to convert the if-then rules in the XML Repository to executable simulations of the rules. To do this it would have a rule compiler that accepts XML rules together with logical expressions provided by a trained logic engineer and compiles them into executable form. It would also have a rule execution engine that can correctly execute rule sets when provided initial state and input stream scenario. Such a fully formalized rule set might well increase the level of automation by augmenting, or even eliminating, the manual discovery and implementation of test paths. In such a form, it would constitute the MIL-STD Executable Reference Model — an oracle, fully capable of answering any and all questions about MIL-STD compliant system behavior.

The ***Test Generation Component*** provides the tools to generate actual test drivers that will operate in a simulation

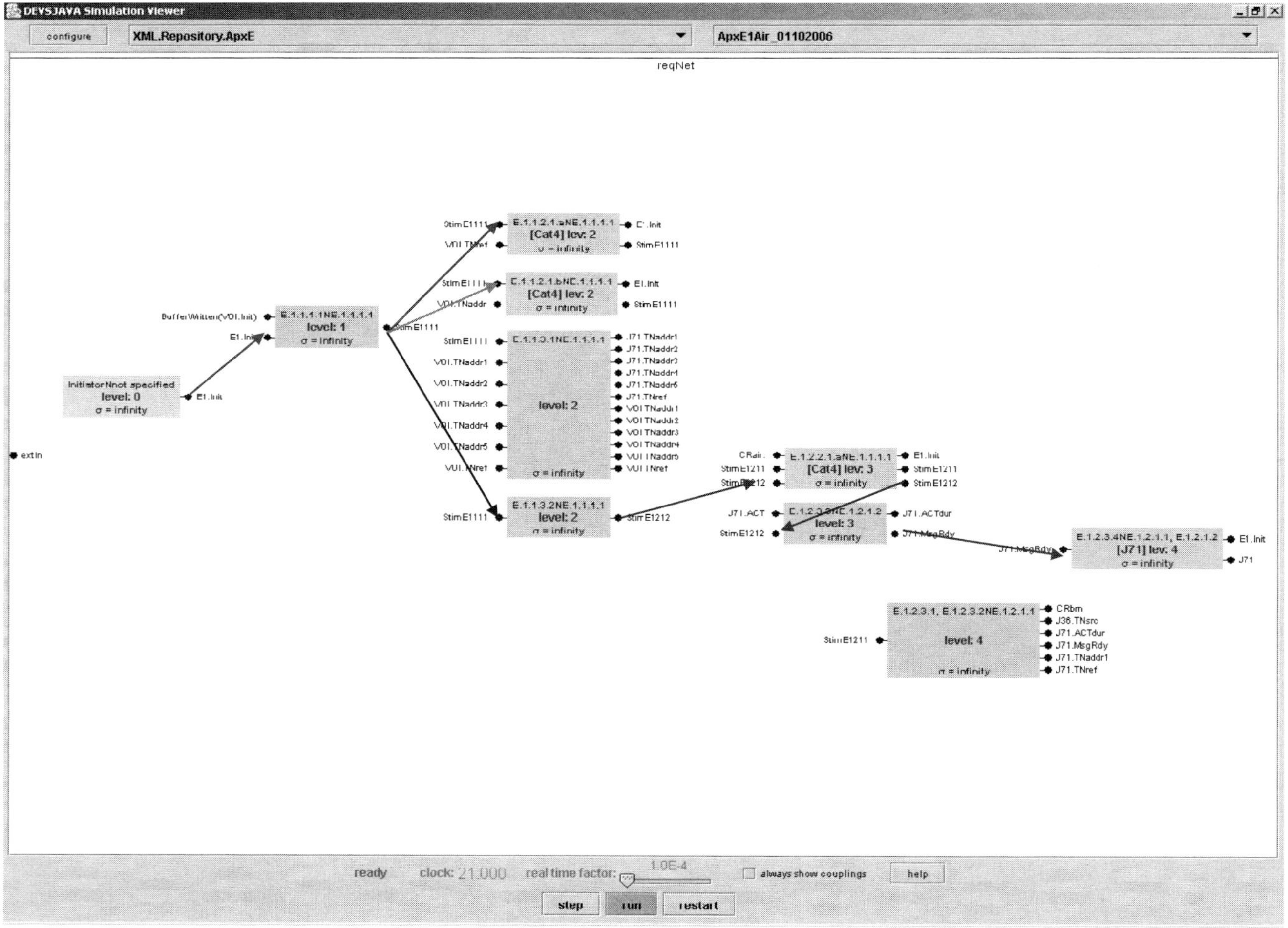
DEVSJAVA Simulation Viewer
configure
XML.Repository.ApxE
ApxE1Air_01102006
reqNet
InitiatorNnot specified
level: 0
σ = infinity
E.1.1.1.1NE.1.1.1.1
level: 1
σ = infinity
E.1.1.2.1.aNE.1.1.1.1
[Cat4] lev: 2
E.1.1.2.1.bNE.1.1.1.1
[Cat4] lev: 2
E.1.1.3.1NE.1.1.1.1
level: 2
E.1.1.3.2NE.1.1.1.1
level: 2
E.1.2.2.1.aNE.1.1.1.1
[Cat4] lev: 3
level: 3
E.1.2.3.4NE.1.2.1.1, E.1.2.1.2
[J71] lev: 4
E.1.2.3.1, E.1.2.3.2NE.1.2.1.1
level: 4
ready
clock: 21.000
real time factor:
1.0E-4
always show couplings
help
step
run
restart

Figure 17.7

platform to stimulate the SUT and iterate through the test cases developed earlier in the process. This component must supply a real-time environment with discrete event simulations that can execute the test cases according to the models developed. This component contains a Test Model Generator and the Test Driver infrastructure. It generates test models for MIL-STD compliance testing according to given test criteria while providing traceability back to the source document to ensure every test is justified by designated portions of the MIL-STD document.

Semi-automation in the ATC-Gen

As indicated earlier, much of the progress achieved in ATC-Gen development can be attributed to a judicious decomposition of the test generation process into phases whose potential level of automation could be understood and managed. Figure 17.8 diagrams this process and annotates its phases with their current level of automation. Rule capture and translation into XML is inherently manual, depending for its success on the collaborative knowledge and abilities of analysts, experts, and standards authorities. Nevertheless, such a process can be supported by appropriate tools that render the task easier to do and more efficient. Phases of the process that involve certain kinds of syntactic checking, or translation from one form of expression to another, offer instances of full automation. Dependency analysis is shown as semi-automated in that, although a dependency graph is automatically generated from a given rule set, selection of rule sets and interpretation of dependency relations depend on human understanding of the underlying standard. Similarly, test generation is semi-automated in that the choice of paths and their data instantiation (assigning values to variables that render the path executable) require human intervention. However, once a path is instantiated, it is mapped automatically into a DEVS model suitable for execution by a simulator. As suggested earlier, were the rule formalization component fully implemented, discovery and data instantiation of test paths might well be automated. However, such formalization would require a high degree of training in logic and would require considerable time and effort (until now, such formalization in other contexts has been fairly limited in scale). In a workforce environment where such training is rare, and where budgets are severely constrained, the dependency abstraction proved to be the key enabler to support a process that, albeit not fully formalized or automated, nevertheless, has proved measurably superior to traditional methods, as will be described next.

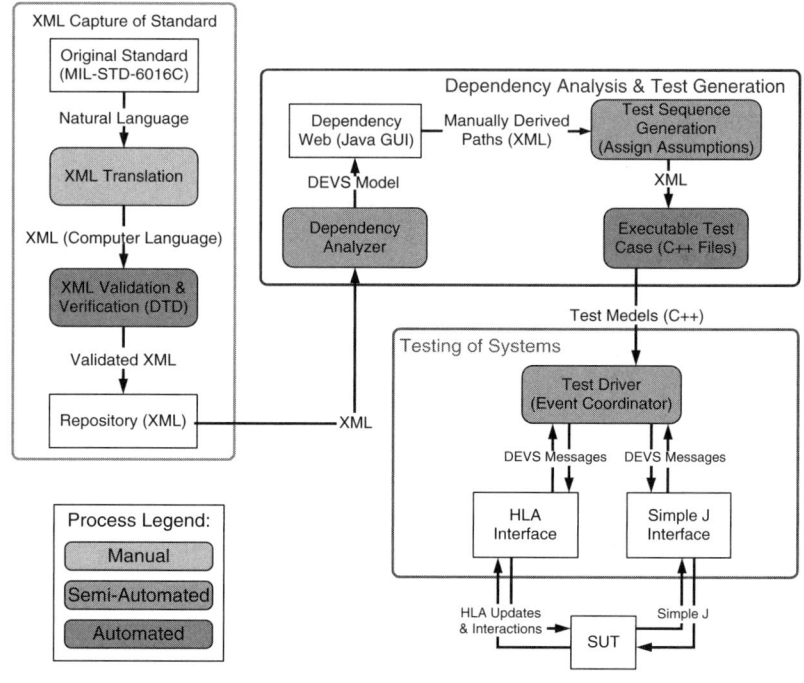

Figure 17.8

ATC-Gen Process Flow Diagram Showing Levels of Automation

Measuring Success: ATC-Gen Test Cases

Test cases were developed for some of the functional areas covered by Link-16 listed in Figure 11.2. Focus was on such tests for correlation and decorrelation. These are very difficult to develop without simulation-based methods because they have intrinsic dynamics and state dependence. In actual test exercises, ATC-Gen was able to complete more test cases in less time than allocated and the test cases were effective in verifying both correct and incorrect behaviors. Importantly, the ATC-Gen process provided assurance that results of test cases could be trusted. Traceability to source requirements was a major factor in uncovering and explaining the reasons that failed test cases signaled true departures and how they should be corrected in subsequent development. Based on these results, and subsequent validation tests, the ATC-Gen was certified as a test artifact. The interested reader can find more detail on the theory underlying the ATC-Gen methodology in Refs. 7–9.

Automation in Standards Conformance Testing: Summary and Conclusions

Introducing automation into standards conformance testing increased the depth of testing while reducing the time needed to prepare test exercises. Critical areas in which a new approach could have a demonstrable effect were selected for initial development. The ATC-Gen project developed a tool-set and methodology that were successful in meeting the requirements of the critical application and established general agreement that the approach was ready for adoption and integration into testing paradigms. Systems theory and DEVS were critical to the successful development of the automated test case generator. The dependency abstraction was key to coming up with a workable, albeit semi-automated, methodology and tool-set.

Employing Agent-based Technology for Web-Services Testing[1]

An effective software test framework should be able to automatically construct test models and dynamically configure them according to different test scenarios for the system under test. Software agents, with characteristics such as autonomy, modularity, on-demand creation, and self-destruction, provide a promising technology to support such a framework. This paper presents an agent-based test framework where test agents are automatically created and dynamically configured according to test scenarios derived from the behavior specification of a system under test. We present this framework in the context of web services and focus on testing input/output (I/O) behavior of complex defense information technology systems.

Web service is becoming one of the enabling technologies for the emerging large-scale net-centric systems, such as distributed supply chain management systems, networked crisis response systems, and joint simulation and training systems. Requirements for testing such systems include those derived from their decision-making, net-centric, large-scale, and dynamic behaviors. The new paradigm of service-oriented computing requires research and technology developments from all aspects of computer science.[10,11] Among them a challenging and increasingly

---

[1] This section was extracted from "A Dynamic Agent-Based Framework for Web-Services Testing," Xiaolin Hu, Bernard P. Zeigler, Moon Ho Hwang, Eddie Mak, working paper, 2006.

important issue is how to support web service testing in a systematic and effective manner. Two of the factors underlying the challenges posed by testing web services are:

1. The increasing complexity of service-oriented computing results in sophisticated service interactive patterns. In particular, stateful web services, i.e., services that access and manipulate state and/or stateful resources in a timely fashion are widely deployed. A stateful service may implement a series of operations such that the result of one operation depends on a prior operation and/or prepares for a subsequent operation.[17] Indeed, in session-based conversational message exchanges, the timing and sequencing of messages in a stream may determine how services process data and interact with each other. Testing the dynamic behavior of stateful web services is difficult because of the temporal and logical relationships among input/output messages.
2. Service-oriented applications are by nature evolutionary. This nature implies that supporting the sustainable long-term evolution of service applications should play a central role in developing quality assurance and testing techniques and tools.[15] Service-oriented computing emphasizes service integration and composition where new services are dynamically discovered and integrated. This evolutionary nature calls for new test mechanisms such as "on-demand" test development to support this paradigm.

A software agent is considered a promising and powerful technology for supporting service-oriented computing in the open distributed grid environment.[18] When applied to web service testing, the autonomy and modular nature of software agents makes it possible to automatically and dynamically construct test agents. Such agents have the potential to "intelligently" exploit the semantics and ontology of services and service applications to achieve systematic and effective testing. In applying agent technology to testing web services and web applications, one focus has been on testing at the function level by exploiting the internal structure and/or behavior of a web application.[13] Test plans are derived from data flow analyses of web applications and test agents are constructed to carry out these test plans.[14] Another focus is at the software component level. For example,[12] utilizes "validation agents" to automate the testing of software components. These validation agents depend on the "component aspects" that specify the functional and non-functional cross-cutting concerns of software components to construct and run automated tests. The Agentcities testbed[16] is an agent-based test environment developed to support large-scale dynamic service synthesis testing. In relation to the objectives of this paper, two themes emerge from current research: 1) automating the construction of test agents, and 2) deploying multiple agents in test federations to collaborate in testing complex distributed interaction patterns.

Our discussion explicitly focuses on these themes in the application of agent-based technology to testing web services orchestrations at the I/O behavior level. We consider this a preferred approach in a service-oriented environment where web services are typically perceived by their interfaces and I/O behaviors. Systematic testing of these I/O behaviors, i.e., the logical and/or temporal relationships between sequences of inputs and outputs, can provide direct measurements of how well a service will interact with others. The I/O behavior test also has the advantage that a test agent need not know about the internal structure of a web service under test. This removes the dependency on web services' internal representations that are inaccessible to testing in a service-oriented environment based on service description interfaces. To carry out a web service test at the I/O behavior level, we take the system theory point of view that an online service (e.g., software components, process models, or service organizations) is a system with I/O behaviors. Based on this system's theory foundation, we develop techniques to extract *testable behaviors* of the system/services under test (SUT). Testable behaviors are information exchanges among system components or services that are publicly observable over a network and do not require inspection into the components or services themselves. Later we will show how development of such testables should be constrained by desiderata such as level of coverage of critical interactions, support for determination of appropriate measures of effectiveness (MOE), and measures of performance (MOP). It is worthy to note that this approach does not aim to cover all possible combinations of a web service's inputs and outputs. Instead, it derives test scenarios according to an SUT's behavior specification and constructs test agents to carry out the tests. Constructing and deploying test agents in this work is supported by the DEVS modeling and simulation/execution framework. The event-driven nature and modular model construction of DEVS prove to be essential to develop test agents for system tests at the I/O behavior level. DEVS is not just a theoretical framework, as it has been operationalized to serve as a practical simulation and execution tool in a variety of implementations.

System Specification and Testable I/O Behaviors

A web service at the I/O behavior level can be viewed as a system that receives inputs (messages) and generates outputs through its interfaces. These inputs/outputs and their logical/temporal relationships represent the I/O behavior of the system. From the design point of view, given such a system behavior specification, the developers' job is to realize a system to fulfill the required behavior; and the testers' job is to verify that the developed system conforms to the behavior that has been specified. This view of system design and verification based on system specification is rooted in systems theory and serves as a theoretic foundation of this work.

Recall the hierarchy of system specifications which defines levels at which a system may be known or specified (Chapter 11). Although this hierarchy provides a good framework for structuring the design and testing processes, a number of its aspects have to be considered in greater depth to enable it to support actual testing. One of these aspects is the development of a more compact form of the I/O Function Specification that is usable for development of semi-automated test suites. Thus in this section we turn our discussion to this issue by first reviewing the concepts of input/output behavior and their relation to the internal system specification in greater depth. This will put us in a position to understand the complexities that arise in specifying test sequences and approaches to their management.

Testable Form of I/O Specification

For a more in-depth consideration of input/output behavior, we start with the top of Figure 17.9 which illustrates an input/output (I/O) segment pair. The input segment represents message types X, with content x and Y, with content y arriving at times t_1 and t_2, respectively. Similarly, the output segment represents message type Z occurring twice with content z and z', at times t_3 and t_4, respectively. The bottom of the figure shows a more compact representation of this same information, in which arrows above the time line represent inputs received and those below represent outputs sent.

To illustrate the specification of behavior at the I/O level we consider a simple system — an adder. All it does is add values received on its input ports and transmit their sum as output. However simple this basic operation is, there are still many

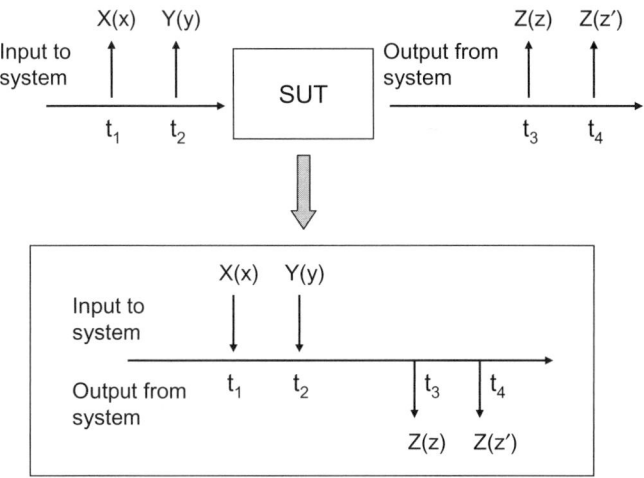

Figure 17.9

Representing an Input/Output Pair

possibilities to consider to characterize its I/O behavior such as which input values (arriving at different times) are paired to produce an output value and the order in which the inputs must arrive to be placed in such a pairing.

Figure 17.10 portrays three possibilities, each described as a DEVS model at the I/O System Level of the Specification Hierarchy. In (a), after the first inputs of types X and Y have arrived, their values are saved and subsequent inputs of the respective types refresh these saved values. The output of message type Z (represented as an open end arrow) is generated some time after the arrival of an input and its value is the sum of the saved values. In (b), starting from the initial state, the output can only be computed when a message of type Y is the next message to arrive after a message of type X. In this case the output is computed from the latest values of X and Y and the system is reset to its initial state. In (c), the order of arrival constraint is removed but the reset requirement is retained — from the initial state, both types of messages must arrive before an output is generated (from their most recent values) and the system is reset to its initial state after the output is generated. This example shows that even for a simple function, such as adding two values, there can be considerable complexity involved in the specification of behavior when the temporal pattern of the messages bearing such values is considered. Two implications are immediate. One is that there may be considerable incompleteness and/or

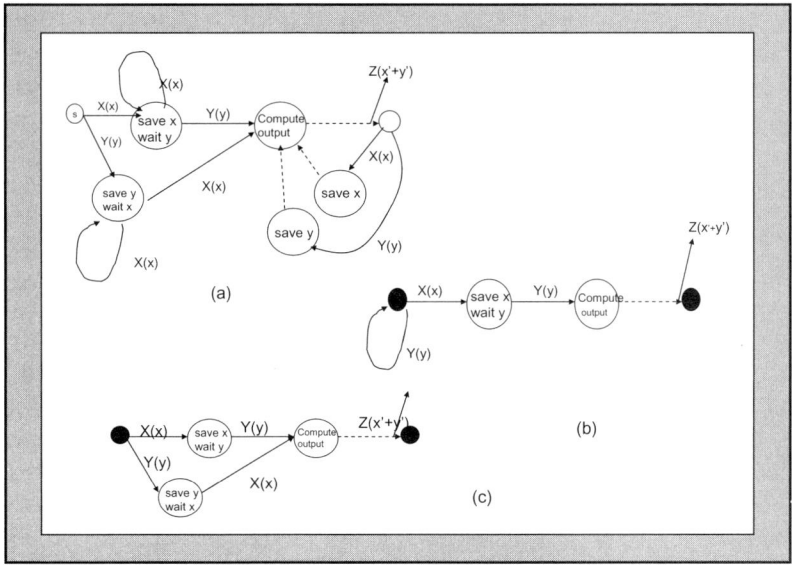

Figure 17.10

Variants of a DEVS Model for Processing Messages

ambiguity in a semiformal specification where explicit temporal considerations are often not made. The second implication follows from the first: an approach is desirable to represent the effects of timing in as unambiguous a manner as possible and in such a way that a discriminating test suite can be derived from this representation. We outline such an approach in the following.

Figure 17.11 shows some I/O pairs that serve to distinguish between the triplet of behaviors in Figure 17.10. In Figure 17.11(a), after initial arrivals of message types X and Y, with subsequent output of type Z, successive arrivals of either message type, X or Y, results in an output of type Z with values shown. In Figure 17.11(b), however, a second Y message in a row does not result in an output, nor does an X following it. To explicitly specify this, a dashed arrow is used to indicate that an output should *not* be generated. In contrast, for Figure 17.11(c), while a second Y message still does not result in an output, the latter will be produced when the next arrival is an X message. Note that these patterns are discernible at the Input/Output behavior level and form the basis for testing where access to the SUT is through input/output interfaces. Development of these patterns

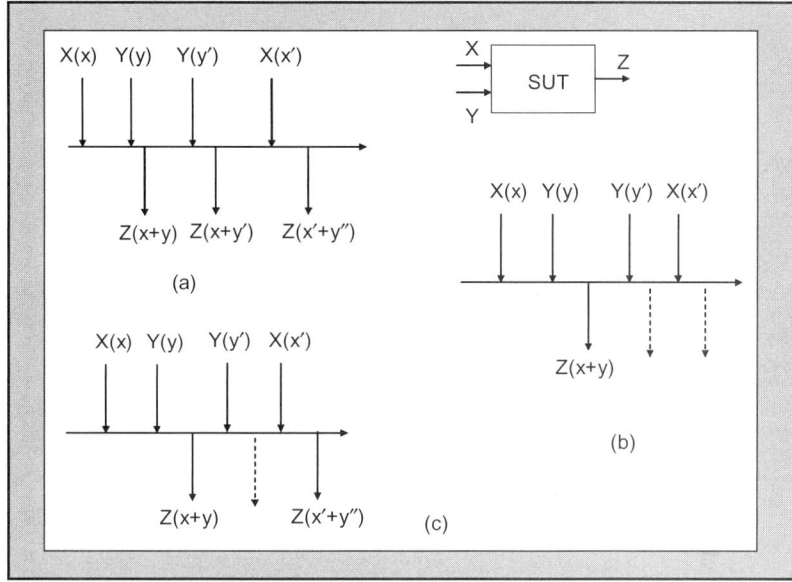

Figure 17.11

Input/Output Pairs Corresponding to the Variants of Figure 17.10

follows from understanding and manipulation of the Level 3 state-based specification.

Minimal Testable Input/Output Pair

An I/O segment pair may have any finite number of input messages and output messages in its interval of definition. It is much easier to deal with segments of limited complexity and synthesize tests from such segments. Thus we introduce the concept of *minimal testable I/O pair*. This is an I/O pair in which the output segment has at most one event at the end of the segment. Such segments are illustrated in Figure 17.9. We can easily extract such pairs from an I/O specification at Level 2, by going from a given state to the next one that results in an output event, because each input segment produces a unique output segment if the initial state is given. Figure 17.12(a) and Figure 17.12(b) illustrate this process for the corresponding variants given in Figure 17.10.

An I/O Specification at Level 2 for which all pairs are minimally testable is said to be a *minimal testable* I/O representation.

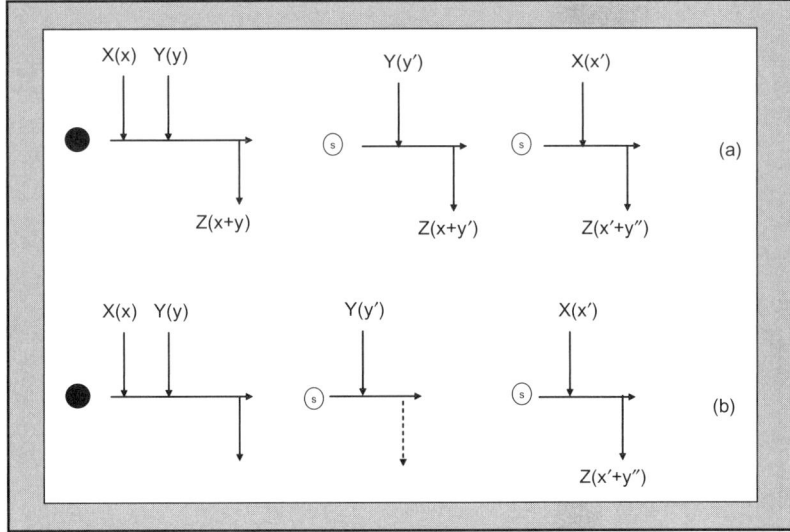

Figure 17.12

Some Minimal Testable Input/Output Pairs

A minimal testable I/O representation can be captured by the System Entity Structure (SES) and documented in XML format. SES is a structural knowledge representation scheme that systematically organizes a family of structures that characterizes decomposition, coupling, and taxonomic relationships among entities (Chapter 11). Describing SES in XML format is straightforward. SES uses nodes to express the structure of a testable I/O representation, and XML uses elements to break up the structure and represent the SES nodes. Figure 17.13 shows an SES example and its corresponding XML elements for representing a minimal testable I/O representation. More information about representing minimal testable I/O representations using SES and XML can be found in Zeigler.[7]

The concept of minimal testable I/O pairs and their representation in SES and XML provide the foundation for automated construction of test models. We can go from an I/O Behavior Specification at Level 2, and in particular from its minimal testable representation, to a test model that interacts with an SUT to assess whether it satisfies the specification. The test models can be synthesized from the segments because each I/O pair has a finite number of input messages and output messages in its interval. Based on this foundation, the next section elaborates on the agent-based framework for web service testing.

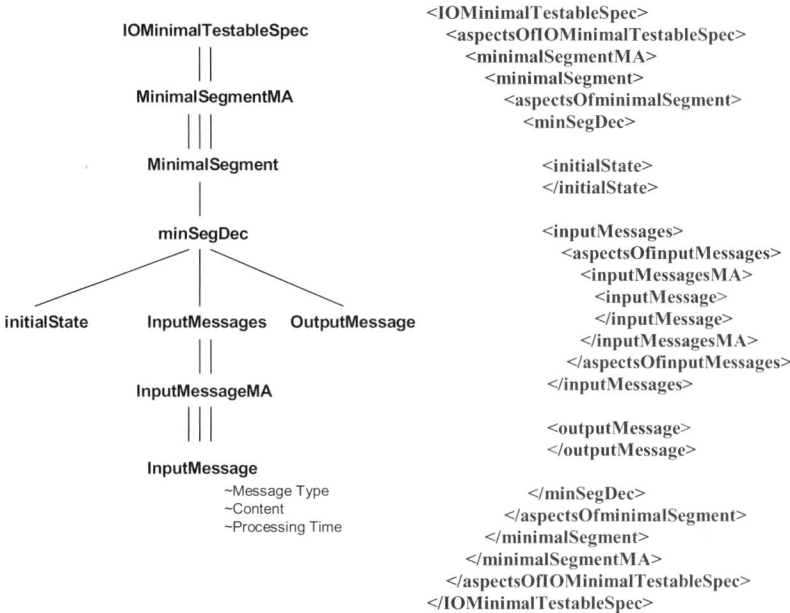

Figure 17.13

SES and XML Tags for Minimal Testable I/O Representation

Overview of the Agent-based Testing Framework

In order to achieve systematic and productive tests, one of the main goals of the agent-based test framework is to support (semi-) automated agent construction from system behavior specification. Motivated by this, a design process (shown in Figure 17.14) has been developed to guide the development of agent-based testing. This process starts from the system behavior specification of the SUT and goes through several stages, where each stage is dependent on the artifacts from its previous stage and works on them to generate new models. This process ends when test agents for specific test scenarios are created.

The first step of this process is to extract minimal testable I/O pairs from the system behavior specification of the SUT. The second step is to sequence such I/O pairs into test scenarios to be embodied by test agents. The third step is to apply a model mapping concept to implement the minimal testable I/O pairs that are included in a test scenario as DEVS primitives. The DEVS primitives are simple atomic models such as: *processDetect*, *waitReceive*, and *waitNotReceive* that will be described in section 3.3 (Test Agent Construction). These primitives act as the

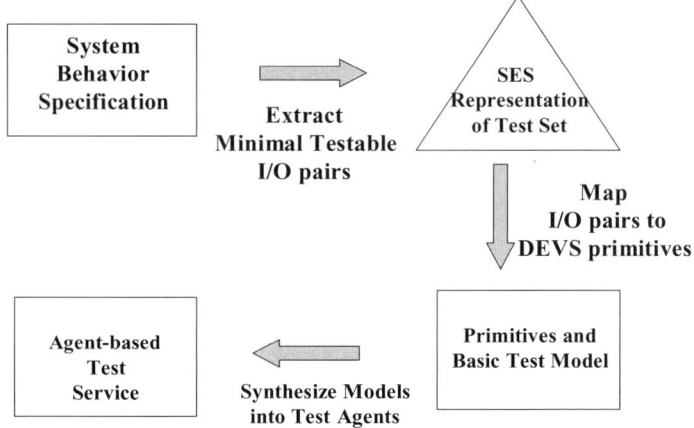

Figure 17.14

Framework for Agent-based Testing of Web Services

building blocks to form basic test models corresponding to the minimal testable I/O pairs. In the fourth step, the basic test models are coupled together according to the test scenarios and their compositions become test agents with specific characteristics such as autonomy, modularity, on-demand creation, and self-destruction. Finally, the created test agents are deployed in test federations to inject or observe inputs sent to, and outputs emitted by, the SUT.

Controlled and Opportunistic Test Modes

Before delving into the details of agent construction, we describe the two types of test configurations that this test framework intends to support: controlled test and opportunistic test. Figure 17.15 uses an example scenario that includes one minimal testable I/O pair to illustrate the interactions between a test agent and the SUT for controlled testing and opportunistic testing, respectively. This minimal testable I/O pair has inputs $Mx1(data1)$, $Mx2(data2)$ and output $Mx3(data3)$, denoted as $\langle Mx1(data1), Mx2(data2) \rightarrow Mx3(data3)\rangle$, and is shown by the grey box in the figure. We note that for both controlled testing and opportunistic testing, in order to test a web service in a service-oriented environment, a test agent will interact with the web service through SOAP. A DEVS test server is responsible for providing the runtime environment for the test agent as well as providing the communication support on the network.

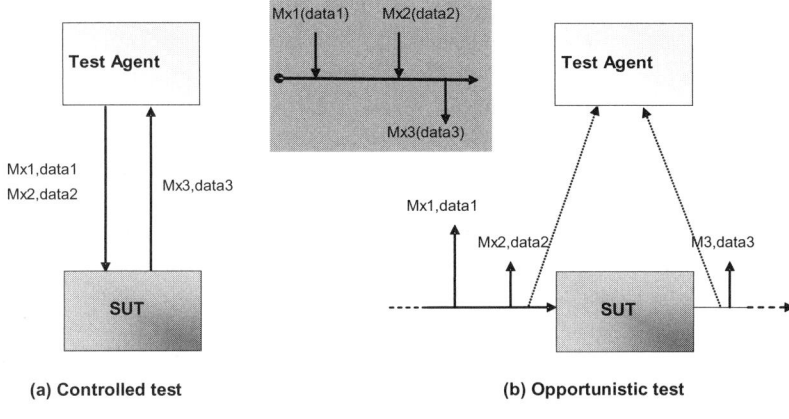

Figure 17.15

Controlled Test and Opportunistic Test

In the controlled test as shown in Figure 17.15(a), the SUT is coupled solely to the test agent, which fully controls the test scenario by sending inputs to the SUT and checking if the output responses of the SUT are correct. The sequence of inputs injected by the test agent and the kind of output that the test agent expects are derived from the scenario (which includes one minimal testable I/O pair for the example of Figure 17.12) of the SUT. Specifically, based on the minimal testable I/O pair $\langle Mx1(data1)$, $Mx2(data2) \to Mx3(data3)\rangle$, the test agent first generates message type *Mx1* with value *data1*, then message type *Mx2* with value *data2* and sends these to the SUT in timed, sequential order. After that, the agent waits for the output generated by the SUT and checks if it is *Mx3* with value *data3*. The test agent issues a Pass or Fail for this test scenario according to whether *Mx3(data3)* was received within an allowed time window. As can be seen, the controlled test corresponds to the tests that are typically conducted before a system is deployed to its executing environment, because the test agent assumes full control of the SUT.

The opportunistic mode, Figure 17.15(b), extends the controlled mode and complements it. In contrast to the controlled test, a test agent in the opportunistic test does not control the test scenario. Instead, it observes the I/O behaviors of an operating SUT and looks for opportunities to carry out the test. Thus a test agent in opportunistic mode is also called *test detector*. In the opportunistic mode, the SUT is only loosely coupled to the test detector in the sense that its operation is not driven by the test detector. Because of this, opportunistic mode can be employed after a SUT is deployed to its executing environment where it interacts with other systems/services. In the opportunistic test, a test detector works with a *test deployer* (not shown in Figure 17.15(b)), which dynamically creates and deploys the test detectors based on the test objectives. Once a test detector is deployed, it starts to observe the inputs and outputs of the SUT, and will activate a test when its conditions are met. For example, the test agent (detector) in Figure 17.15(b) is set up to test the minimal testable I/O pair $\langle Mx1(data1), Mx2(data2) \rightarrow Mx3(data3) \rangle$. If this agent observes input messages $Mx1(data1)$ and $Mx2(data2)$ in the right timed, sequential order (this means the agent's test condition is met), it will check if the right output $Mx3(data3)$ will be generated by the SUT within the allowed time. Otherwise, e.g., if the agent notices the first input message is $Mx1(data1)$ but the following message is not $Mx2(data2)$, the agent will withdraw the test by reinitializing itself and will not issue any result for the abandoned test. The re-initialization prepares the agent to continue to look for other opportunities to conduct the test. As can be seen from this example, opportunistic testing has the advantage that the test agents do not interfere with the operation of the SUT. This loose coupling nature makes it possible to dynamically create multiple test agents for the same SUT. These agents work concurrently and can have unique objectives to test a specific aspect of the SUT. This feature of loose coupling and dynamic deployment of test agents is especially useful in service-oriented environments where services evolve and change dynamically.

The functional difference between a test agent in controlled testing and an agent in opportunistic testing implies that the constructions of these agents will also be different. In controlled mode, an agent needs to send inputs to the SUT and wait for output from the SUT. Thus the agent is constructed from primitives: *holdSend*, *waitReceive*, and *waitNotReceive*. In opportunistic testing, an agent needs to detect inputs sent to the SUT and wait for output from the SUT. Thus the agent is constructed from primitives: *process Detect*, *waitReceive*, and *waitNotReceive*. Despite the many operational differences that stem from the modes of execution, both controlled and opportunistic modes share the same underlying design processes, i.e., mapping minimal testable I/O pairs to DEVS primitives and then automatically synthesizing models into test agents. Next we discuss in detail how a test agent in opportunistic mode is automatically constructed. The description can be easily adapted to cover the case of controlled mode.

Test Agent Construction

The starting point of agent construction is the minimal testable I/O representation derived from the system behavior specification of the SUT. Based on the minimal testable I/O representation, a test detector has to watch for messages arriving to, and departing from, the SUT. Correspondingly, there are three atomic models (also referred to as primitives): *processDetect*, *waitReceive*, and *waitNotReceive*. A *processDetect* primitive waits for a message to arrive to the SUT within a prescribed interval. A *waitReceive* primitive waits for a SUT response within a pre-defined time interval, and determines the pass-fail condition by comparing it with the expected response. On the other hand, a *waitNotReceive* primitive watches for messages that are not supposed to be sent by the SUT. This is for the case of the I/O minimal test pair that does not end in an output. This primitive holds for a pre-defined time interval and determines failure if the SUT produces a message on the watched list in this interval. Each primitive is implemented as a DEVS atomic model. To give an example, Figure 17.16 shows the input/output ports and the state transition of the *waitReceive* primitive. The constructor of a *waitReceive* primitive consists of three fields: model name, the message that the model expects, and wait time. As can be seen, the model has two input ports: *start* and *In_msg*, and one output port: *pass*. The behavior of the model is described as follows. The model begins at the "Passive" state. When it receives a start

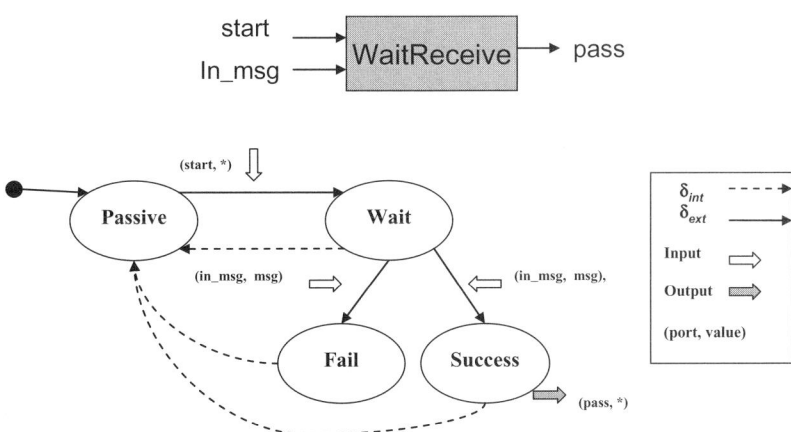

Figure 17.16

Interface and State Transition of the *WaitReceive* Primitive

event, the external transition function dictates a new state, "Wait," with the wait time provided by the constructor. If the model receives a message event before the specified time expires, it compares the received message with the message that it expects. If they are identical, it will log a successful outcome, and the state will change to "Success," where the output function will generate a "pass" message, and then the internal transition function will change the model's state to "Passive." Otherwise, a failure will be logged and the state will change to "Passive."

The three primitives described above act as the building blocks to compose basic test models according to the minimal testable I/O pairs derived from system behavior specification. Specifically, for each minimal testable I/O pair, a basic test model is constructed by coupling one or more *processDetect* primitives in sequential order corresponding to the inputs, and then one *waitReceive* (or *waitNotReceive*) primitive corresponding to the output. Figure 17.17 illustrates how to construct basic test models using these primitives. The behavior of the SUT is described by

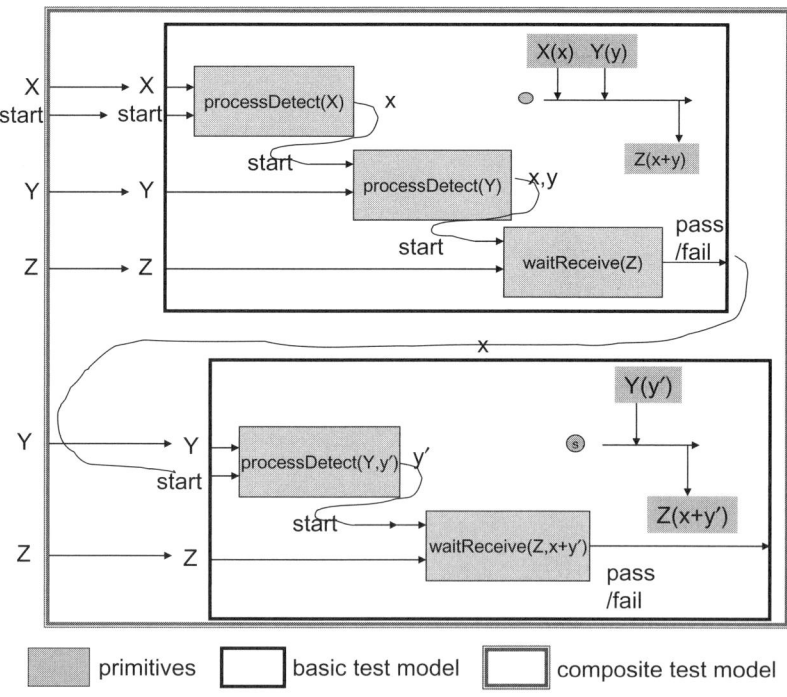

Figure 17.17

Basic and Composite Test Models

the minimal I/O pairs shown in Figure 17.12. Corresponding to the first minimal testable I/O pair ⟨X(x), Y(y) → Z(x + y)⟩, the initial *processDetect(X)* watches for the arrival of a value on port X; when such a value, x, arrives, it sends a start message containing x to the second *processDetect(Y)* which looks for a value coming on the Y port. When such a value, y, arrives, it sends the pair of values x, y to the *waitReceive(Z)*, which waits for a value on the Z port. When such a value, z, arrives, it compares it to x + y and emits a pass or fail decision accordingly. Note that communication from one primitive to the next establishes the context or information needed to maintain the state information needed for the SUT. This first level composition of primitives is called a Basic Test Model.

When the basic test models derived from the minimal testable I/O pairs are cascaded together, the second level composition is called a composite test model and implements a test scenario (defined by one or more minimal testable pairs, in sequential order). The atomic models and the coupled models are triggered by a start event through the start port. When the condition of the basic test model is met, it will trigger another basic test model. For example, the basic test model shown in the lower part of Figure 17.17 is for the I/O minimal segment shown which checks for arrival of a message on the Y port subsequent to processing from the initial state. This model consists of a *processDetect(Y)* that is started with receipt of a value y'. When receiving a value y' it sends the pair (x, y') to the *waitReceive(Z)* which operates as just described. Consider now the SUT described in Figure 17.10(c). Here the second basic model in the cascade is a *waitNotReceive(Z)*, rather than a *waitReceive(Z)*, since the SUT is not supposed to respond unless both inputs are refreshed.

We can couple sets of composite test models together in hierarchical fashion to form complex test scenarios. It is important to automate the transformation from the minimal testable pairs to the test models. Such automated model transformation also provides traceability support between the required SUT behavior and the DEVS test models. In our approach, both the minimal testable I/O pairs and test models are described in SES and written in XML format. Transformation from minimal testable pair to test models is based on the mapping relationship between the corresponding nodes in the two models.

Figure 17.18 illustrates the transformation from minimal testable I/O pairs to test models. For example, the initial state entity in a minimal testable I/O pair is mapped to the initial state requirement variable of the *BasicTestModel* entity. The *InputMessages* entity is mapped to the *processDetect_Components*

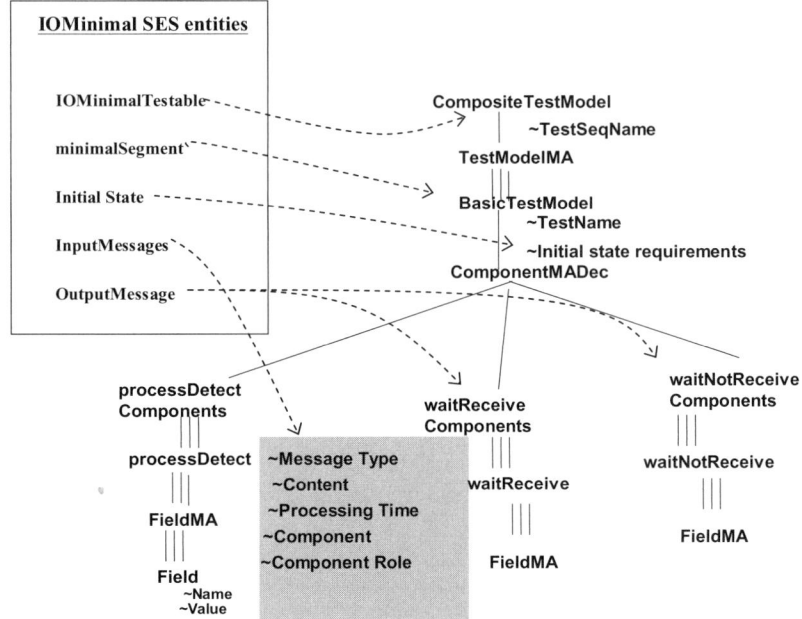

Figure 17.18

Mapping Minimal Testable Pairs into Test Model SES

entity. The *OutputMessage* entity is mapped to either *waitReceive_Components* or *waitNotReceive_Components* depending on whether it is required to be, or not to be, produced, respectively. Details can be found in Mak.[9]

Dynamic Configuration of Test Agents

A test agent, after its creation, is added to a test suite to execute the test. A mechanism is needed to support dynamical inclusion/removal of these test agents and/or reconfiguration of agents' connections at runtime. Compared with using a fixed set of test agents, dynamic inclusion/removal of agents will result in cohesive test suites, thus making it easier to manage the active test cases. It allows for changing from one test focus to another in real-time, e.g., going from format testing to correlation testing once the first has been satisfied. Furthermore, dynamic configuration of the agent's interconnections makes it possible to re-organize multiple cooperative test agents to accomplish complex test scenarios. Motivated by the advantages mentioned above, our test framework supports on-demand inclusion of test agent instances, and allows an agent to remove itself after finishing its

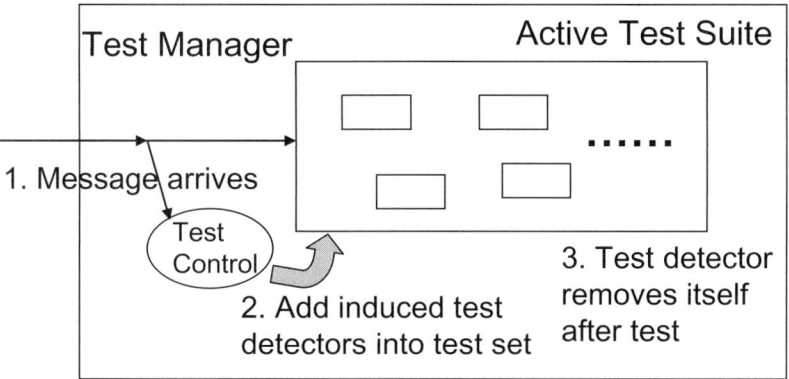

Figure 17.19

Dynamic Configuration of Test Agents

test. Realization of the dynamic configuration is based on DEVS' variable structure modeling capabilities, which allow models and their connections to be dynamically added/removed.

Figure 17.19 illustrates a setup of an opportunistic test that employs dynamic configuration to manage an active test suite. In this example, a Test Manager (a DEVS coupled model) hosts a Test Control (a DEVS atomic model) and an Active Test Suite (a DEVS coupled model) that includes multiple test detectors. The Test Control has the intelligence to decide when and how to create test detectors and add them into the active test suite. For example, as shown in the figure above, the Test Control observes some interested messages of the SUT, and when such a message arrives, it dynamically constructs one or more test detectors and adds them to the active test suite. Using the variable structure operations of DEVS, this can be accomplished as follows (pseudo code).

TestDetector = Create_testDetector( ); // create the test detector
addModelToChildOf (TestDetector, "Active Test Suite");
addCoupling("Test Manager", "message", TestDetector,
  "message");

The *addcoupling( )* operation establishes a connection between the Test Manager's message port to the added test detector's message port. This allows the test detector to receive the input/output messages of the SUT, thus to observe its I/O behavior for carrying out the test. After the test is finished, i.e., when the test case result is known (either pass or fail), the test detector removes itself from the active test suite. This can be accomplished using the following code:

removeModel("test detector"); //based on the name of the model

To support the dynamic configuration needed in this work, we extended the variable structure operations developed in Ref. 19 into a new set of operations. The extension results in more allowed variable structure operations and more relaxed operation boundaries for those operations. With the new variable structure operations, a model can add or remove other models, and add or remove couplings between pairs of models anywhere in the model tree. The new operations are necessary in an agent-based framework because agents typically have knowledge about other agents even if they do not belong to the same peer group in the model tree. Special attention was paid in implementing these variable structure operations to ensure a structure change will not cause violation to the hierarchical property of the model tree. An example of test agent application is given in Appendix B.

Summary

On page 337 we concluded that introducing automation into standard conformance testing could increase the depth of testing, while reducing the time needed to prepare test exercises. Systems theory and DEVS were critical to the successful development of the automated test case generation methodology.

We went on to present a dynamic agent-based framework that enables automatic constructing and configuring software agents to test web services' I/O behavior. The framework starts from the I/O behavior specification of a web-service collaboration and derives minimal testable I/O pairs that are mapped to basic test models. These primitives are then synthesized into test agents for specific test scenarios and deployed in test federations. A prototype was demonstrated in a net-centric testbed that exhibits characteristics of a collaborative web-services environment. The correctness and effectiveness of such web-service collaborations are heavily dependent on the dynamic behaviors of individual services or service organizations. The agent-based test framework presented in this paper provides a starting point to develop tools for automatic and dynamic behavior testing of such collaborations.

References

1. NATO STANAG 5516 TACTICAL DATA EXCHANGE — LINK 16 http://en.wikipedia.org/wiki/Linkv16 (accessed May 4, 2007).
2. Jacobs, Robert W. "Model-Driven Development of Command and Control Capabilities For Joint and Coalition Warfare," Command and Control Research and Technology Symposium, June 2004.
3. Zeigler, B. P., H. Praehofer, and T. G. Kim, "Theory of Modeling and Simulation: Integrating Discrete Event and Continuous Complex Dynamic Systems," 2nd ed., New York: Academic Press, 2000
4. Dahmann, J. S., F. Kuhl, and R. Weatherly, Standards for Simulation: As Simple As Possible But Not Simpler — The High Level Architecture for Simulation. *Simulation*, 71(6), pp. 378–387, 1998
5. Modeling and Simulation in Manufacturing and Defense Acquisition: Pathways to Success, Washington, DC: National Academy Press, 2002
6. Technology for the United States Navy and Marine Corps, 2000–2035: Becoming a 21st-Century Force: Volume 9: Modeling and Simulation, Washington, DC: National Academy Press, 1997
7. Zeigler, B. P., D. Fulton, P. Hammonds, and J. Nutaro, "Framework for M&S–Based System Development and Testing in a Net-Centric Environment," *ITEA Journal of Test and Evaluation*, Vol. 26, No. 3, 21–34, 2005
8. Nutaro, James, and Phil Hammonds, "Combining the Model/View/Control Design Pattern with the DEVS Formalism to Achieve Rigor and Reusability in Distributed Simulation," *Journal of Defense Modeling and Simulation: Applications, Methodology, Technology*, pp. 19–28, Vol. 1, No. 1, 2004
9. Mak, E., "Automated Testing Using XML and DEVS," MS Thesis, Department of Electrical and Computer Engineering, The University of Arizona, May 2006
10. Manes, A. T., *VantagePoint 2005–2006 SOA Reality Check*: Version 1.0, Burton Group Publication, June 29, 2005
11. Hull, R., and J. Su, "Tools for Composite Web Services: A Short Overview," ACM SIGMOD Record, Vol. 34, No. 2, June 2005
12. Grundy, J. C., G. Ding, and J. G. Hosking, "Deployed Software Component Testing Using Dynamic Validation Agents," *Journal of Systems and Software: Special Issue on Automated Component-based Software Engineering*, vol. 74, no. 1, pp. 5–14, January 2005
13. Yu Qi, David Kung, and Eric Wong, "An Agent-Based Testing Approach for Web Applications," *compsac*, pp. 45–50, 29th Annual International Computer Software and Applications Conference (COMPSAC'05) Volume 2, 2005
14. Huo, Qingning, Hong Zhu, and Sue Greenwood, "A Multi-Agent Software Environment for Testing Web-based Applications," *COMPSAC*, p. 210, 27th Annual International Computer Software and Applications Conference, 2003

15. Hong Zhu, "Cooperative Agent Approach to Quality Assurance and Testing Web Software," *COMPSAC*, pp. 110–113, 28th Annual International Computer Software and Applications Conference — Workshops and Fast Abstracts (COMPSAC'04), 2004
16. Willmott, Steven, Simon Thompson, David Bonnefoy, Patricia Charlton, Ion Constantinescu, Jonathan Dale, and Tianning Zhang, "Agent Based Dynamic Service Synthesis in Large-Scale Open Environments: Experiences from the Agentcities Testbed," *aamas*, pp. 1318–1319, Third International Joint Conference on Autonomous Agents and Multiagent Systems — Volume 3 (AAMAS'04), 2004
17. Foster, I., J. Frey, S. Graham, S. Tuecke, K. Czajkowski, D. Ferguson, F. Leymann, M. Nally, T. Storey, and S. Weerawaranna, *Modeling Stateful Resources with Web Services*, Globus Alliance, 2004, www.globus.org (accessed Nov. 2006)
18. Foster, N., R. Jennings, and C. Kesselman. "Brain meets brawn: Why grid and agents need each other." Autonomous Agents and Multi-Agent Systems (AAMAS'04), 2004
19. Hu, X., B. P. Zeigler, and S. Mittal, "Variable Structure in DEVS Component-Based Modeling and Simulation," *SIMULATION: Transactions of The Society for Modeling and Simulation International*, Vol. 81, No. 2, pp. 91–102, 2005

Appendix A: Infusion of Modeling and Simulation into Defense Acquisition

In the 1990s, the U.S. Army had undertaken to reorganize itself and to use simulation to enhance and speed up the training of its soldiers. The extensive use of simulation in training spurred advances in distributed simulation — in which computers, both general purpose processors and specialized ones such as tank simulators, are connected together using a data communication network. This in turn led to DoD-wide standards such as the Distributed Interactive Simulation (DIS) protocol and later the High Level Architecture (HLA) for data exchange and coordination of simulated events[4] (see text references). Spurred by successes in training, and cognizant of increasingly extensive use of modeling and simulation (M&S) in system development, the defense department published several directives that encouraged the adoption of M&S throughout its agencies and activities. By 2001, this trend became codified in a set of revised policies for acquiring new systems, called *simulation-based acquisition*, which required the use of M&S in all phases of the system development life cycle, especially starting early on before "metal has been bent"[5,6] (see text references). Although it took longer to be fully appreciated, a major implication of simulation-based development was the concomitant need for simulation-based testing. As became clear in the Single Integrated Air Picture (SIAP) project, you need a simulation-capable infrastructure to test a system that is first formulated as an abstract model — one that can be executed only as a software prototype. Since it is divorced of real-world inputs and outputs, such an object can be communicated with only through certain designated channels, a situation ideally suited for distributed simulation in which both the System Under Test (SUT) and the test device are coupled by exchanging data packets on a network.

Further, the dawning of the information age brought with it new plans for weapons and defensive systems whose very operation depended on the collection and processing of information. Such C4ISR systems, as they became known, augmented Command and Control (the first two Cs) with Communication and Computers (the next two Cs) and were informed by satellites and other sensors providing Intelligence, Surveillance, and Reconnaissance (ISR). To deal with the increasing complexity and advanced decision capabilities of such systems, testing methodology had to become more rigorous, in-depth, and thorough. At the same time, to keep up with the rapid change and short development life cycles expected from the system builders, testing had to also become more nimble itself. Where traditional ways of designing and deploying test exercises could take months to accomplish, new demands required tests to be ready to conduct in time scales compatible with the agile development strategies of new systems. Such demands became increasingly insistent as the DoD started to move all of its operations on to the Global Information Grid (GIG), its adoption of the World Wide Web to a new high speed infrastructure structured like the Internet.

Appendix B: Opportunistic Message Tester

The agent-based test framework is being developed in the context of testing service collaborations on network implementations of service-oriented architectures. Based on this context, this section provides an illustrative example in a net-centric testbed that exhibits characteristics of a web services environment. This testbed emulates a collaborative network of sensor and decision nodes acting as web services that exchange information and decisions to achieve given objectives (correctly detecting and identifying objects). The protocols for collaboration are given in a specification that dictates the manner in which messages are received, processed, and transmitted to accomplish system objectives.

The Opportunistic Message Tester (OMT) is intended to demonstrate opportunistic testing at the I/O behavior level following the basic principles of the agent-based framework discussed above. The objective of the OMT is to passively observe the message traffic among nodes and, when possible, test for conformance to the specified collaborative behavior, particularly relating to object matching. The goal of matching is to determine whether multiple sensors are detecting the same object or not.

Figure 17B.1 illustrates the overall configuration of the agent-based testbed environment whose SUT consists of one local sensor/decider unit and multiple remote units. The OMT listens in to messages exchanged between the local unit under observation and remote units. It activates agents to look for testable message patterns. Similar to the setup shown in Figure 17.15, the OMT is a coupled DEVS model that consists of a Test Control and a set of MatchSequence test models representing the active test suite.

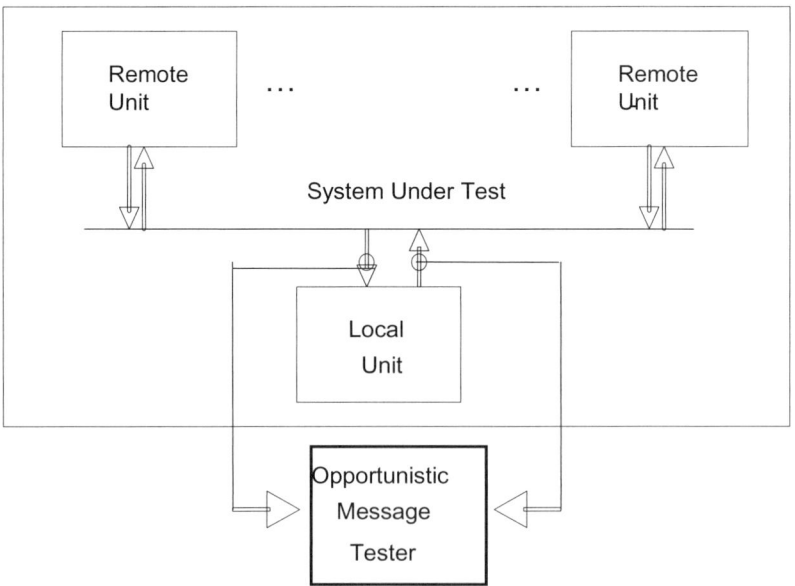

Figure 17B.1

Overall Configuration of Agent-based Testbed

Appendix B: Opportunistic Message Tester

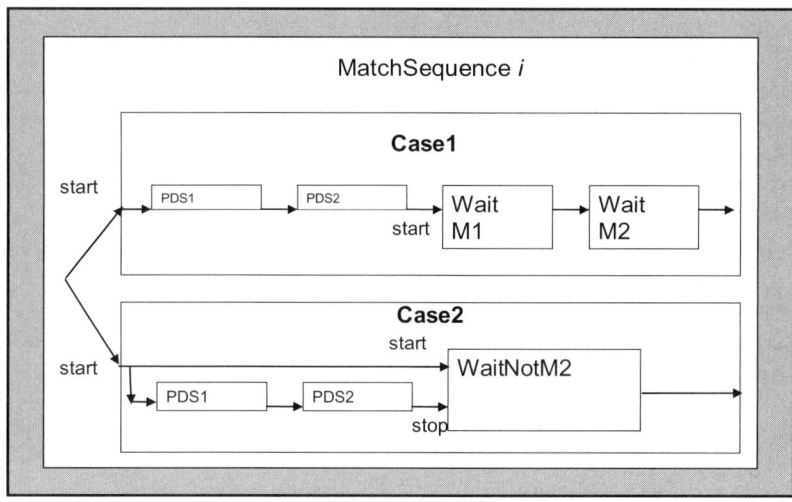

Figure 17B.2

DEVS Hierarchical Modular MatchSequence Model

A MatchSequence implements agent behavior in determining whether the received message stream is allowable in terms of the rules governing object matching.

As shown in Figure 17B.2, there are two test cases working in one *MatchSequence*: Case1 and Case2. Once testControl assigns a remote object signature to a specific MatchSequence, it sends an activating message through the "start" port to both Case1 and Case2. So PDS1 of Case1 and PDS1 and WaitNotM2 of Case2 are activated at that time. ProcessDetect primitives PDS1 and PDS2 listen for messages of type S bearing locally and remotely generated signatures arriving to the local unit. Two consecutive successful matches between remote and local signatures by PDS1 and PDS2 activate WaitM1 of Case1. At the same time the WaitNotM2 is disabled because identical PDS1 and PDS2 are used in both Case1 and Case2. In other words, activation of WaitM2 of Case1 and WaitNotM2 of Case2 behave in mutually exclusive fashion.

An opportunity to test occurs when such pairs of successful matched signatures are detected. As described earlier, subsequently, *waitReceive* models (here WaitM1 and WaitM2) will be activated to test whether messages M1 and M2 are subsequently correctly handled by the local system. If both tests complete successfully, a pass is logged. Otherwise, a fail is logged. Failure is also logged when an M2 message is received, in the absence of

Table 17B.1. Cases for Testing by OMT

		SUT	
		Matches	**Does not match**
O M T	Match expected	Pass	Fail
	Match not expected	Fail	Not tested in OMT

prior consecutive S message sequences which is done by Wait-NotM2 of Case2. The two fail cases as well as a pass case are summarized in Table 17B.1.

A test was performed in which traffic was injected into the network using a playback of a prerecorded live scenario. This prerecorded scenario had 1 local and 4 remote units, and approximately 60 remote and local signatures. MatchSequence models are assigned individual remote signatures, while local signatures are broadcast to all the MatchSequence models. This allows multiple remote signatures to be simultaneously tested for matching against the same local signature, exploiting the fact that remote signature messages are more frequent than local ones. In this prototype system experiment, fewer than 100 *MatchSequence* agents were active at any time and dynamic structuring was not used. Subsequent review of the logged test results showed that the 3 testable cells of Table 17B.1 were correctly populated by the test agents thereby verifying the logical behavior of the tester. To assess its performance attributes, especially the advantage of dynamic structure, we must scale up to large numbers of sensor/decider units, heavy message traffic, and a large suite of opportunity-based test cases.

18

Testing in a Net-Centric Environment: Multiple Levels

Interoperability, or lack thereof, is the characteristic that highlights the fact that systems that perform well individually typically do not work well when brought together. Inability to exchange information is often at the heart of interoperability problems. For example, we have seen that in geospatial sensing, an enormous number of single purpose "stove-piped" processing chains have emerged in which very little of the available information from all the systems can be fused together (Chapter 3). In military and business organizations, lack of interoperability between stove-piped systems prevents rapid, seamless, and collaborative exchange of information. Unsharable information is actually detrimental in that it accumulates in storage, clogs up processing, and puts the burden on the end user to make sense of. As we have seen, such stove-piping is supposed to be eliminated by the transition to the Service Oriented Architecture (SOA), in which services are designed to be accessed without knowledge of their internals through well-defined interfaces that are readily discoverable and composable (Chapter 15). However, such a transition is easier said than done.

Having emphasized the distinction between syntactic, semantic, and pragmatic concepts, it is time to put these into a unified structure that can be used to design and test for improved interoperability. This chapter presents a layered architecture for information exchange that blends together existing concepts of Levels of Information Systems Interoperability and Levels of Conceptual Interoperability. Based on a stratification along the lines of the linguistic levels of pragmatics, semantics, and syntax, we discuss an approach to multilevel testing in net-centric environments including that of the SOA over the Global Information Grid (GIG).

Levels of Information Systems Interoperability

Introduced in 1998, the Levels of Information Systems Interoperability (LISI)[1] is a set of models and associated processes based on five levels of interoperability described in Table 18.1.

Levels of Conceptual Interoperability Model

Although the LISI models are used successfully to determine the degree of interoperability between information technology systems, they do not provide a systematic formulation of the underlying properties of information exchange. To remedy this situation,[5] the Levels of Conceptual Interoperability Model (LCIM) in which there are various levels of interoperability between participating systems was introduced.[5] The current version of LCIM, outlined in Table 18.2 was developed to become a bridge between conceptual and technical design for implementation, integration, or federation.[4,6]

The last column lists key conditions that are required to reach an interoperability level from the one below. Of course, the conditions accumulate as the level increases. We note that the conditions given in the LCIM for pragmatic interoperability require that the *use* of data be mutually understood, where the term "use" is interpreted as the context of its application. In this book, we have developed more definitive concepts for pragmatic interoperability including the concepts of pragmatic frames and pragmatic equivalence. Moreover, the definition of the semantic level requires the use of a single reference semantic model as a hub for information exchange among participants in collaboration.[7,8] However, as discussed in Chapter 13, such a hub and spokes approach, while desirable, is not always feasible.[1] In the stratifi-

---

[1] Turnitsa et al.[9,10] evaluated a common information exchange model, C2IEDM, as an interoperability-enabling ontology for command and control. The

Table 18.1. Levels of Information Systems Interoperability

Level of information system interoperability	Characteristic	Information exchanges at the level
Enterprise	Shared data and applications	Global Information Grid (GIG) web services, service-oriented architecture, advanced collaboration
Domain	Shared data, separate applications	Access to common databases, sophisticated collaboration
Functional	Minimal common functions, separate data and applications	Annotated images, maps with overlays
Connected	Electronic connection	Tactical data links, email, file transfer
Isolated		

cation to be introduced below, we propose a more streamlined and extended account of information exchange levels based on the concepts and approaches developed in this book.

Linguistic Levels

The definitions given earlier (Chapter 2) agree in general, but differ substantially, with those used in the LCIM. Recall that we defined:

- *Pragmatics:* Data used in relation to data structure and context of application (Specialized; Implementation Context: Authority, Urgency/Consequences, Relationship with other systems)
- *Semantics:* Low-level semantics focus on definitions and attributes of terms; high-level semantics focus on the combined meaning of multiple terms (Generalized Context). Note this allows both a many-to-many and many-to-one approach to harmonization of ontologies. In contrast to the LCIM requirement for semantic interoperability, this definition focuses on the underlying requirement for achieving shared meanings rather than how this requirement is achieved.

conclusion is that even if there is room for improvements, the model supports almost all basic needs for such a semantic bridge. Lasschuyt et al.[11] claim that in its current form, the model is unbalanced in its levels of detail and too large to be practical.

Table 18.2. Levels of Conceptual Interoperability

Level of conceptual interoperability	Characteristic	Key condition
Conceptual	The assumptions and constraints underlying the meaningful abstraction of reality are aligned	Requires that conceptual models be documented based on engineering methods enabling their interpretation and evaluation by other engineers
Dynamic	Participants are able to comprehend changes in system state and assumptions and constraints that each is making over time, and are able to take advantage of those changes	Requires common understanding of system dynamics
Pragmatic	Participants are aware of the methods and procedures that each is employing	Requires that the use of the data — or the context of their application — is understood by the participating systems
Semantic	The meaning of the data is shared	Requires a common information exchange reference model
Syntactic	Introduces a common structure to exchange information	Requires that a common data format be used
Technical	Data can be exchanged between participants	Requires that a communication protocol exist
Stand-alone	No interoperability	

- *Syntax:* Syntax focuses on a structure and adherence to the rules that govern that structure, e.g., XML (Rules and Structure)

The authors of LCIM associate the lower layers with the problems of simulation interoperation while the upper layers relate to the problems of reuse and composition of models.[2,3] They conclude "simulation systems are based on models and their assumptions and constraints. If two simulation systems are combined, these assumptions and constraints must be aligned accordingly to ensure meaningful results."[4] This suggests that levels of

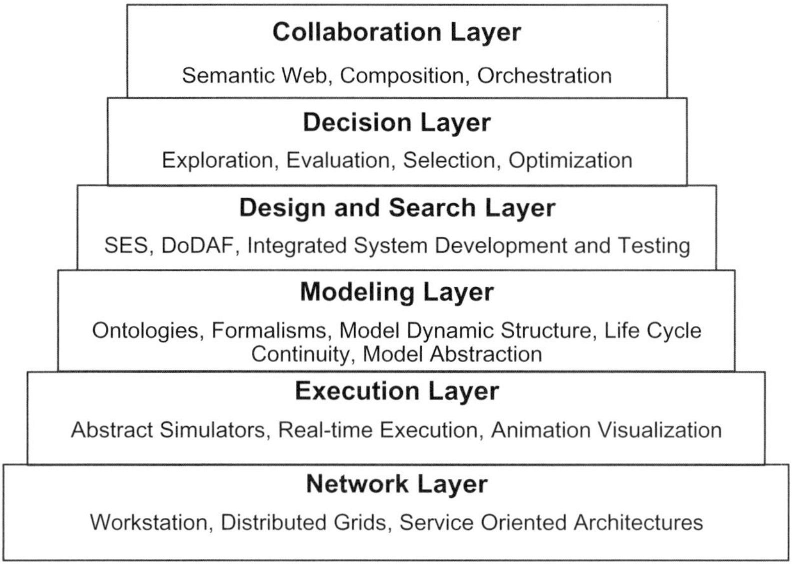

```
┌─────────────────────────────────────────────────┐
│              Collaboration Layer                │
│     Semantic Web, Composition, Orchestration    │
├─────────────────────────────────────────────────┤
│                Decision Layer                   │
│  Exploration, Evaluation, Selection, Optimization │
├─────────────────────────────────────────────────┤
│            Design and Search Layer              │
│  SES, DoDAF, Integrated System Development and Testing │
├─────────────────────────────────────────────────┤
│                Modeling Layer                   │
│ Ontologies, Formalisms, Model Dynamic Structure, Life Cycle │
│         Continuity, Model Abstraction           │
├─────────────────────────────────────────────────┤
│               Execution Layer                   │
│ Abstract Simulators, Real-time Execution, Animation Visualization │
├─────────────────────────────────────────────────┤
│                Network Layer                    │
│ Workstation, Distributed Grids, Service Oriented Architectures │
└─────────────────────────────────────────────────┘
```

Figure 18.1

Architecture for Modeling and Simulation

interoperability that have been identified in the area of modeling and simulation (M&S) can serve as guidelines to the discussion of information exchange in general. Therefore, we consider an earlier developed conceptual layered architecture for M&S.[19] We'll correlate the above linguistic definitions with the layers outlined below and shown in Figure 18.1.

- *Network Layer* contains the actual computers (including workstations and high performance systems) and the connecting networks (both LAN and WAN, their hardware and software) that do the work of supporting all aspects of the M&S life cycle.
- *Execution Layer* is the software that executes the models in simulation time and/or real-time to generate their behavior. Included in this layer are the protocols that provide the basis for distributed simulation (such as those that are standardized in the High Level Architecture (HLA). Also included are database management systems, software systems to support control of simulation executions, visualization, and animation of the generated behaviors.
- *Modeling Layer* supports the development of models in formalisms that are independent of any given simulation layer

implementation. The HLA just mentioned also provides object-oriented templates for model description aimed at supporting confederations of globally dispersed models. However, beyond this, the formalisms for model behavior, whether continuous, discrete, or discrete event in nature) as well as structure change, are also included in this layer. Model construction and, especially, the key processes of model abstraction and continuity over the life cycle are also included. We'll also add ontologies to this layer; this is consistent with the definitions given earlier (Chapter 1) where they are understood as models of the world for a particular conceptualization intended to support information exchange.
- *Design and Search Layer* supports the design of systems, such as in the DoDAF where the design is based on specifying desired behaviors through models and implementing these behaviors through interconnection of system components. It also includes investigation of large families of alternative models, whether in the form of spaces set up by parameters or more powerful means of specifying alternative model structures such as those provided by the SES methodology.[20-22] Artificial intelligence and simulated natural intelligence (evolutionary programming) may be brought in to help deal with combinatorial explosions occasioned by powerful model synthesizing capabilities.
- *Decision Layer* applies the capability to search and simulate large model sets at the layer below to make decisions in solving real-world problems. Included are course-of-action planning, selection of design alternatives, and other choices where the outcomes may be supported by concept explorations, "what-if" investigations, and optimizations of the models constructed in the modeling layer using the simulation layer below it.
- *Collaboration Layer* enables people or intelligent agents with partial knowledge about a system, whether based on discipline, location, task, or responsibility specialization, to bring to bear individual perspectives and contributions to achieve an overall goal.

Using the definitions for linguistic levels above, we correlate such levels with the layers just discussed. As illustrated in Figure 18.2, at the syntactic level we associate network and execution layers. The semantic level corresponds with the modeling layer — where we have included ontology frameworks as well as dynamic system formalisms as models. Finally, the pragmatic level includes use of the information such as that identified in the upper layers of the M&S architecture. This use occurs, for example, in design

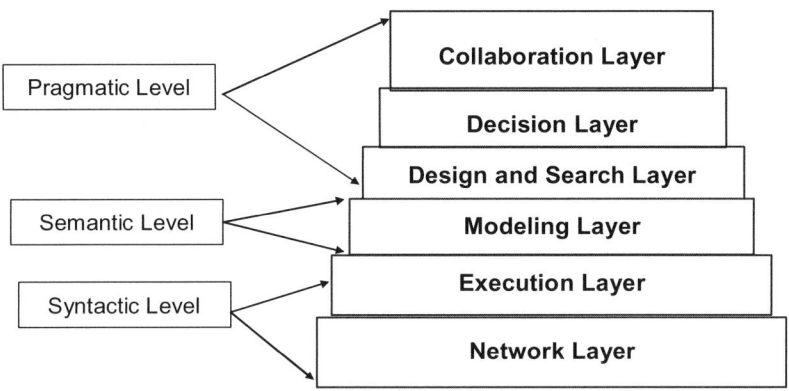

Figure 18.2.

Associating Linguistic Levels with Layers of Modeling and Simulation

and search, making decisions and collaborating to achieve common goals. Indeed, such mental activities, along with real-world physical actions that they lead to, provide the basis for enumerating the kinds of pragmatic frames that might be of interest in particular applications — the context of use.

The resulting stratification leads us to propose Table 18.3 for defining effective interoperation of collaborating systems or services at the identified linguistic levels (first and second columns).

Note that the stratification employs the traditional three levels of linguistics to represent much of the same ground as the seven levels of the LCIM. This happens because we have given the levels broader interpretation than in traditional linguistics. This allows the stratification to collapse the technical and syntactic levels into the syntactic level, the semantic and conceptual levels into the semantic level, and the pragmatic and dynamic levels into the pragmatic level. More specifically, the claim is that:

- Technical interoperation can be considered syntactic in the sense that rules of one sort or another underlie the operation of computer and network devices, but such rules are blind to the larger context in which they operate. They provide the computational and communicational substrate for, and under the command of, higher level activities.
- The conceptual level concerns shared model features such as abstractions and assumptions that can be incorporated in the model-based semantic level as we have defined it.

Table 18.3. Interoperation of Services at Identified Linguistic Levels

Level	A collaboration of systems or services interoperates at this level if:
Pragmatic (includes LCIM dynamic level)	The systems or services use the information received in a manner that is consistent with the shared agreements about the use of information exchanged. Roughly, the receiver reacts to the message in a manner that the sender intends (assuming nonhostility in the collaboration).
Semantic (includes LCIM conceptual level)	The systems or services employ common ontologies and other models to interpret/generate the messages received/sent, or if not, then translators exist to correctly reinterpret messages on the basis of one ontology/model to another. Roughly, the receiver assigns the same meaning as the sender did to the message.
Syntactic (includes LCIM technical level)	The systems or services employ common formats and protocols for communicating message data frames, or, if not, gateways and adapters exist to correctly map from one protocol to another. E.g., they employ TCP/IP, HTTP, and SOAP at successively higher levels.

- The dynamic level, with its concern for evolution over time, refers to the more advanced form of pragmatics that we have called dynamic pragmatics in Chapter 16.

Simultaneous Testing at Multiple Levels

Implications for simultaneous testing at the enhanced linguistic levels are shown in Table 18.4. Here we use the locution "testing that everything is as it should be" as a place holder to be filled in detail in subsequent analysis. The formulation recognizes the fact that we can't properly test a level without first gaining assurance that everything is as it should be at lower levels. Due to combinatorial complexity, it is extremely difficult to diagnose the source of a deviance at a higher level without first being assured that everything below is working as it should. We call the approach to testing that takes account of this fact, "testing at all levels simultaneously."

At the Pragmatic level, mission threads are given for testing of collaboration among participants and to establish use contexts

Table 18.4. Testing at Identified Linguistic Levels

Level	Testing that everything is as it should be at this level involves:	Testing details
Pragmatic	1) Establishing that everything is as it should be at the semantic and syntactic levels 2) Examining whether there are shared agreements about the use of information exchanged 3) Observing whether or not the messages exchanged among the systems or services result in the uses that are specified in the pragmatic agreements	Mission threads are given for testing of collaboration among participants Pair-wise translations between participant ontologies or to common information exchange model are evaluated for equivalence within pragmatic frames that are involved in end-to-end mission threads (Chapter 13) Test agents are deployed to observe end-to-end mission thread performance and evaluation of measures of effectiveness of information exchanges Tests whether mission threads are executed to completion within allowed timings
Semantic	1) Establishing that everything is as it should be at the syntactic level 2) Examining whether there are common or harmonized ontologies and models to support consistent interpretation and use 3) Observing whether or not the messages sent/received are interpreted or re-interpreted consistently with the ontologies/models.	Test whether SES expressed in XML is valid Test whether typical PESs conform to SES (Chapter 8) Test whether typical PESs satisfiy pruning rules (Appendix A) Generate XML instances from PESs to stimulate inputs to participants
Syntactic	1) Examining whether there are common formats and protocols are being used or gateways if needed 2) Observing whether the computers and communications networks are functioning as they should be	XML Schema are well-formed XML Schema are valid XML Instance documents are well-formed XML Instance documents are valid with regard to Schema

for information exchange. Such mission threads are intended to capture how effectively the net-centric capabilities are being used in real-world situations. Examples of such mission threads are:

- Direct action in an urban environment
- Joint close air support
- Noncombatant evacuation operations
- Coordination of first responders to natural disaster

We must evaluate the existence, and adequacy, of pragmatic frame agreements about how information that is exchanged will be used in the end-to-end mission threads to be tested. On a pair-wise system-to-system basis, we can test information exchange and interoperability. To do this, we must determine whether translations exist between participants' ontologies or to a common information exchange model. If such translations exist, we must ascertain whether they result in pragmatic equivalence within the uses contexts that arise during the course of mission threads execution. In terms of the SES, for pragmatic equivalence, we can check whether ontologies expressed in the SES framework share the common entities required in the pragmatic frames (Chapter 13).

Pair-wise evaluation is followed by test agent-mediated observation and end-to-end evaluation of measures for effectiveness of information exchanges among the systems or services as specified in the pragmatic frame agreements. In a combat context, such measures evaluate how effectively participants can receive role-appropriate command and control information and share a current and accurate understanding of the overall operational situation. In other words, we must be able to *measure to what extent participants receive the right information at the right time* (Measures of Performance) and to what extent such information sharing increases the quality and speed of decision making (Measures of Effectiveness).

The ultimate test is whether mission threads are executed to successful completion within allowed time and resource constraints.

At the Semantic level, for data exchange we can test whether typical PESs conform to the governing SES (Chapter 8), and whether a PES satisfies the given pruning rules (Appendix A).

At the Syntactic level, we can test whether the XML Schema are well-formed, and whether the XML Schema are valid. Like-

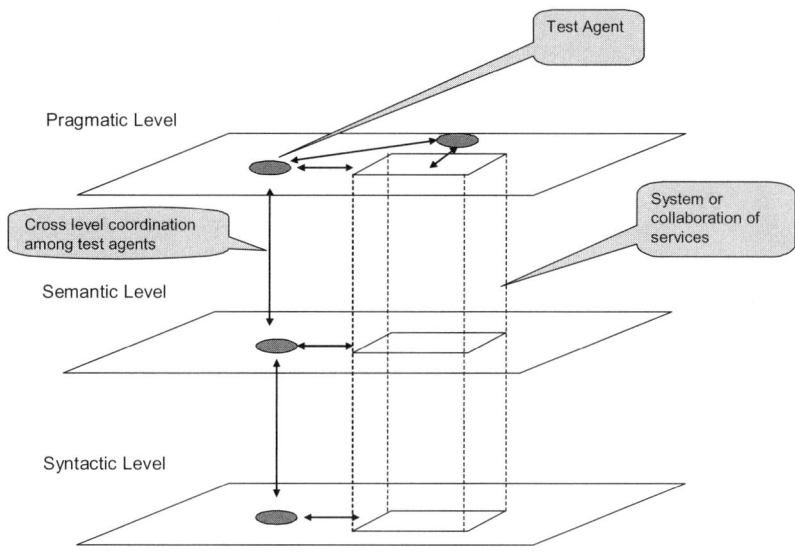

Figure 18.3

Simultaneous Testing at Multiple Levels

wise, we check that the XML Instance documents are well-formed and are valid with respect to Schema.

Building on the test agents for web services introduced in Chapter 17, Figure 18.3 depicts a concept of how such testing might be structured. Test agents at each level are designed according to the I/O Behavior-based approach discussed earlier. Such agents interact with the system or services under test at their assigned levels. This implies access to the message traffic and ability to interpret message contents within the assigned level. In addition, agents communicate and coordinate with other agents at other levels. This is where the term "simultaneous" applies. For example, an agent that detects an anomaly at the syntactic level should inform colleagues in the semantic level that all is not well at the lower layer. This should cause agents at the semantic level to stop attempting to conduct tests.

In the next section, we discuss how the information needed for testing at the various levels can be derived from an architectural system design specification. Indeed, in principle, there should be a common specification that drives both development and testing.[12]

Integrated Architectures for System Development and Testing

Having established a set of levels of information exchange for interoperability as well as a framework for simultaneous testing at these levels, we proceed to a more in-depth consideration of how such testing can be implemented.

Multiple System Views

The Department of Defense Architectural Framework (DoDAF) is mandated for expressing high level system and operational requirements and architectures that cross organizational and national boundaries.[13,14] Its objective is to provide a common denominator of understanding, comparing, and integrating these Family of Systems (FoSs), System of Systems (SoSs), and interoperating and interacting architectures. It comprises three major views:

1. *Operational View (OV)*: This view provides information on what needs to be accomplished and who should be doing it. It deals with the functional capabilities of the architecture.
2. *Systems View (SV)*: This view provides information on which systems are employed to provide the functionalities expressed in OV. It provides the bridge between the conceptual functionalities and real systems that would provide them.
3. *Technical View (TV)*: This view provides information on what standards are being used to employ the systems required in SV and what standards are under development to address the future needs of the current architecture.

These general view categories are elaborated into distinct specifications or documents, some of which are listed in Table 18.5.

DoDAF requires an extensive set of specification documents, each of which may have several replications applicable to different system aspects. We'll use the SES to get a holistic perception of how the DoDAF views fit together to make a whole system specification.

Test Development Approach

End-to-end interoperability is defined to include both the technical exchange of information and operational effectiveness of that

Integrated Architectures for System Development and Testing

Table 18.5. Some Views of the DoD Architectural Framework

Overview and Summary Information	AV-1
High-Level Operational Concept Description	OV-1
Operational Node Connectivity Description	OV-2
Operational Information Exchange Matrix	OV-3
Organizational Relationships	OV-4
Operational Activity Model	OV-5
Operational Event — Trace Description	OV-6b,c
Systems Interface Description	SV-1
Communication Description	SV-2
Systems to Systems Matrix	SV-3
Functionality Description	SV-4
Operational Activity to Function Traceability Matrix	SV-5
Data Exchange Matrix	SV-6
Technical Standards Profile	TV-1

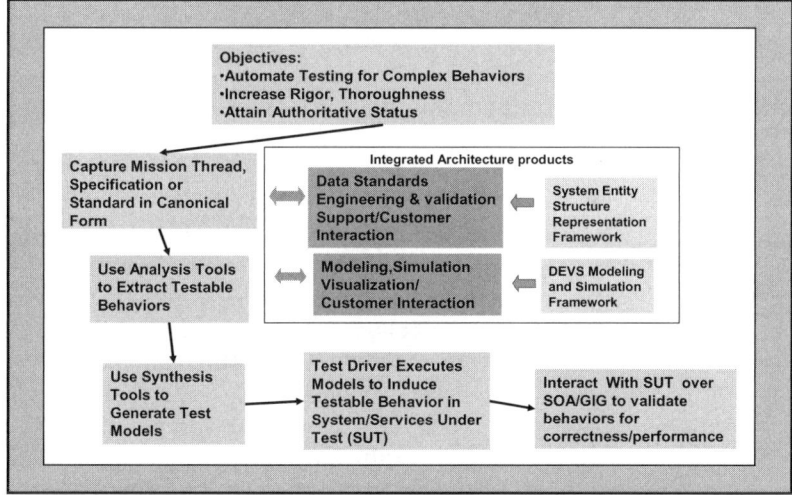

Figure 18.4

Approach to Developing Tests for Joint Interoperability

exchange, as required for mission accomplishment.[17] Accordingly, the strategy discussed next has the overarching goal of assessing end-to-end information exchange in critical joint mission threads that are stated within DoDAF by systems developers and users. Figure 18.4 provides an overview of the approach illustrating how it operationalizes the guidance for testing Joint Interoperability development and assessment.[17]

The process depicted has the following phases:

INITIALIZATION Define objectives of testing for this particular application. General objectives are to increase the automation level of testing, especially for complex behaviors while increasing the levels of rigor and thoroughness. Additionally, an essential goal is to develop tests that are traceable to the governing integrated architecture (DoDAF) products thereby attaining authoritative status with all concerned stakeholders.

CAPTURE MISSION THREADS AND APPLICABLE SPECIFICATIONS OR STANDARDS Recall that Joint Critical mission threads capture how the system's capabilities will be used in real-world situations and are documented in operational event-trace descriptions. Expressed in various forms compliant with DoDAF, such threads and associated artifacts will be converted into an equivalent standardized form that supports rapid test generation and deployment. To enable Communities of Interest to express their data and threads in testable forms, the test agency offers support for data standards engineering and validation.

USE ANALYSIS TOOLS TO EXTRACT TESTABLE BEHAVIORS The SES representation supports analysis tools to extract testable behaviors of the system/services under test (SUT). Testable behaviors are information exchanges among system components or services that are publicly observable over a network that do not require inspection into the components or services themselves. Development of such testable behaviors is informed by desiderata such as level of coverage of critical interactions, support for determination of appropriate measures of effectiveness (MOE) and measures of performance (MOP).

USE SYNTHESIS TOOLS TO GENERATE TEST MODELS The testable behaviors derived above are converted to simulation models that can interact with live and other virtual components or services over the GIG. Such test models are expressed in the Discrete Event Systems Specification (DEVS) formalism that provides a rigorous framework for automating the generation of models and their deployment as web services capable of sending and receiving messages to participate in a mission thread under test. Such test federations can be animated prior to deployment to allow Community of Interest personnel to visualize mission threads and adjust them according to the goals of the test.

DEPLOY TEST DRIVER TO EXECUTE TEST MODELS Various forms of test drivers have been developed to execute test models and thereby to induce testable behavior in the SUT. Such forms involve modes of control over the mission thread development, levels of intrusiveness into an ongoing interaction over the network, and modes of monitoring network traffic for opportunities to test.

TEST DRIVER INTERACTS WITH SUT Deployment of the test federation over the GIG, in compliance with its Service Oriented Architecture (SOA) is the end goal intended to validate the behaviors of the SUT for technical correctness of its information exchanges and to measure the performance and operational effectiveness of that exchange, as required for mission accomplishment.

This process was employed successfully in the automated development of test models and test federations for application to testing of Link 16 implementations (Chapter 17).

In the following sections we provide greater detail for the phases just enumerated.

Representing DoDAF within the System Entity Structure: Multiple Aspects

The System Entity Structure (SES) is a high level ontology framework targeted to modeling, simulation, systems design, and engineering. Its expressive power, both in strength and limitation, derive from that domain of discourse. An SES is a formal structure governed by a small number of axioms that provide clarity and rigor to its models. The structure supports hierarchical and modular compositions allowing large complex structures to be built in stepwise fashion from smaller, simpler ones. Tools have been developed to transform SESs back and forth to XML allowing many operations to be specified in either SES directly or in its XML guise. The axioms and functionally based semantics of the SES promote pragmatic design and are easily

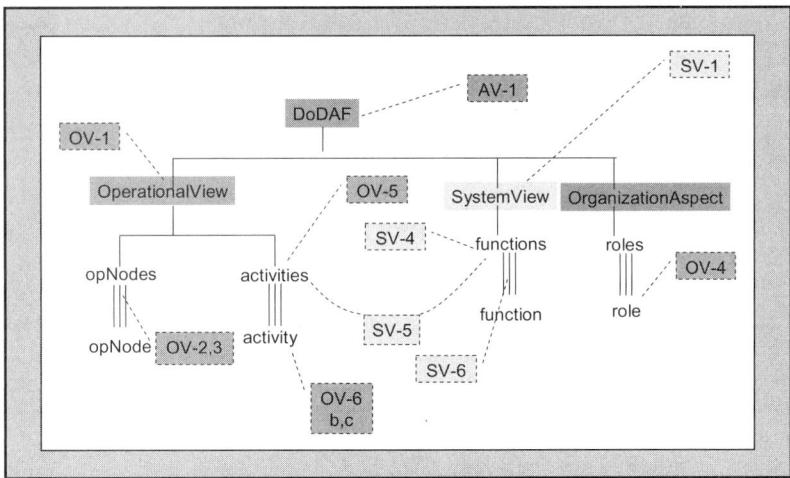

Figure 18.5

Representing DoDAF within the SES Framework

understandable by data modelers. Together with the availability of appropriate tool support, this makes development of XML Schema transparent to the modeler. Finally, SES structures are compact relative to equivalent Schema, and automatically generate associated executable simulation models.

Figure 18.5 shows the various DoDAF views mapped into the SES framework. Operational and System perspectives are considered two different decompositions of the system under consideration. They are represented by corresponding nodes called aspects labeled by the names, OperationalView and SystemView, respectively. The OperationalView aspect has entities labeled opNodes (operational nodes) and activities. The various operational views of DoDAF (other than OV-4) are easily interpreted as describing the entities and their interactions. Likewise, the SystemView aspect has entities labeled functions with DoDAF views that are associated with the functions and their interactions. The one exception is SV-5 which is a relation between the functions of the SystemView and the activities of the Operational-View. This view describes how the activities are implemented via executable functions supplied by the system. To accommodate OV-4 we have added another aspect, the OrganizationalAspect, which represents the decomposition of the system into the roles played by participating personnel.

EXAMPLE: Net-Enabled Command Capability (NECC)

The Net-Enabled Command Capability (NECC) is the DoD's principal command and control (C2) information technology. The NECC capability integrates existing and emerging C2 capabilities supported by the GIG Enterprise Services (GES) and is based on small modular components (i.e., services) organized into Capability Modules that are meant to be agile and adaptable.[18] NECC capabilities are defined by joint mission capability packages such as Force Projection, Force Readiness, Intelligence, Situational Awareness, etc. These capabilities are to be implemented by web services such as planning support, readiness information provision, task force analysis, etc.

A DoDAF specification for such a system has to specify how the desired capabilities web-services as well as how information will be exchanged among these services. Since DoDAF was developed before the emergence of the SOA it must be tailored to accommodate the greater flexibility that web-service implementation platforms offer.[15,16]

Deriving Testable Behaviors from DoDAF Specification

So far the SES has been shown to provide a means of pigeonholing the various DoDAF views. The power of this representation, however, lies in the support it provides for deriving system behaviors that can be transferred in semi-automated fashion to executable test federations. In Figure 18.6, the SystemView is further refined by explicitly adding messages as entities to it. For simplicity, components represent both the functions and their decomposition into services. The coupling associated with the component's aspect specifies how messages are routed among the components. This information is what is required to automatically map the SystemView to a simulation model, that is, in this case, a test federation. To obtain such information, we develop a process for deriving it from specifications associated with the OperationalView and mappings between the OperationalView elements and their realizations in the SystemView. Figure 18.6 describes such a derivation in the form of a logical deduction. It states roughly that if an opNode is engaged in an activity which requires a certain information exchange and the opNode is mapped to a component that executes the function implementing the activity, then this component must be observed to receive and send messages associated with that information exchange.

Figure 18.7 illustrates how relation-based pruning of the SES (Chapter 9) can support implementation of the process depicted in Figure 18.6.

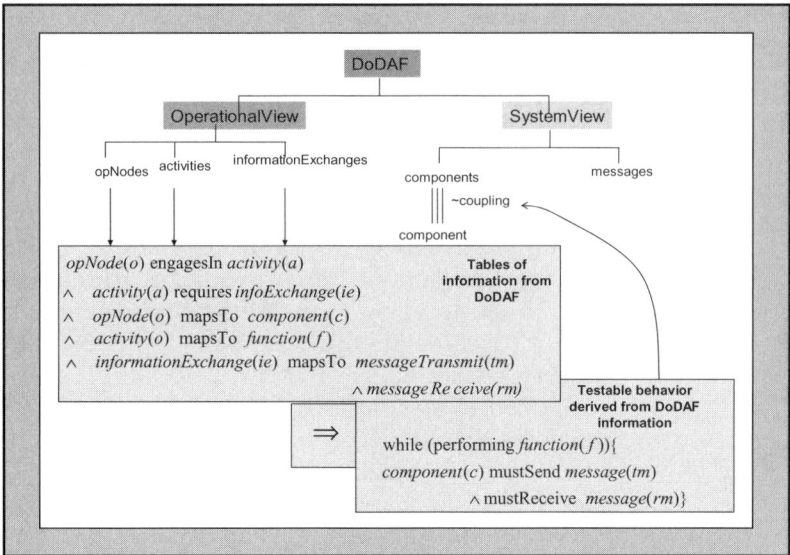

Figure 18.6

Extracting Testable Behaviors from DoDAF Representation in SES

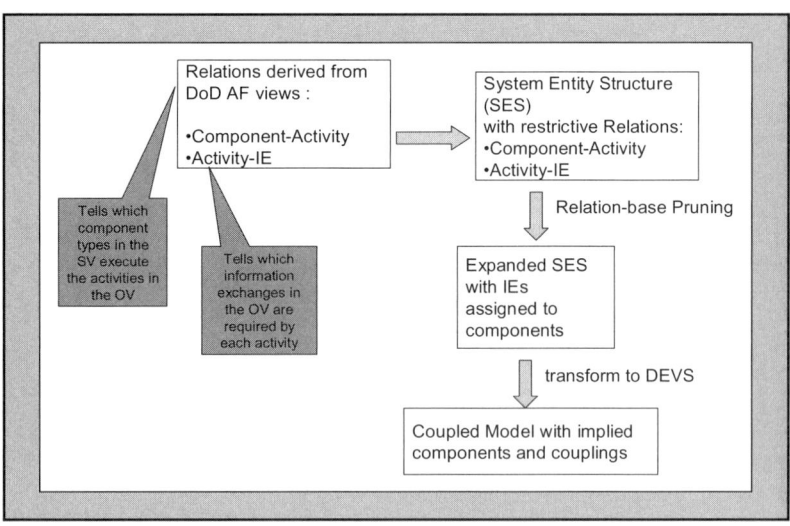

Figure 18.7

Relation-based Pruning Implementation of the Test Derivation Process

Integrated Architectures for System Development and Testing

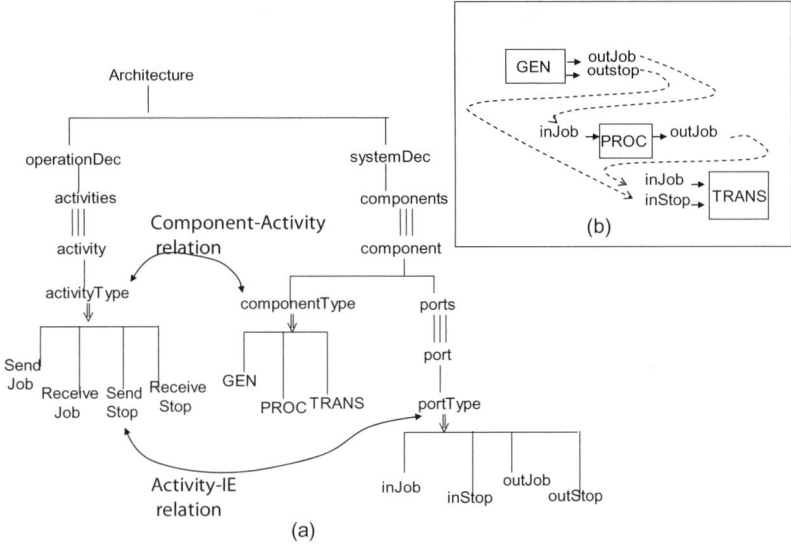

Figure 18.8

An SES Architecture and Its Pruning

We start with the following relations derived from DoDAF views:

- Component-Activity — tells which component types in the System View execute which activities in the Operational View
- Activity-IE — For each activity, tells which information exchanges in the OperationalView are required to execute that activity

These relations are attached to a System Entity Structure so that they can constrain pruning. For example, consider the SES in Figure 18.8(a). Table 18.6 gives the activities performed by the components and Table 18.7 gives the ports needed to engage in activities. First, the SES is expanded and pruned for the activities needed for a given mission thread. Next, using Table 18.6, relation-based pruning selects the components needed to execute the selected activities. Using Table 18.6, relation-based pruning then continues to assign each component the information exchanges associated with its activities. Information exchanges are actually assigned as input and output ports. A component with an output port of the form, outX can send a message X on this port that will be received by any component with an input port, inX. For example, the generator has an output port outJob

Table 18.6. Component-Activity Relation

componentType/ activityType	Send Job	Receive Job	Send Stop	Receive Stop
GENerator	x		x	
PROCessor		x		
TRANSducer		x		x

Table 18.7. Activity-IE Relation

activityType/ portType	inJob	outJob	inStop	outStop
SendJob		x		
ReceiveJob	x			
SendStop				x
ReceiveStop			x	

so it can send a Job message to a processor with an input port of inJob. The end result is a coupled model in which components have the right ports and couplings to engage in their assigned activities. For example, the DEVS model in Figure 18.8(b) supports a mission thread containing all of four types of activities.

Deploying Test Agents over the GIG/SOA

Figure 18.9 depicts a logical formulation test federation that can observe an SUT to verify the message flow among components as derived from information exchange requirements. In this context, a mission thread is a series of activities executed by operational nodes. In playing out this thread, DEVS test models are informed of the current activities (or see to detect their onset) as well as the operational nodes that execute these messages. These test models watch messages sent and received by the components that host the participating operational nodes. The test models check whether such messages are the ones that should be sent or received under the current function.

Implementation of Test Federations

A test federation observes an orchestration of web services to verify that the message flow among participants adheres to information exchange requirements. As derived from DoDAF inputs, a mission thread is a series of activities executed by operational nodes and employing the information processing functions of web

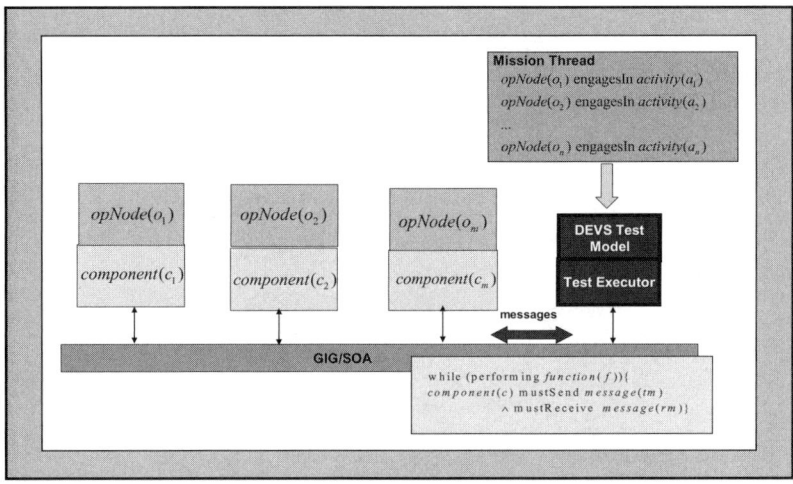

Figure 18.9

Logical Formulation of Test Agent Federation

services. As discussed in Chapter 17, test agents watch messages sent and received by the services that host the participating operational nodes. Depending on the mode of testing, the test architecture may, or may not, have knowledge of the driving mission thread under test. If thread knowledge is available, DEVS test agents can be aware of the current activity of the operational nodes it is observing. This enables it to focus more efficiently on a smaller set of messages that are likely to provide test opportunities. A DEVS distributed federation is a DEVS coupled model whose components reside on different network nodes and whose coupling is implemented through middleware connectivity characteristic of the environment, e.g., SOAP for GIG/SOA. The federation models are executed by DEVS simulator nodes that provide the time and data exchange coordination as specified in the DEVS abstract simulator protocol.

To help automate setup of the test we use a capability to interconvert between DEVS and XML. DEVSML allows distributing DEVS models in the form of XML documents to remote nodes where they can be coupled with local service components to compose a federation.[12] The layered middleware architecture capability is shown in Figure 18.10.

At the top is the application layer that contains models in DEVS/JAVA or DEVSML. The second layer is the DEVSML layer itself that provides seamless integration, composition, and

Chapter 18 Testing in a Net-Centric Environment: Multiple Levels

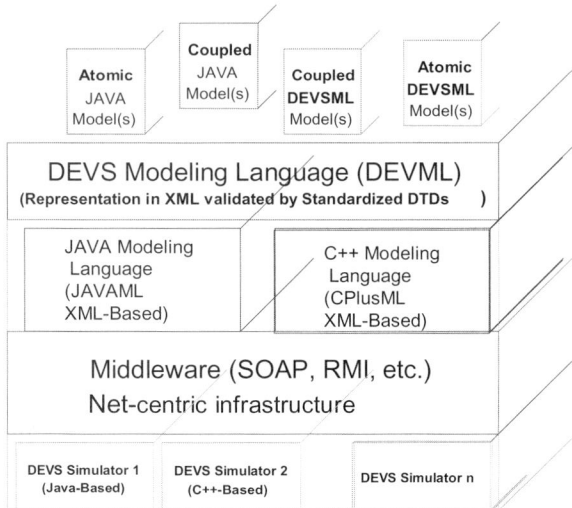

Figure 18.10

DEVSML Layered Architecture Providing Simulator Transparency

dynamic coupled model construction resulting in portable models in DEVSML that are complete in every respect. These DEVSML models can be transmitted in SOAP messages to remote DEVS simulators using the web-service infrastructure.

The simulation engine is totally transparent to model execution over the net-centric infrastructure. The DEVSML model description files in XML contain metadata information about its compliance with various simulation "builds" or versions to provide true interoperability between various simulator engine implementations. Such run-time interoperability provides great advantage when models from different repositories are used to compose models using DEVSML seamless integration capabilities. Finally, a test federation is illustrated in Figure 18.11 where different models (federates) in DEVSML collaborate for a simulation exercise over GIG/SOA.

Summary

When examined closely, lack of interoperability is not a trivial matter to do away with. Stove-piping exists because individual applications gain efficiency by, in effect, employing ontologies whose scope is restricted to support the problem at hand. Trying

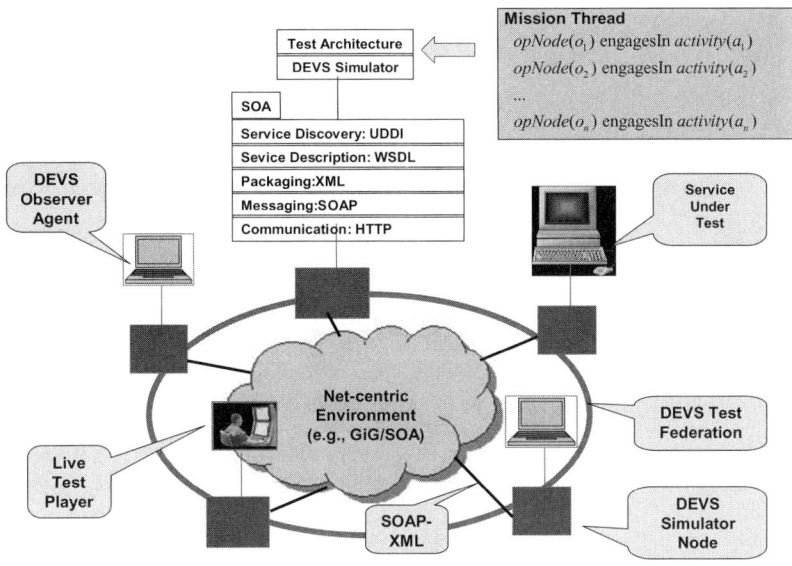

Figure 18.11

Prototypical DEVS Test Federation

to interoperate such systems is difficult because their idiosyncratic ontologies have to be reconciled. We have seen that such harmonization needs to be done at the syntax, semantics, and pragmatics levels. Likewise, testing for effective information exchanges across such ontologies must be done simultaneously at the three levels. We have seen how systems with architectural designs and mission thread specifications can be evaluated using test agents and federations that can be rapidly generated from the design and thread specifications.

References

1. Levels of Information Systems Interoperability (LISI), http://www.sei.cmu.edu/isis/guide/introduction/lisi.htm (accessed Nov. 2006)
2. Hoffmann, M., Challenges of Model Interoperation in Military Simulations. *SIMULATION*, Vol. 80, pp. 659–667, 2004
3. Chaum, E., M. R. Hieb, and A. Tolk, "M&S and the Global Information Grid," Proceedings Interservice/Industry Training, Simulation and Education Conference (I/ITSEC), 2005

4. Muguira, James, and A. Tolk, "Applying a Methodology to Identify Structural Variances in Interoperations," *JDMS: The Journal of Defense Modeling and Simulation*, Vol. 3, No. 2, 2006
5. Tolk, A., and J. A. Muguira, The Levels of Conceptual Interoperability Model (LCIM). Proceedings Fall Simulation Interoperability Workshop, 2003
6. Turnitsa, C., "Extending the Levels of Conceptual Interoperability Model," *Proceedings IEEE Summer Computer Simulation Conference,* 2005
7. Tolk, A., "XML Mediation Services Utilizing Model Based Data Management," *Proceedings IEEE Winter Simulation Conference,* pp. 1476–1484, IEEE CS Press, 2004
8. Tolk, A., and S. Y. Diallo, "Model Based Data Engineering for Web Services." *IEEE Internet Computing,* Volume 9, Issue 4, pp. 65–70, 2005
9. Turnitsa C., and A. Tolk, "Evaluation of the C2IEDM as an Interoperability-Enabling Ontology," Proceedings of Fall Simulation Interoperability Workshop, 2005
10. Turnitsa C., S. Kovurri, A. Tolk, L. DeMasi, V. Dobbs, and W. P. Sudnikovich, "Lessons Learned from C2IEDM Mappings within XBML," Proceedings of Fall Simulation Interoperability Workshop, 2004
11. Lasschuyt E., M. van Henken, W. Treurniet, and M. Visser, "How to Make an Effective Information Exchange Data Model," RTO-IST-042/9, 2004
12. Mittal, S. "Extending DoDAF to Allow DEVS-Based Modeling and Simulation," *JDMS: The Journal of Defense Modeling and Simulation,* Vol. 3, No. 2, 2006
13. Office of the Assistant Secretary of Defense (Networks & Information Integration, NII, "Department of Defense Architecture Framework (DODAF)," http://www.defenselink.mil/nii (accessed Nov. 2006)
14. Kobryn, C., and C. Sibbald, Modeling DoDAF Compliant Architectures: A Telelogic Approach for Complying with the DoD Architecture Framework, http://www.telelogic.com/download/paper/Modeling-DoDAF-WhitePaper.pdf (accessed Nov. 2006)
15. Dandashi, F., H.-W. Ang, and C. Bashioum, Tailoring DODAF to Support a Service Oriented Architecture. OMG Press, 2004
16. Trbovich, S., and R. Reading, Simulation and Software Development for Capabilities Based Warfare: An Analysis of Harmonized Systems Engineering Processes. *Proceedings Spring Simulation Interoperability Workshop*, 2005
17. Joint Interoperability Directives & Instructions, CJCSI 6212.01D http://jitc.fhu.disa.mil/jitc_dri/jitc.html (accessed Nov. 2006)
18. The NECC Provisional Technical Transition Architecture Specification, http://www.ditco.disa.mil/News/Documents/File/NECC_ptta_v0_5_7.pdf (accessed Nov. 2006)
19. Zeigler, B. P., T. G. Kim, and H. Praehofer, *Theory of Modeling and Simulation*, 2d ed., New York: Academic Press, 2000

20. Couretas, Jerry M., Bernard P. Zeigler, and George V. Mignon, "SEAE-SES Enterprise Alternative Evaluator: Design and Implementation of a Manufacturing Enterprise Alternative Evaluation Tool," Proceedings of SPIE — Volume 3696, *Enabling Technology for Simulation Science III*, Alex F. Sisti, ed., pp. 136–146, 1999
21. Couretas, Jerry M., Bernard P. Zeigler, and U. Patel, "Automatic Generation of System Entity Structure Alternatives: Application to Initial Manufacturing Facility Design." *Transactions of the SCS*, 16(4), pp. 173–185, 1999
22. Couretas, Jerry, System Architectures: Legacy Tools/Methods, DoDAF Descriptions & Design through System Alternative Enumeration, *JDMS: The Journal of Defense Modeling and Simulation: Applications, Methodology, Technology*, Vol. 3, No. 2, 2006

Appendix A: Testing for Pruning Rule Satisfaction

Consider testing the rule:

If select entNm from specNm then must select entNm2 from specNm2

The form of the rule is $p \Rightarrow q$

where:

- p = select entNm from SpecNm
- q = select entNm2 from specNm2

From propositional logic, we have

$$p \Rightarrow q \text{ is equivalent to } \neg p \vee q$$

The latter form is useful for testing whether the rule is satisfied. Most programming languages have the primitive boolean functions for negation $\neg$, conjunction $\wedge$, and disjunction $\vee$.

For example, in the C family of languages,

```
! p       returns true ⇔ p == false, (¬)
p & q     returns true ⇔ p == true and q ==
  true,   (∧)
p | q     returns true ⇔ p == true or q ==
  true,   (∨)
```

Notice that the form $\neg p \vee q$ says that if p is false then $\neg p$ is true, and since one argument's truth is enough to make the $\vee$

true, we don't have to test q. However, if p is true then $\neg p$ is false and we do have to test if q is true to make the $\vee$ true.

The sequential forms of the logic tests in the C family of languages,

p && q returns false if **p == false;** otherwise tests q
p || q returns true if **p == true;** otherwise tests **q**

are especially suited to our needs, since they will only test the second argument if needed.

Using these forms, following the pattern $\neg p \vee q$, the SES/JAVA test predicate for the rule is:

```
public static boolean checkSpecChoiceOf(String
   entNm, String specNm, String entNm2, String
   specNm2) {

return !isChoiceOf(entNm,
   specNm)||isUniqueChoiceOf(entNm2,specNm2)
```

where

isChoiceOf(entNm, specNm) **returns** true $\Leftrightarrow$ entNm is a member of the entities in specNm

and

isUniqueChoiceOf(entNm2, specNm2) $\Leftrightarrow$ entNm is the one and only member of the entities in specNm

Three other forms of the selection rule can be obtained by changing a positive selection to its negated version. For example, "select entNm" is changed to "do not select entNm." Test patterns for each of the three selection rule forms can be obtained by correspondingly changing the basic pattern $\neg p \vee q$.

Thus, we have:

- If do not select entNm from specNm then must select entNm2 from specNm2

   has the form $\neg p \Rightarrow q$ that is equivalent to $p \vee q$

- If select entNm from specNm then must not select entNm2 from specNm2

  $p \Rightarrow \neg q$ that is equivalent to $\neg p \vee \neg q$

- If do not select entNm from specNm then must not select entNm2 from specNm2

  $\neg p \Rightarrow \neg q$ that is equivalent to $p \vee \neg q$

Exercise

Write the test predicates corresponding to each of the rule forms in Java.

19

Bringing It All Together: Modeling and Simulation-based Data Engineering

Two levels, those of ontology and implementation, constitute the overall architecture for our simulation-based ontology engineering methodology. As depicted in Figure 19.1, at the ontology level, the modeler develops one or more SESs that are merged to create an ontology to satisfy the pragmatic frames of interest in a given application domain. An SES can be specified in XML, via restricted natural language, through a GUI, or directly in SESJAVA. It is then automatically encoded to an XML schema/document type definition (XSD or DTD) at the implementation level. Such automation is an important advantage, since other ontology developments based on UML currently lack the combination of automation and a broad range of tools that the SES framework supports. The XML instance documents specified by a schema are formally represented by the family of pruned entity structures at the ontology level. In the context of dynamic data engineering, each completely pruned PES specifies a DEVS simu-

Bringing It All Together: Modeling and Simulation-based Data 389

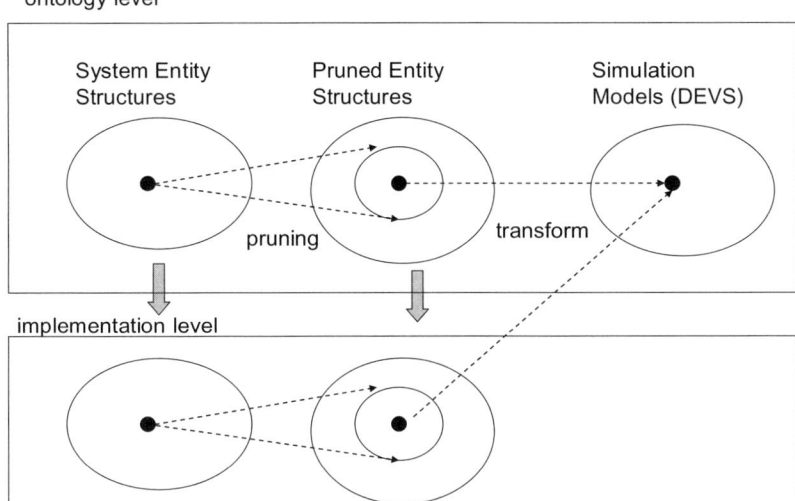

Figure 19.1
Overall Architecture for Simulation-based Data Engineering Methodology

lation model that constitutes a capability to describe world states that evolve in time. When limited to static data engineering, the family of PES represents a logically possible set of world states. Each such state is like a snapshot of the world as depicted by the ontology in service of the application contexts (the pragmatic frames). An XML document instance is the concrete encoding of the abstract PES. It encodes data in an information exchange that either directly represents a world state in the static case, or can be transformed to a simulation model in the dynamic case.

For example, DEVSML is a DTD that provides a lossless representation of DEVSJAVA models (Chapter 18). As discussed in Chapter 11, these models can be synthesized to satisfy user simulation objectives, i.e., to support a given pragmatic frame (also called experimental frame in this context). In contrast to the static case, where a document carries *passive* information only, DEVSML instances also carry *active* information, i.e., information that becomes executable by a DEVS simulator.

In this chapter, we place the simulation-based ontology engineering framework into a larger context that includes application of the Unified Modeling Language (UML) as an ontology framework. A discussion of this context has recently appeared in a very enlightening book to which we refer the interested reader.³ The International Standards Organization approaches to ontology development in the geospatial context have been referenced several times in the book (Chapters 3, 13, 14). We briefly explicate the ISO concepts at the ontology and implementation levels and go on to compare and contrast these with the methodology in Figure 19.1. Since the ISO concepts are framed within UML, in its earlier form, and XML, this sets the stage to view the SES framework within the UML context, using its most recent definition (UML 2.0).

A software environment that supports the simulation-based ontology engineering methodology, called the Scalable Entity Structure Modeler (SESM), is briefly described in Appendix A.

Revisiting the SES Framework

We begin by addressing some of the unique features of the SES ontology that set the basis for later comparison.

Vertical and Horizontal Relationships

The SES emphasizes *vertical hierarchical* relationships. That is to say, it views all entities as components in either decomposition (aspect) or variation (specialization) hierarchies. These hierarchies interact in well-specified formal terms (axioms) and are visualized as tree-like graphs. We'll discuss these formulations in relation to UML soon. Referring to the vertical hierarchical relationships as *primary*, we can designate as *secondary* all other relationships among components. In other words, the basic SES axioms, given in Chapter 5, determine the primary vertical relationships. They also set up the framework to consider the secondary relationships that are not directly mentioned in the axioms. Such secondaries include *horizontal* relationships, i.e., relationships among peer components at the same level in a hierarchy, and *cross-level* associations that relate components at different levels. The pruning relations discussed in Chapter 9 fall into the latter category. They can constrain selections from a specialization based on selections made from other specializations in a manner that is independent of their locations in the hierarchy. We'll now discuss the horizontal relations.

As an ontology framework for simulation models, the SES can rely only on the coupling specifications associated with aspects to determine horizontal relationships. This is true because cou-

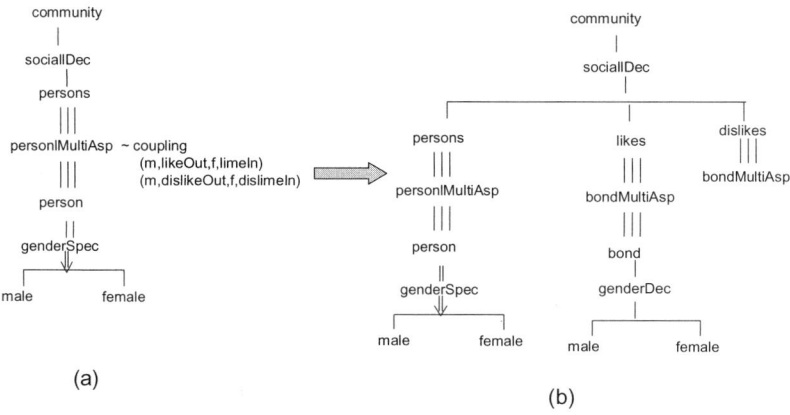

Figure 19.2

Extending the Concept of Coupling So That Horizontal Relations Can Be Expressed

pling determines the interactions among peer and parent components whose internal behaviors are specified in simulation models synthesized under control of the pruned entity structure. For example, consider the SES in Figure 19.2 for the island states of affection (Chapter 1). Note that the coupling associated with the multiAspect that decomposes the community into persons can specify how persons exchange assertions of likes and dislikes, while the dynamics underlying these exchanges are determined within the simulation models themselves. In this book, we have expanded the scope of the SES ontology to include description of world states whether generated by simulation or not. Consequently, we extend the concept of aspect to include horizontal relationships among peers. For example, in Figure 19.2(b), we have added likes and dislikes as the entities to the socialDec aspect, each of which are collections of bonds, a bond being a male-female pair.

Exercise

Allowing all three pairings is actually a little harder to specify because the valid brothers axiom prohibits children with the same name under genderDec. Show how expressing bond as a multiAspect solves the problem.

In general, an n-ary relationship among the children of an aspect can be expressed by a multiAspect whose generic entity is decomposed into n-entities representing the n-tuples of the relation.

Include a representation of love triangles involving members of the community. *Hint:* Such a triple (x,y,z) occurs when both x and y like z.

Living with the Strict Hierarchy Axiom

The strict hierarchy axiom has no correlate in UML where the equivalent self-looping is allowed. Figure 19.3(a) depicts an SES for the DEVS formalism which violates strict hierarchy since the entity, DEVS, occurs more than once within a path from the root to the leaf entities. However, this SES does describe the fact that components of composite models can themselves be composite. This pattern shows up often in indefinitely hierarchical or nested software elements such as windows (where windows may be included within other windows). However, in practice such systems will rarely exceed a fixed depth. Figure 19.3(b) shows how an SES for a given depth can be developed by unfolding the basic relations. The DEVS specialization, compSpec, can be augmented with successive generations of unfoldings, reflecting the use of earlier generated models in the current stage. Indeed, the first layer has only Atomic components, the second layer has Atomic and composite Models of layer 1; the third layer has Atomic and composite Models of layers 1 and 2, and so on. Figure 19.3(b) shows the result of stopping at layer 3, allowing only hierarchical models of depth 3. The resulting SES satisfies the strict hierarchy axiom.

Completing the Representation

A common modeling approach in UML is to develop a generalization hierarchy for a class of objects and to employ the more specialized classes in compositions. Figure 19.4(a) illustrates the equivalent SES representation of this approach. It shows a specialized type of building, house, used as the generic entity of a multiAspect to create a residential street. A limitation of this approach is that it does not provide an explicit generic construction process even though it is implied by the special case. In the absence of such a generalization, each case of construction, such as a commercial street, will be treated as an *ad hoc* case. In contrast, the representation in Figure 19.4(b) shows how the

Revisiting the SES Framework

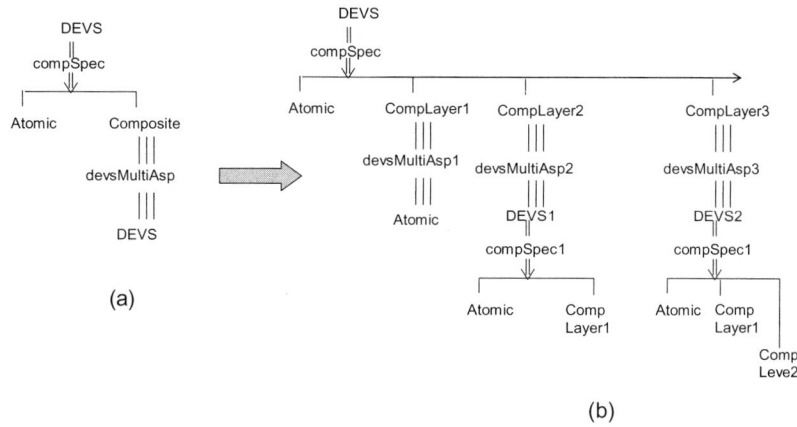

Figure 19.3

Limiting the Generation of Hierarchical Models to a Maximum Depth

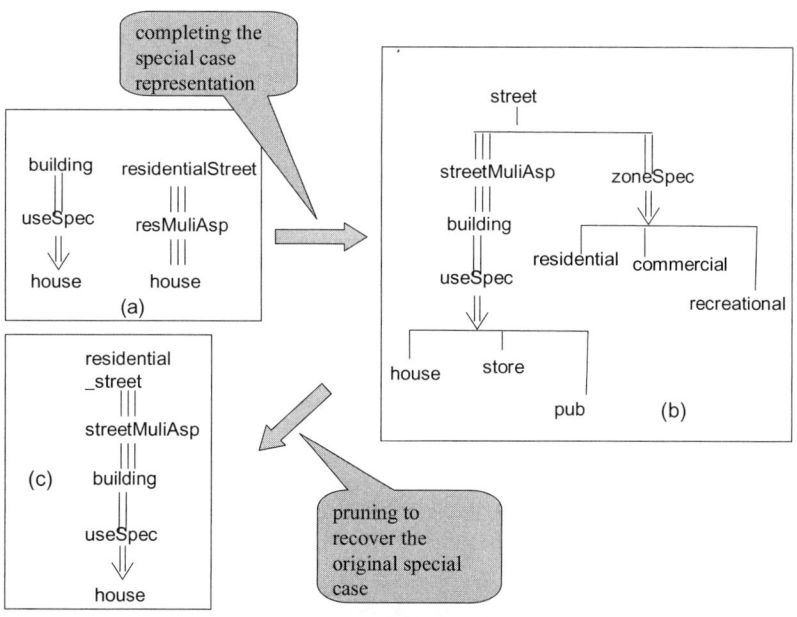

Figure 19.4

Illustrating How a Partial Specification Can Be Completed to the General Case

particular case can be subsumed within a systematic pruning process in which streets of various compositions can be constructed. In particular, a residential street can be obtained by expanding streetMuliAsp so that only houses are employed in the expansion, as in Figure 19.4(c). In fact, a pruning rule can constrain selection from useSpec under building based on the selection from zoneSpec (Chapter 9). In general, a partial representation based on a specialization of components should be completed to explicitly create a generic construction process. This can be achieved by creating a specialization for the top level composition (e.g., zoneSpec) that is parallel to the existing low level specialization (e.g., useSpec). A beneficial side effect: this also allows the process to be controlled in the pruning stage using pruning constraint rules/relations.

Geographic Information Representation in UML and Encoding in XML

A report of the ISO committee on Geographic Information/Geomatics[1] provides guidance for encoding geographic application schemata (ontologies) represented in UML into XML. The application schemata must be expressed in the UML schema language (UML version 1.4) according to the rules specified in ISO/TS 19103 and ISO 19109.[2] A profile is discussed in which only certain UML concepts are employed, and these are given more detail using extensibility mechanisms called stereotypes.[1] The profile does not use the other allowed extensibility mechanisms, viz., constraint and tagged value. The profile introduces a generic instance model (IM) that is capable of representing data described by application schematas. The attributes and associations defined by a class are mapped to ordered sequences of name-value pairs. Five value types are defined as follows:

- IM_BasicValue represents a value of a basic type.
- IM_ObjectReference represents a link or reference to a target object.
- IM_Object represents an object.
- IM_StructuredValue represents an attribute group as a sequence of name-value pairs.

---

[1] Stereotype is a UML modeling element that extends the semantics of the metamodel. Stereotypes must be based on certain existing types or classes in the metamodel. Stereotypes may extend the semantics, but not the structure of pre-existing types and classes. Certain stereotypes are predefined in the UML, others may be user defined.

- IM_CollectionValue represents collection types, such as Set, Sequence, Bag, and Dictionary.

The IM profile uses class stereotypes: BasicTypes and DataTypes, Unions, Enumeration, CodeList, and Type to provide greater detail by interpreting them as IM value constructs. (UML constructs: interfaces, abstract classes, operations, and constraints are not used.) The UML concepts: Attribute, Association, Aggregation, Composition, and IM_Property are interpreted using the IM property constructs. Schema conversion rules define how to produce an XML Schema Document (XSD) from an application schema. A summary of the rules is presented in Table 19.1.

Representation of Geographic Application Schemata in SES

We now discuss the structure development concepts of the IM and show how they are represented using the SES framework. We will see that the primary constructs of the IM are covered by the SES concepts of aspect and specialization, where the latter provide for increased representation capability because of the labeling that supports finer specification and greater combinatorics and reusability.

As indicated in Table 19.1, association, aggregation, and composition are the UML binary relation constructs employed in the IM profile. Figure 19.5 illustrates the coding rules for these relations.

The primary relations, aggregation and composition are employed to create composites of basic types, data types, and previously created composites. Figure 19.6(a) shows a data set expressed as a composition of objects. The corresponding SES representation starts at the top of the hierarchical decomposition in Figure 19.6(c) and uses a multiAspect to obtain the multiplicity of objects. As in Table 19.1, a multiple inheritance hierarchy is mapped into an XML sequence of elements that are copied down from parents (Figure 19.6(b)). The corresponding SES representation is shown in Figure 19.6(c) where T4 is equivalent to the product of multiple specializations, each with a single entity. Further, the IM introduces the substitution type described in Table 19.1. As illustrated in Figure 19.6(a), this construct allows a modeler to define an object as a family of alternatives that map to a choice group in XML. The equivalent representation in the SES is the specialization called ObjSpec whose entities are the members of the substitution type.

Based on the above discussion, Figures 19.7, 19.8, and 19.9 provide a side-by-side comparison of the development of a complex

Table 19.1. Summary of Encoding Rules (see Reference 1)

Instance model construct	is converted to XML Schema as
class stereotyped `<<BasicType>>` Including CharacterString, Integer, Decimal, Boolean, Date, Time, Probability, Logical, Binary, UnlimitedInteger	simpleType
class stereotyped `<<DataType>>` Including Mutiplicity, UnitsOfMeasure	complexType
class stereotyped `<<Enumeration>>` Including Sign	simpleType that restricts a text string to a number of enumerated values
class stereotyped `<<Union>>`	complexType encapsulating a choice element
Object types: A class with no stereotype or stereotyped `<<Type>>`	complexType with same name to include identification attributes either inherited from IM_Object or through a reference to the IM_ObjectIdentificaton attribute group
Single inheritance	Use XML's extension mechanism for complexTypes; use restriction mechanism for redefinitions
Multiple inheritance	Copy down the attributes/associations to target with standard traversal (starting with the left supertype and copy its attributes and associations, then continue with the next supertype to the right until the rightmost supertype is reached. The subtype's attributes are added last. A conflict occurs if a supertype or subtype defines an attribute or association with the same name as previously copied. In case of a name conflict the latter attribute or association shall take precedence and replace the previously copied one.
Substitution types	The use of a supertype as an attribute type means that an instance of the attribute can be of one of the concrete subtypes defined by the inheritance hierarchy of the supertype. XML Schema does not support this dynamic type mechanism directly.
Association, Aggregation, Composition	complexTypes for one or both of the end classes which encapsulate sequence elements and references to the other class

object as developed by ISO in UML with the corresponding representation in SES. One salient difference is the hierarchical structures of the SES that are visualized as tree-like graphs. These result from the top-down interwoven decomposition and specialization development approach that the SES encourages.

Geographic Information Representation in UML and Encoding in XML 397

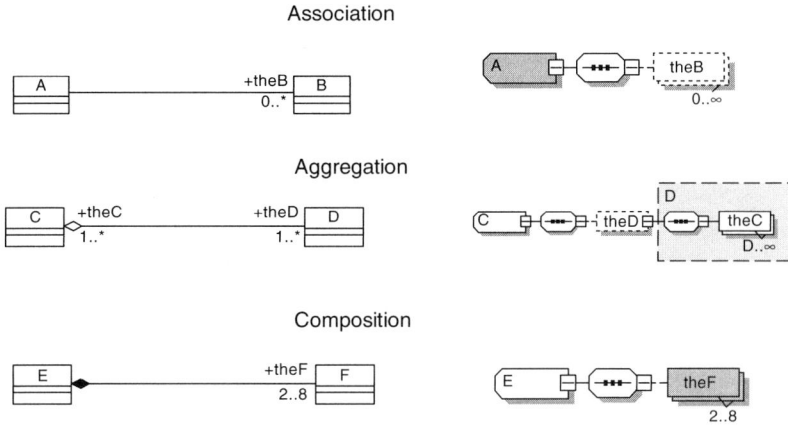

Figure 19.5

Encoding UML Binary Relations into XML

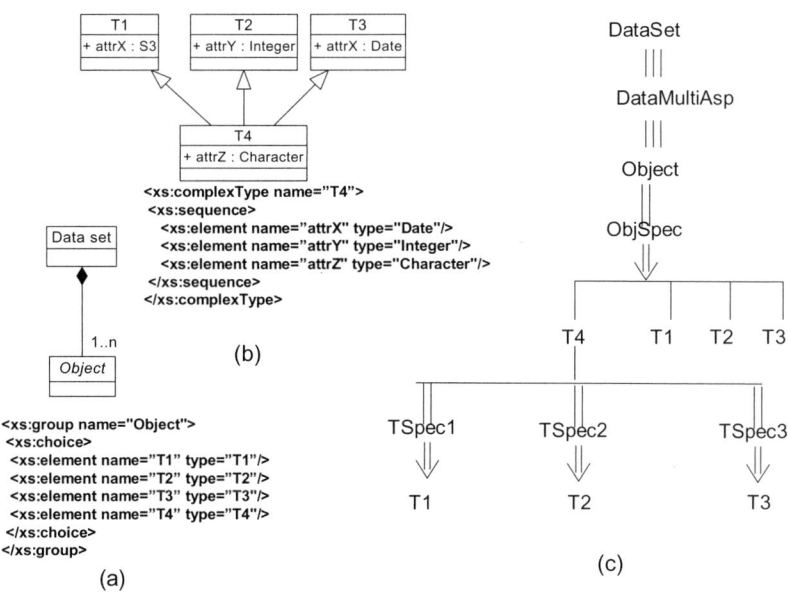

Figure 19.6

Illustrating Multiple Inheritance, Substitution Groups, Composition and Their SES Representation

398 Chapter 19 Bringing It All Together: Modeling and Simulation-based Data

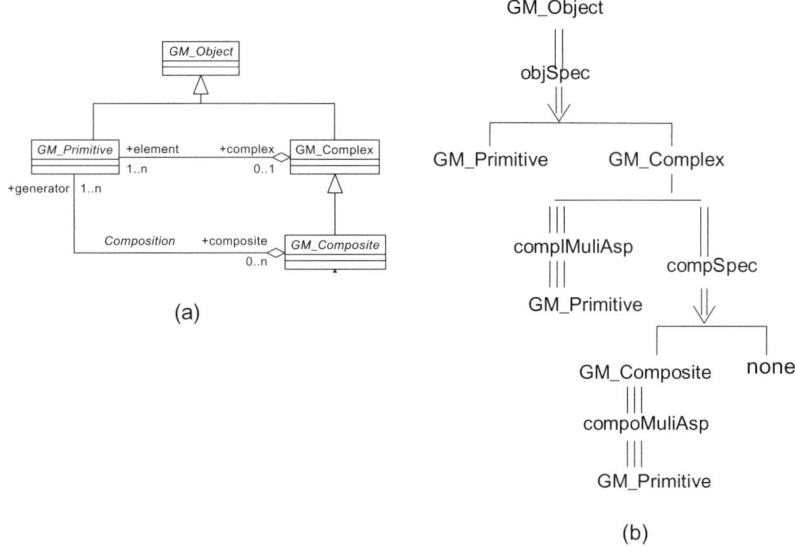

Figure 19.7

Composition Hierarchy for GM_Object and Its SES Representation

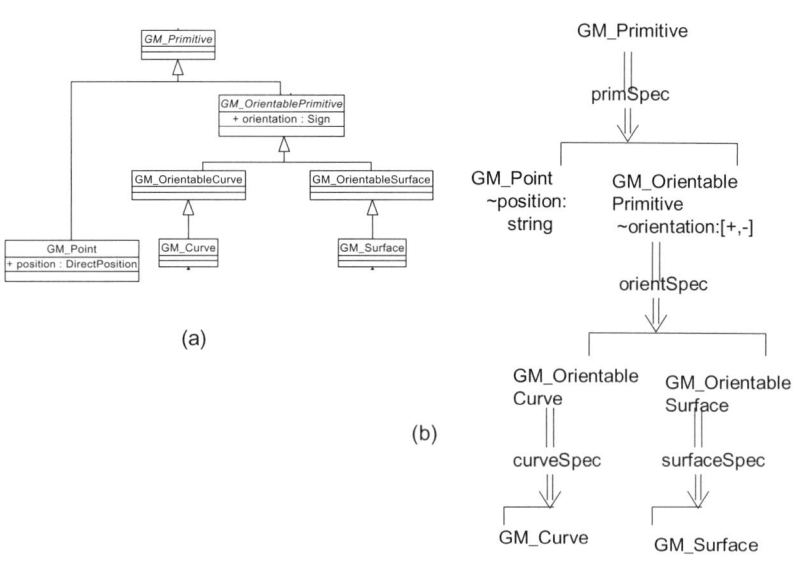

Figure 19.8

Generalization Hierarchy for GM_Primitive and Its SES Representation

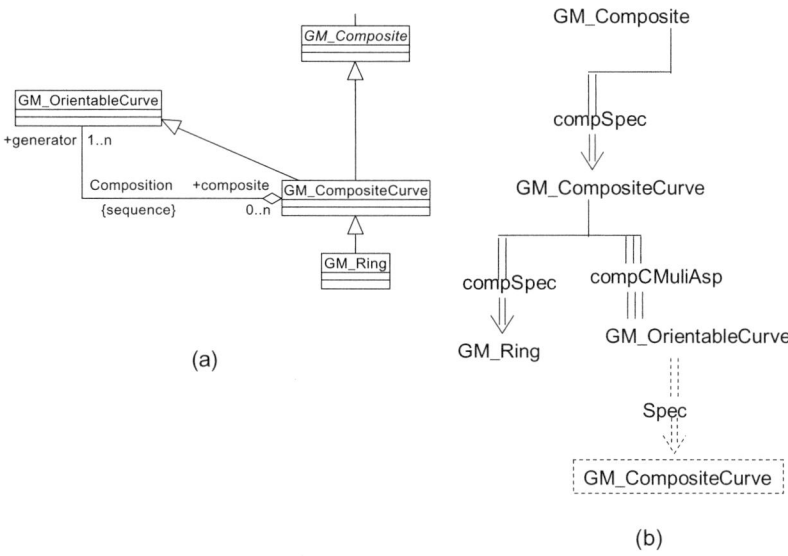

Figure 19.9

Composition Hierarchy Continuation

In contrast, the UML diagram is a general graph that has no inherent directional development except for its generalization hierarchies.

Figure 19.7 shows how a UML diagram unfolds to a tree structure in the SES.

Figure 19.9(a) shows a cycle in the UML diagram that manifests as a violation of strict hierarchy in the SES in Figure 19.9(b). The meaning of such strict hierarchy violation in the SES and its resolution were discussed earlier. Strict hierarchy checking in the SES brings these violations out explicitly and allows the modeler to decide on the level of recursion really needed for the application.

Figure 19.10(a) illustrates the problem of incomplete representation discussed above. The problem in the construction of a surface boundary is given only for the special case of GM_Ring. The equivalent partial SES construction is shown in Figure 19.10(b) and the completion of the representation is illustrated in Figure 19.10(c) where a pruning rule might state that the selection of GM_RingGenerated entity from the GM_Object-BoundarySpec specialization must constrain the selection of GM_Ring from GM_Object specified earlier at the top of the hierarchy (see Figure 19.7(a)). This representation would then allow uniform generation of other possible surface boundary constructions.

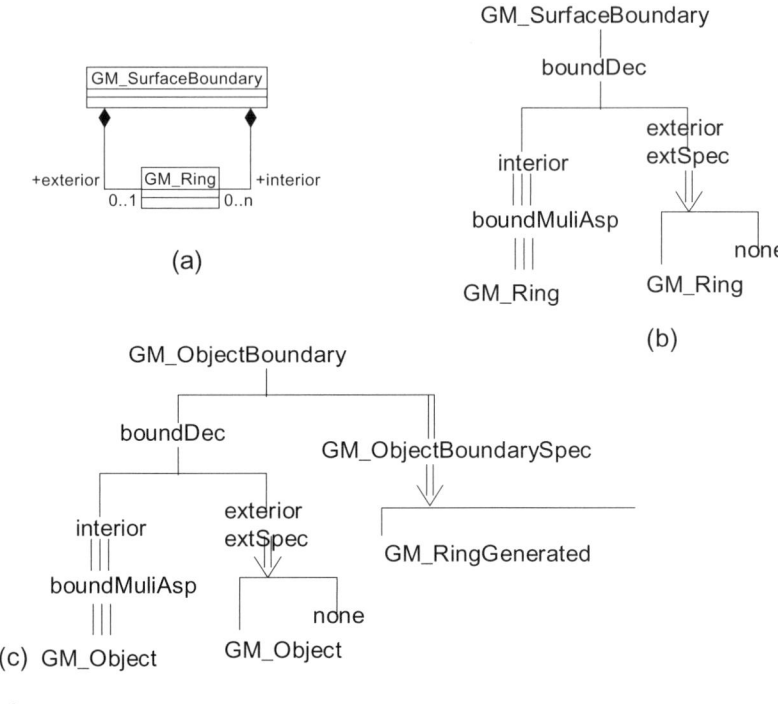

Figure 19.10

Completing the Representation of Boundaries

Web Ontology Language (OWL)

Recall that the Web Ontology Language (OWL) is a W3C recommended ontology standard under development to support intelligent queries.[5] The basic ontology framework employs classes, class hierarchies, and binary relations called properties. We'll use the example of a Wine ontology, similar to that developed by Ref. 3 to explain some of its capabilities of OWL and compare them with those of the SES framework.

The following is a fragment of an OWL description of a class Wine:

```
<owl:Class rdf:ID="Wine">
    <owl:equivalentClass>
        <owl:Class>
            <owl:UnionOf rdf:
              parseType="Collection">
```

```
            <owl:Class rdf:about =
              #WineTaste"/>
            <owl:Class rdf:about =
              #WineColor"/>
         </owl:UnionOf>
      </owl:Class>
   </owl:equivalentClass>
</owlClass>
```

We note that it is declared to be equivalent to an unnamed class that is the union of two classes, WineTaste and WineColor. The WineColor class is described as a subclass of Wine and also as equivalent to a class in which an instance can be Red, White, or Rose:

```
<owl:Class rdf:ID="WineColor">
    <rdfs:subClassOf rdf:resource=#Wine"/>
    <owl:equivalentClass>
        <owl:Class>
            <owl:OneOf rdf:
                parseType="Collection">
                <wineColor rdf:about= #Red"/>
                <wineColor rdf:about= #White"/>
                <wineColor rdf:about= #Rose"/>
            </owl:OneOf>
        </owl:Class>
    </owl:equivalentClass>
</owl:Class>
```

The WineTaste class can be similarly described.[2]

A corresponding SES is displayed in Figure 19.11. It uses labeled specializations for the color and taste families to express the fact that wine is described by both its color and taste. Thus specializations represent the effect of the union of classes in OWL. Further, similar to the subclass construct of OWL, the entities of a specialization such as WhiteWineColor inherit from Wine to accumulate features of both (see Chapter 5).

An example of the property feature in OWL employs the property construct to express the relation between regions and the wines they grow. The binary relation, regionGrows, is defined as a property with domain equal to class Region, and range equal to class Wine.

---

[2] OWL is an extension of an earlier web language RDFS, for describing web resources. This explains the occurrences of tags from both sources.

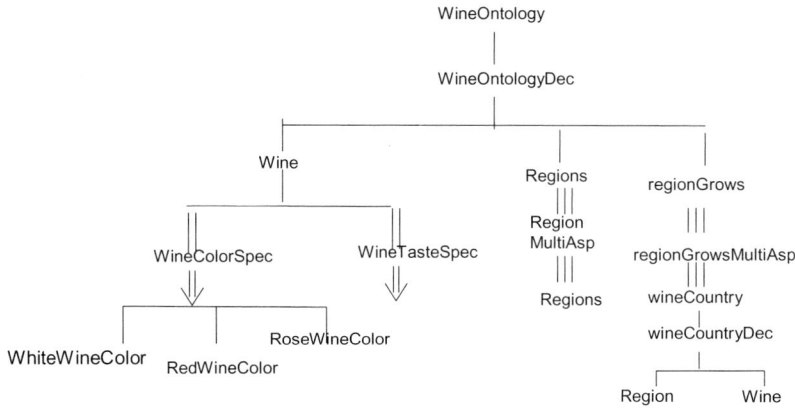

Figure 19.11

SES Representation of the Wine Ontology in OWL

```
<owl:ObjectProperty   rdf:ID="regionGrows">
    <rdfs:domain rdf:resource=#Region"/>
    <rdfs:range rdf:resource=#Wine"/>
</owl:ObjectProperty>
```

The class Region is defined to be a subclass of an unnamed class that restricts the regionGrows property to have values coming from class Region:

```
<owl:Class rdf:ID="Region">
    <rdfs:subClassOf>
        <owl:Restriction>
          <owl:onProperty rdf:
             resource="regionGrows"/>
          <owl:someValuesFrom rdf:
             resource=#Region"/>
        </owl:Restriction>
    </rdfs:subClassOf>
<owl:Class>
```

The equivalent of the regionGrows property is represented by the multiAspect with the same name in the SES in Figure 19.11. We note that this is a horizontal relation in the SES framework as discussed earlier.

In the next section, we discuss OWL and SES in the context of UML, specifically the most recent version of UML 2.0 that has capabilities related to ontology development.

UML and Ontology Development

As we have seen, Unified Modeling Language (UML)[4] is a software development language and environment that has found application in geographic information modeling and other data engineering. There is a natural affinity between software and data engineering — both work with software and with modeling. Although UML was primarily developed for modeling software components and their interrelations, its object-oriented modeling concepts proved attractive to real-world ontology developers. Likewise, recently software developers have realized that there are significant benefits from including application domain ontologies in their architectural designs. In this light, the Object Modeling Group (OMG) has initiated a process for developing an Ontology Definition Metamodel (ODM) for modeling Semantic Web ontology languages within the context of OMG's Model-Driven Architecture (MDA). A proposal for such a definition is presented in Ref. 3 which develops an Ontology UML Profile (OUP) capable of representing the OWL. As illustrated in Figure 19.12, both ODM and OUP are expressed in the MetaObject Facility (MOF),[7] which supports definition of metamodels similar to the instance model discussed earlier. The bottom layer of Figure 19.12 illustrates the development of a capability to capture UML diagrams within the OWL profile and convert them into OWL documents. This approach uses a processing chain in which

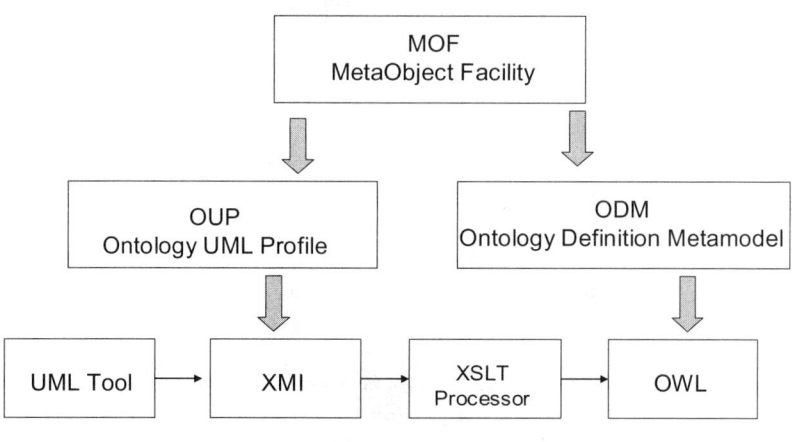

Figure 19.12

Architecture of Gašević et al.'s UML Representation of OWL

the captured UML specification is converted to XML using the XML Metadata Interchange (XMI)[6] standard followed by transformation of the XML into OWL using a processor developed in XSLT, a W3C language for transforming XML documents into other XML documents.

Limitations of UML Ontology Development Support

Figure 19.12 actually represents an advanced formulation that is supported by few tools available today. Although UML was developed to support interoperability of tools, few can actually exchange UML models without information loss. More to the point of this book, UML XMI does not impose a standard set of encoding rules so that the XML encodings of UML ontologies developed in one tool are not likely to make sense when imported to another tool.

The SES Framework within the UML Context

The Geographic Information ontology was developed in UML 1.4, a version that has been superseded by a new, more capable version, UML 2.0 (UML2 for short). Although current UML tools do not support many of the new features, they may evolve in these new directions. Therefore we close by examining the possibility that a profile of the SES may be developed in UML2 that might parallel the approach of Figure 19.12. Two relevant extensions introduced in UML2 are GeneralizationSet and AssociationClass.

GENERALIZATIONSET — allows grouping subclasses into categories. Also called subtype partitions, the concept graduates the IM substitution type (see Table 19.1) into a full-fledged class. Figure 19.13(a) illustrates the grouping of subclasses FemalePerson and MalePerson into a GeneralizationSet named gender, thus separating these subclasses from other subclasses such as Employee. The generalizationSet concept brings UML closer to the concept of specialization in SES. Figure 19.13(b) illustrates the SES concept that employs two specializations, genderSpec and statusType. However, there appear to be both similarities and differences in the semantics of GeneralizationSet and SES specialization. The uniformity axiom in SES allows the reuse of a specialization under any entity (so long as strict hierarchy is not violated). Thus, genderSpec might be hung under electrical connectors in which such distinctions are made. While the class

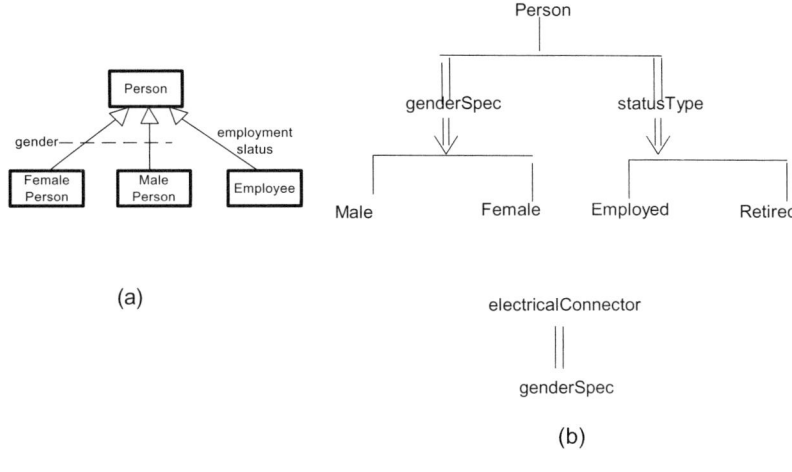

Figure 19.13

The UML GeneralizationSet in Relation to the SES Specialization

status of Generalization might allow such reuse, it is not clear whether other constraints prohibit such reuse since current UML specification is not explicit on this and some of the other interactions involving GenerationSet with other UML concepts. For example, in SES, inheritance works from the bottom up, so that both Person and electricalConnector can inherit properties of Male to become Male_Person and Male_electricalConnector, respectively. However, in typical object oriented programming interpretations, the stand-alone Male class would inherit properties from Person and electricalConnector.

ASSOCIATIONCLASS — generalizes the association feature in old UML in two ways: 1) it is no longer restricted to representing binary relations and 2) it now has the status of both a class and an association. The associationClass might be a vehicle to represent the SES concept of aspect. Recall that, as illustrated in Figure 19.14(a), an entity may have several aspects representing different decompositions or views. Aspects need to be labeled to distinguish such differing decompositions just as specializations need to be labeled to represent different categorizations. In old UML, the corresponding concepts of aggregation and composition represented unlabeled binary whole-part relations that could not be treated as a single decomposition in the manner of an SES aspect. In new UML, the associationClass presents the possibility of doing this, as illustrated in Figure 19.14(b).

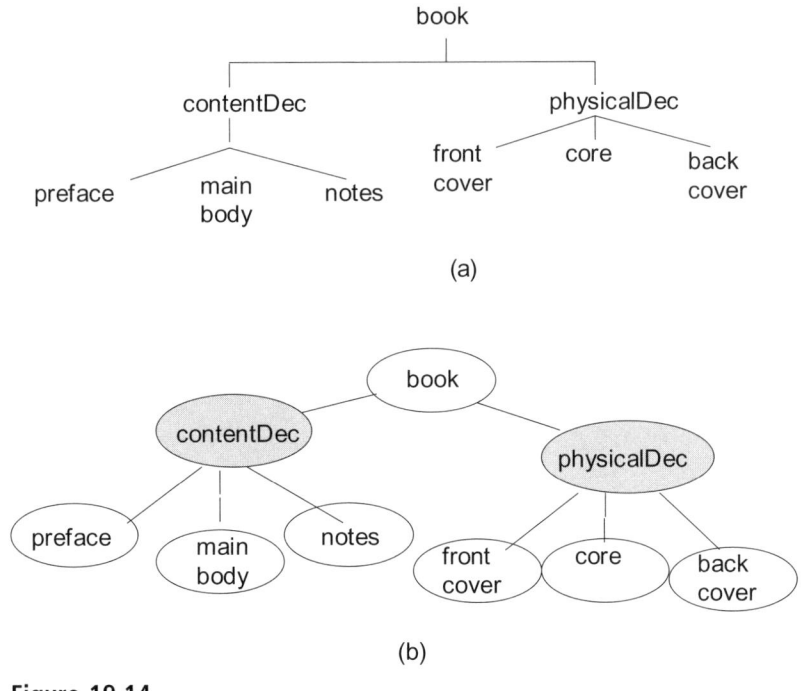

Figure 19.14

Association Classes in UML2 in Relation to SES Aspect

Here contentDec and physicalDec are association classes with multiple end classes. Unfortunately, the interpretation of such associationClasses as aggregations or compositions conflicts with UML2's concept of aggregation since it explicitly insists that aggregation can only be an associationClass with at most two end classes (see Constraint [4] in Section 7.3.3, Ref. 8). The consequence is that standard UML2 tools might not interpret the associationClass as an aspect in the manner intended for XML generation, code generation, and other applications.

Tables 19.2 and 19.3 spell out a relationship between the SES framework and UML2 that might support development of an SES profile in MOF. Table 19.2 makes a correspondence between SES items and UML classes, while Table 19.3 does the same thing for SES axioms. The axioms would have to be expressed within the Object Constraint Language (OCL).[8]

An SES-specific profile in MOF would raise the level at which SES models can be integrated with those of other ontologies. Unfortunately, UML vendors have not been eager to support such integration in general. However, recall from Figure 19.1 that the SES structure specification itself is only one part of the complete

Table 19.2. Relating the SES Items to UML2 Concepts

Item	Denotes	Relation to UML 2.0 conceptual mapping
entity	A thing in the real or modeled world	A class without methods
aspect	The relationship between a thing and its components when decomposed from a certain perspective	Composition or Aggregation are binary and do not support labeling of aspects. Could be formulated as an AssociationClass to obtain labeling and uniformity properties (however, UML2 currently restricts Aggregation to a binary relation)
MultiAspect	A special kind of aspect in which all the components are homogeneous in nature	Composition or Aggregation — with multiplicity specified at the unique end. Same difficulty as with aspect
specialization	The relationship between a thing and its variants from a given family	GeneralizationSet a PowerType with constraints ={disjoint, complete}
variable	A property, quality, or attribute of an entity to which it is attached	attribute of the class
aspect coupling	Internal, external input, and external output port-to-port range translation mappings involving the parent entity and the entities in the composition	Relations on classes and ports attached to the AssociationClass representing the aspect (however, see aspect)

framework. For example, pruning is a critical behavior feature and there is no equivalent concept to pruning of SES instances in the UML *per se*. It is important to note that while UML provides a framework for software architectures, it does not explicitly deal with the manner in which such specifications are construed as alternative variations of the same architecture. Of course, UML, together with its metalevel model, MOF, is part of an ambitious project to support all of object technology. To the extent that it can serve as a high-level specification for any object-oriented architecture, UML will be able to specify behavioral aspects of the SES framework such as pruning as well as subsequent transformation of pruned entity structures to DEVS models. However, here a major gap appears. DEVS is a formalism that intrinsically controls time advances in arbitrarily complex

Table 19.3. Relating the SES Axioms to UML2 Concepts

Axiom	Informal statement	Relation to UML 2.0 conceptual mapping. Restrictions assume capabilities of OCL
uniformity	Any two nodes that have the same labels have identical attached variable types and isomorphic subtrees	Uniformity is a consequence of representing items as Classes
strict hierarchy	No label appears more than once down any path of the tree	Restricts Classes not to have recursive definitions involving their associations
alternating mode	Each node has a mode that is either entity, aspect, or specialization; if the mode of a node is entity, then the modes of its successors are aspect or specialization; if the mode of a node is aspect or specialization, then the modes of its children are entity. The mode of the root is entity.	Restricts the participating classes in GeneralizationSets and CompositeAggregation (as AssociationClasses) to be classes that are not GeneralizationSets nor CompositeAggregations
valid brothers	No two brothers have the same label.	Restriction on multiplicity on Classes associated through the representation of an aspect and specialization
attached variables	No two variable types attached to the same item have the same name.	Same
inheritance	The parent and any child of a specialization combine their individual variables, aspects and specializations when pruning is activated.	Subclasses inherit properties from superclasses. There is no pruning concept to support alternative design selections.

models. On the other hand, UML runtime semantics explicitly state that UML does not dictate how time elapses between its events (see Section 6.3.1 of Ref. 4). Therefore, while UML may express a software architecture for implementing the framework of Figure 19.1, it is not able to incorporate the framework as a profile within MOF. (See ref. 9 for an in-depth comparison of time management and port-coupling in UML and DEVS.)

Table 19.4. Summarizing SES and PES Operations

Operation	Summary description	Applications	Key axiom or feature
SES merging	SESs can be merged as components to form larger SES (Chapter 12)	Synthesize large Schema such as UPHD (Chapter 14), Test Agents (Chapter 17)	Hierarchical modular structure of SES and DEVS
SES substructuring	Substructures can be extracted as stand-alone SESs (Chapter 6)	Decompose large Schema into manageable parts (Chapters 13, 14)	Strict hierarchy
SES path labeling	All items have unique path identifiers (Chapter 6)	Provide context and access for elements in large Schema such as UPHD (Chapters 10, 14)	Uniformity, valid brothers, strict hierarchy
SES comparing	Set theory-based tests can be done to test pairs of SESs for inclusion, equality, etc. (Chapter 13)	Harmonization of legacy formats and checking of profiles (Chapter 14); metadata-based process chain management (Chapter 15)	Set theoretic formulation
SES restructuring	Change in structure from one form to another (Chapter 7)	Improving representation efficiency, reducing size, etc. Reverse engineering applied to harmonization (Chapter 13)	

Continued

Table 19.4. Summarizing SES and PES Operations—cont'd

Operation	Summary description	Applications	Key axiom or feature
SES pruning	Assign values to variables, select from specializations, and choose aspects. Allows partial pruning (Chapter 8)	Generates Schema instances (Chapter 8); supports transformation to DEVS (Chapter 11)	
SES validation	Check that SES satisfies all axioms (Chapter 6)	Validation checking supports error discovery and location; assure consistency of SES synthesized from multiple source SESs; testing of standards and collaborations (Chapter 18)	Set theoretic axioms
PES merging	PESs can be merged as components to form larger PES (Chapter 12)	Synthesize instances of large Schema such as UPHD (Chapter 14), Test Agents (Chapter 17)	Hierarchical modular structure of PES and DEVS
PES substructuring	Substructures can be extracted as stand-alone PESs (Chapter 8)	Decompose large Schema instances into manageable parts (Chapter 14)	Strict hierarchy

PES path labeling and value retrieval	Extract minimal information needed to disambiguate multiple occurrences and access values (Chapter 10)	Provide context and access to values for elements in large Schema instances such as UPHD (Chapter 14)	Uniformity, valid brothers, strict hierarchy
PES differencing	Identify differences and their locations in a pair of PESs (Chapter 10)	Provide basis for change-based information exchange (Chapter 10)	
PES validation/ conformance	Validation applies to completely pruned structure; conformance accepts partial pruning stages (Chapter 6)	Validation checking supports error discovery and location; testing of standards and collaborations (Chapters 17, 18)	
PES transformation to DEVS	Synthesize DEVS simulation models from PES (Chapter 11)	Synthesize distributed Test Agents (Chapter 17)	

Summary

Table 19.4 summarizes the various System Entity and Pruned Entity Structure operations that have been introduced in the book.

Further Research and Development

As we come to the end of this book, we note that our formulation of the System Entity Structure as a relational structure has enabled us to characterize and implement the operations summarized in Table 19.4. These operations support the framework introduced in Chapter 1 for net-centric information exchange and its extension of ontology concepts with pragmatic considerations. This is the good news. The bad news is that although some of the theory has been applied to real-world problems, much of it has not. But this is not so bad after all. As the transition to Service Oriented Architecture proceeds in the next few years, the kinds of problems in harmonization and testing that we have identified will become more pressing and the kinds of solutions that we have sketched here will become more critical. The book's framework and approach will then serve as a starting point for software and services that implement such solutions. Experience with such implementations will no doubt encounter complexities in harmonization and testing that we have only been able to hint at here. This should stimulate further research and lead to better understanding of the issues involved. Perhaps at that stage and looking back, this book will be seen as the first step in a long journey toward better appreciation of the proper roles of syntax, semantics and pragmatics in effective net-centric information exchange.

References

1. International Standards Organization, "Geographic Information — Encoding," sent to the ISO Central Secretariat for publication ISO/TC 211 Geographic information/Geomatics ISO reference number: 19118
2. ISO/TS 19103:2005 Geographic information — Conceptual schema language http://www.iso.org/iso/en/CatalogueDetailPage.CatalogueDetail?CSNUMBER=37800 (accessed Nov. 2006)

3. Gašević, D., D. Djurić, V. Devedžić, "Model Driven Architecture and Ontology Development," Berlin: Springer-Verlag, ISBN: 3-540-32180-2, 2006
4. Unified Modeling Language (UML) Superstructure, version 2.0 http://www.omg.org/technology/documents/formal/uml.htm (accessed Nov. 2006)
5. http://www.w3.org/TR/2004/REC-owl-guide-20040210/ (accessed Nov. 2006)
6. MOF 2.0/XMI Mapping Specification, v2.1, http://www.omg.org/technology/documents/formal/xmi.htm/ (accessed Nov. 2006)
7. Meta Object Facility (MOF) Core Specification, http://www.omg.org/technology/documents/formal/mof.htm/ (accessed Nov. 2006)
8. Object Constraint Language, http://www.omg.org/technology/documents/formal/ocl.htm/ (accessed Nov. 2006)
9. Huang, D., H. S. Sarjoughian, "Software and Simulation Modeling for Real-time Software-intensive System," The 8th IEEE International Symposium on Distributed Simulation and Real Time Applications, pp. 196–203, Budapest, Hungary, Oct. 2006
10. Hessam Sarjoughian, Robert Flasher, "System Modeling with Mixed Object and Data Models," DEVS Symposium, Norfolk, VA, 2007
11. Mittal, Saurabh, *DEVS Unified Process For Integrated Development and Testing Of Service-Oriented Architectures*, Doct. Diss. ECE Dept. U. of Arizona, Tucson, AZ, 2007

Appendix A: Scalable Entity Structure Modeler (SESM)

In this appendix, we briefly overview an environment that supports component-based modeling. In particular the Scalable Entity Structure Modeler (SESM) supports the DEVS and SES framework. Modeling frameworks such as DEVS and UML support modeling dynamics of simulation software models. Each is well-suited for a particular kind of modeling — i.e., DEVS is targeted for simulation modeling and UML for software modeling. A common theme among these approaches is to describe specific system models with strong emphasis on components. In contrast to these, XML is primarily used to model the static view of a system. It can also be used to describe static structures among simple and complex data elements. The System Entity Structure (SES) emphasizes modeling concrete alternative model structures. XML Schemata allow describing arbitrary structures, but they do not have axioms that can establish similarity relationships among different structures of a system. SESM emphasizes a unified logical, visual, and persistent modeling framework for component-based model development.[10]

The principal features of the SESM are scalable multi-aspect/resolution model specification, iterative/incremental model development process, and quantifying complexity metrics for models. The basic concept of the model types and their synthesis facilitate visual and persistent modeling. These capabilities afford automatic creation of well-formed model specifications according to the DTD and XML data language as well as object-oriented programming languages.

The Scalable Entity Structure Modeler is a modeling framework based on Entity-Relation (ER), System Entity Structure (SES), Object-Orientation (OO), visual modeling, simulation modeling, and model transformation. The ER concepts support scalable representation and storage of entities and their relations in databases. The concept for representing a system's structural representation is defined by SES. Its axioms combined with the OO composition, inheritance, and hierarchical structures support organizing alternative structures of a system. The visual modeling concepts offer visual abstractions that are key for systematic handling of tedious, error prone modeling tasks faced by designers and analysts. The simulation concepts are used to account for behavioral modeling of system specifications. Finally, well-formed exchangeable representations play a crucial role for generating alternative models that conform to standardized modeling languages such as XML and programming languages that can be executed with simulation engines. A realization of the SESM approach has been developed using Java and DBMS technologies. SESM's underlying software architecture style is client/server. The first generation of SESM used the Oracle database and subsequently was replaced with MS Access. A modeler can have multiple, independent modeling sessions, each with its own database.

SESM is a component-based modeling approach in which families of system specifications are defined in terms of elementary Template, Instance Template, and Instance model types.[10] Each of the model types is defined to have primitive and composite model components. Every composite model component is a hierarchical tree where its leaf nodes must be primitive model components. A model component can be specialized such that its specializations are distinguishable. SESM defines a set of axioms that characterize compositional and specialization relationships across the elementary model types with support for object-oriented and markup languages. Two kinds of models are defined — i.e., simulatable components are used to define simple and complex dynamic models and non-simulatable models are used to define the static models (models that describe structure, but not behavior).

Appendix A: Scalable Entity Structure Modeler (SESM) 415

The SESM environment supports exporting XML-like models from SESM models and importing XML models for generating SESM models. With forward modeling, XSD and DTD models can be derived from SESM specifications. With backward modeling, XSD and DTD models that are consistent with SESM can be put into the database and thus support visual modeling, analysis of model complexity (i.e., using complexity metrics). These two capabilities offer a round-trip modeling for specifying alternative system designs. Furthermore, component-based simulation code can be generated semi-automatically. In particular, DEVSJAVA atomic and coupled models can be generated from SESM models. The SESM environment supports partial specification of atomic models (i.e., input/output ports, state variables, and the skeletons of the transition functions). Coupled models can be specified completely (i.e., input/output ports, couplings, and hierarchical decomposition).

As shown in Figure A.1, rounded rectangles are used to visually represent primitive components. A rectangle is used to identify composite and specialized components. Components with a multiplicity range are shown as rectangles with dashed lines. The same color coding is used to differentiate model types in the model tree representations. The tree representation uses folder and page visual notations with a letter S to distinguish specialized models. The SESM naming convention tabs for the Template, Instance Template, and Instance models include the SES terms to aid modelers working with the SES concepts. The SES trees can be represented with TM and IM models and

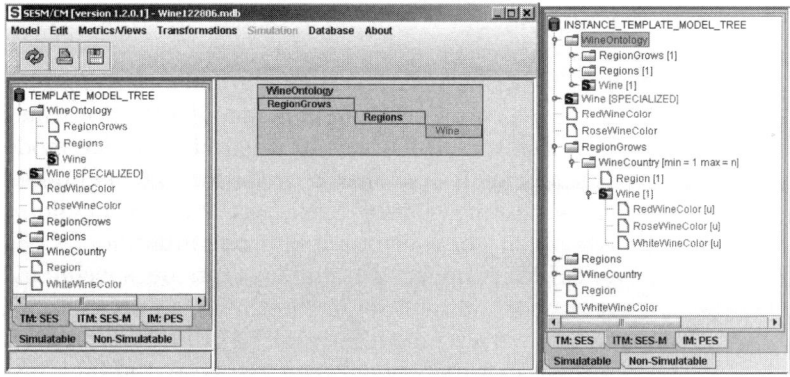

Figure A.1

SESM UI for Tree and Block Model Specification (a) Template Model Tree Structure and Block Model Components (b) Instance Template Model Tree Structure

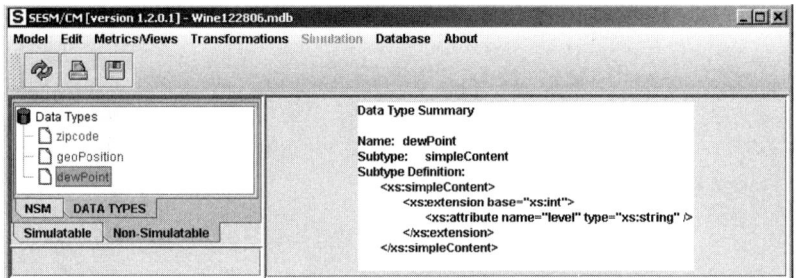

Figure A.2

Data Type Models with XML-S Specification

the pruned entity structure can be represented with IM. The block model components are placed diagonally since the drawing algorithm for (feedforward and feedback) couplings is more efficient and simpler for visual representation and manipulation. Visual modeling of coupling relationships, specification of states (variables and types) and ports (port names, variables, and types) are supported in this panel. Complexity metrics and translation to XML and simulation code are supported in the tree structures, block models, and the menu items (Edit, Metrics/Views, and Transformations lists).

In Figure A.2, three data types are shown. The zip code and geoPosition are XML-S primitive data types. The dewPoint is a simple content element. The simpleContent data type is defined according to the XML-S language. In SESM, a textpad with skeleton tags are provided and the user specifies the rest. These data types can be used as state variables for template models. The Region primitive Template model contains the state variables location and aveDewPoint with geoPosition and dewPoint data types. The Behavioral Metrics for the Region Template model shows input, output, state, and NSM variables. The structural complexity of every template model is also available.

SESM supports the analog of pruning in the SES. For example, it supports direct determination of multiplicities and their combinatorics at multiple levels. Once all multiplicities are determined, instance models can be generated for every primitive and composite Instance Template model. The Instance model for the WineOntology model is shown in Figure A.3.

Given the combined SESM and XML-S specification, every Instance Template model can be transformed into an XSD speci-

Appendix A: Scalable Entity Structure Modeler (SESM) 417

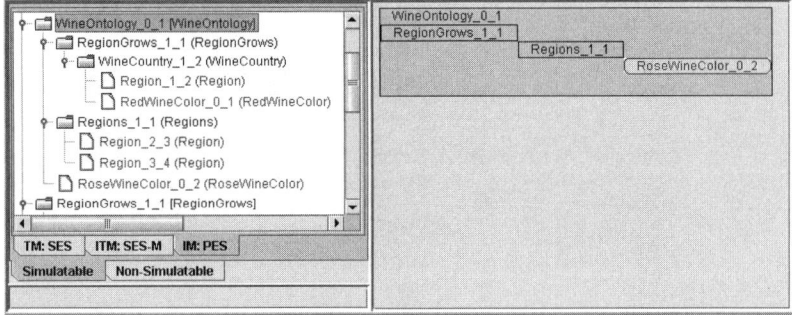

Figure A.3

WineOntology Instance Model

fication. Given the expressiveness of SESM, it can be used to visually specify persistent SES models. The generation of an instance model in SESM results in a model that can have the same representation as those that can be generated with the SES/DTD and SES/XML translators developed for SES/JAVA. This requires (i) there is a correct mapping between SES and SESM and (ii) the representation of XML or DTD data types is also supported in the same way for SES and SESM.

Glossary

ATC-Gen
Automated Test Case Generator generates test cases as federations of test agents to interact with a system under test, exemplifies model-based test automation

Change-based Ontology
Ontology that describes changes in states over time can be implemented by differencing of PESs

Composition
Composing SESs, PESs, or simulation models to create larger hierarchical structures

Constrained Pruning
Constraints on pruning can be stated as rules or relations in the SES framework

DEVS
Discrete Event System Specification formalism describes models developed for simulation; applications include simulation-based testing of collaborative services

Glossary 419

DoDAF
Department of Defense Architecture Framework consists of multiple views broken down into operational, system, and technical categories though which a system is described. DoDAF specifications can be used as input to SES to generate test federations, a form of model-based test automation.

Dynamic Pragmatics
Refers to the change in state of the pragmatic frame due to a change of context occurring in continuous time, discrete time change, or due to a discrete event

GIG
Global Information Grid, DoD's net-centric initiative to construct a SOA on a high bandwidth internet protocol network

Harmonization
Using restructuring and pragmatic equivalence to finding a common ground for, or otherwise reconciling, different data formats or ontologies in order to replace them with a standard for universal and consistent use, e.g., the UPHD standard

I/O Behavior-based test agents
Federations of test agents that embody knowledge of the system behavior under test to observe and assess correctness of observable information exchanges; examples include ATC-Gen and web-service testers

Levels of Interoperability
Levels at which systems can interoperate such as syntactic, semantic, and pragmatic. The higher the level, the more effective is information exchange among participants.

Levels of System Specification
Levels at which dynamic input/output systems can be described, known, or specified ranging from behavioral to structural

Metadata
Data that describes other data; a hierarchical concept in which metadata are a descriptive abstraction above the data it describes

Metadata-based Processing Network Model
Uses the metadata employed by processing nodes to model their organization as layers of processing and to formalize the concept of pragmatic frame

Model-based Test Automation
Automation of test development and deployment that employs models or system specifications, such as DEVS, to derive test artifacts, e.g., I/O Behavior-based test agents

Glossary

Modeling and Simulation Ontology
The SES is interpreted as an ontology for the domain of hierarchical, modular simulation models specified with the DEVS formalism; applications include ATC-Gen and web-service test agents

Net-centric Environment
Network Centered, typically Internet-centered or web-centered information exchange medium

OKBC
Open Knowledge Base Connectivity Knowledge Model is a metalanguage for specifying interoperable ontologies and is specified in Knowledge Interchange Format (KIF), a first-order predicate logic language with set theory

Ontology
Language that describes a state of the world from a particular conceptual view and usually pertains to a particular application domain

PES
Pruned or prunable Entity Structure implemented as instances of the schema generated by an SES

PES Data Extraction
Context-based path identifiers can be employed to access the targeted occurrence of an element

Pragmatic Equivalence
Ontologies are pragmatically equivalent in a pragmatic frame if there is a one-to-one correspondence of their world states descriptions such that corresponding descriptions are used in the same manner within the pragmatic frame of interest

Pragmatic Frame
A means of characterizing the consumer's use of the information sent by a producer; formalized using the concept of processing network model

Pragmatic Framework for Information Exchange
Information exchange occurs when a person or information system reports a new state of the world, or changes in previous states, to another person or information system. The person or system that generates and reports the information is called a *producer* while the user is called a *consumer*. The consumer's *use* of the information should determine the description mechanism, or ontology, used by the producer. The developer of the ontology, also called the data engineer, has the task of tuning the ontology to the pragmatic frame.

Glossary

Pragmatics
Pragmatics is based on Speech Act Theory and focuses on elucidating the intent of the semantics constrained by a given context. Metadata tags to support pragmatics include Authority, Urgency/Consequences, Relationship, Tense, and Completeness.

Predicate Logic
An expressive form of declarative language that can describe ontologies using symbols for individuals, operations, variables, functions with governing axioms and constraints

Processing Network
One or more levels of downstream processing of producer's data by one or more consumers

Restructuring
Mapping whose domain and range are the same, it changes the structure of an object without changing the form in which it is expressed, e.g., an SES can be restructured in the harmonization process

Semantics
Semantics determines the content of messages in which information is packaged. The meaning of a message is the eventual outcome of the processing that it supports.

Sensor
Device that can sense or detect some aspect of the world or some change in such an aspect

SES
System Entity Structure and ontological framework as the basis for modeling and simulation based data engineering. Its pruned entity structures describe static data sets or dynamic simulation models. The SES is formulated in set theory with constraining axioms. The SES generates schemata and its prunings are implemented as schema instances.

SOA
Service Oriented Architecture in which services are designed to be 1) accessed without knowledge of their internals through well-defined interfaces and 2) readily discoverable and composable

Syntax
Prescribes the form of messages in which information is packaged

Testing at Multiple Levels
Federations of agents that can simultaneously observe and test at each level of interoperability

Transformation
Mapping from one representation to another, e.g., from SES to XML

UML
Unified Modeling Language is a software development language and environment that can be used for ontology development and has tools that map UML specifications into XML

UPHD Standard
Universal Phase History Data standard for synthetic aperture radar, developed with the SES framework

XML
eXtensible Markup Language provides a syntax for document structures containing tagged information where tag definitions set up the basis for semantic interpretation

Index

A

Action variables, 331
Active information, 389
Activity-IE relation, 380t
Affection ontology, 23f
Agent-based testing framework, 345-346
Aggregation restructuring, 234-236, 396t
Aligning attributes, 227t
Alignment, 227t
Allowable values, 26
Alternating mode, 71
Alternating mode axiom, 79
Alternative decompositions, 200f
Ambiguity, 258t
Analysis tools, 374
Ancestor contexts, 180
AppendChild, 94
Applicable specifications/standards, 374
Architecture, 365f
  ATC-Gen top-level, 330-331

DEVSML layered, 382
  integrated, 372-382
  model driven, 403
  SOA, 273
Aspect completion ratio, 140
Aspect labels, 114-117
AspectHasEntity, 73
Aspects, 54, 72, 83, 109, 123
Associated specialization, 149
Associating slots, 33-34
Association, 296t
Association classes, 406f
ATC-Gen. *See* Automated Test Case Generator
ATC-Gen process flow diagram, 336f
ATC-Gen top-level architecture, 330-331
Attached new rules, 217f
Attached variables, 72, 110
Attributes, 54
  aligning, 227t
  default, specifications, 148

423

pragmatic frame, 275t
SES, 150f, 153f
Automated Test Case Generator (ATC-Gen), 327
Automation, 324-337
Axes shift, 305f
Axioms for SES, 71-88, 392

B

Basic composite test models, 350f
BasicTestModel, 351
Bit-level format specification, 39
Black box multiplication representation, 67f
Block model components, 415f
Book evolution, 286

C

Capture mission threads, 374
Change-based information exchange, 176-191
  PES, 177-178
  shortening path identifiers, 179-183
Class frames, 32-33
Class relation, 91-93
Classes, 32
Clones, 116
Coherent Data Period, 244
COI. See Communities of Interest
Collaboration layer, 365f
Common entity, 230
Common format approach, 226
Common ontology, 221
Common reference ontology, 221
Common standards-based architecture, 247
Common-activity relation, 380t
Commonality
  measuring, 231, 232f
  renaming, 233f
Communities of Interest (COI), 222
Completeness, 275, 276, 296, 311
Complex data types, 18
Complex parent, 116
ComplexTypes, 105
Component-activity relation, 380t
Composite models, 198f, 199f, 200f
Composition, 396t
  hierarchy, 398f
  hierarchy continuation, 399f
Compounds, 108
Computation sequence, 173, 175
Concatenation, 204
Concatenation operator, 205f

Conceptual interoperability
  levels of, 364f
  model, 362-363
Conceptualization, 5
Conflict-resolution strategies, 228
Conforms, 376
  data engineering, 381, 388-417, 412
  mission thread, 374
  validation, 106, 381
ConformsTo Relation, 138-139
Conjecture, 313
Connective ontology, 25-27
Constrained pruning, 146-175
  implantation, 157-160
  multiAspects interaction, 162-164
  relation-based approach, 154-157
  relation-based pruning for multiAspects, 160-164
  specialization selections, 146-147
Constraining facets, 28f
Context information, 187-190
Continuous range, 124
Controlled test, 347f
Controlled/Opportunistic test modes, 346-348
Converse slot, 158
Copula, 108
Correct comprehension, 15
Corresponding schema, 104
Coupled model, 205f
Cross Product, 204
Cross product operator, 206f

D

Data extraction, 176-191
Data type models, 416f
DDA. See Differential digital analyzer
Decision layer, 365f
Decomposition, 111, 133, 218
Default attribute specifications, 148
Default selection, 153
Department of Defense Architectural Framework (DoDAF), 372
Dependency analysis, 173
Dependency graph, 334f
Dependency-based sequences/cycles, 173-175
Design and search layer, 365f
Determiners, 108
Deterministic corrections, 252
DEVS. See Discrete Event Systems Specification

Index

Differential digital analyzer (DDA), 245
DIS. *See* Distributed Interactive Simulation
Discoverable data, 284-285
Discrete Even Systems Nature, 330f
Discrete Event Systems Specification (DEVS), 328, 374
  DEVSML layered architecture, 382
  interpretation, 199
  models, 204f, 359f
Distributed Interactive Simulation (DIS), 356
Document instance, 96
Document Object Models (DOM), 90, 114
Document type definition (DTD), 16
DoD architectural framework, 373t
DoDAF. *See* Department of Defense Architectural Framework
DoDAF specification, 377
DOM. *See* Document Object Models
Domain(restrictRelationFn), 168
Domain element, 131
Domain space, 5
Downstream processing, 42f
  image, 41-44
  summary of considerations of, 43-44t
DTD. *See* Document type definition
Dynamics, 251
  configuration, 352-354
  data representations, 201-203
  harmonization, 310
  pragmatics, 288-319
    frame comparison, 304-306
    information exchange between humans/systems/services, 290-295
    static pragmatic frame analysis, 303
    static pragmatic frames, 295-301
    service synthesis testing, 339

E

Effective conversations maxims, 45-48
Elementary Logic, 315f
Elements, 95
  domain, 131
  global, 105
  pragmatic frame, 296t
  specialization, 116
  stimulation model, 198t

Encoding rules, 396t
End-to-end interoperability, 372
Engineering-related activities, 249
Enriching semantic equivalence, 302f
Entities, 72, 83
EntityHasAspect, 73, 82, 139-140
Entity-Relation (ER), 414
ER. *See* Entity-Relation
Event set generator, 202f, 204f
Event sets, 202f
Execution layer, 365f
Existing node, 115
Exploitation, 249
Expressive power, 25
eXtended Markup Language (XML), 8, 90, 328
  based characterization, 40f
  parser, 98
  PES, 151f, 152f
  schema, 17-21, 18, 19, 106, 113, 128-145, 370
  SES, 96-98
  specification, 98t
  spy, 100
  syntax, 15-17

F

Facets, 26
  with frames, 33-34
  own, 31-32
  template, 32
Family of Systems (FOSs), 372
FIPA. *See* Foundation For Intelligent Physical Agents
FOSs. *See* Family of Systems
Foundation For Intelligent Physical Agents (FIPA), 5, 25
Frames, 26, 31-32. *See also* Pragmatic frame
  class, 32-33
  compatibility, 306
  facets, 33-34
  general pragmatic, 296f
  slot, 34
  static pragmatic, 295-301

G

General pragmatic frame, 296f
General pragmatics, 292-294
Generalization hierarchy, 398f
Generalizationset, 404-405

Generator, 201
Generic formulation, 248f
Generic instance model (IM), 394
Generic Pruned Entity Structure (GPES), 155
Generic structures, 144
Geographic application schemata, 395-400
Geolocation information, 43t
Georeference, 41, 43t
Geospatial imagery sensors
 background, 37-41
Geospatial sensor data, 35-49, 242-272
GetAttrVal, 177
GetElementOccurrence, 179
GetElementsByTagName, 177
GIG. *See* Global Information Grid
Global elements, 105
Global Information Grid (GIG), 362
Global range definitions, 262-264
Goal-seeking control theory analogy, 310
Gödel numbers, 307f, 313, 314
GPES. *See* Generic Pruned Entity Structure
Graph depictions, 76

H

Harmonization, 113, 220, 292
 of data formats, 225-227
 equivalences, 237f
 increasing commonality via restructuring, 232-233
 metadata, 294
 process, 236
 supported by SES, 228-231
 two approaches to, 221
Harmonizing, 220f
 data, 219-241
 standards, 299-301
HasEntity, 82, 132, 139, 140
Hierarchical event sets
 composite, 207f
 constructing, 207-208
Hierarchical models, 197-199, 393f
Hierarchical structures, 196-197
Hierarchical systems, 195-208
 composing, 214-215
 DEVS model realizations, 203-205
 discourse, 197-199
 dynamic data representations, 201-203

systems specifications, 199-201
why, 196-197
Hierarchy. *See also* Strict hierarchy
 composition, 398f, 399f
 generalization, 398f
 modeling/simulation-based data engineering, 392
 natural, 196f
holdSend, 348
Horizontal relationships, 390-391
HTML. *See* Hypertext Markup Language
Human translators, 291
Hypertext Markup Language (HTML), 15

I

Identified linguistic levels, 368, 369t
IM. *See* Generic instance model
Imagery data, 42, 43t
Impulse Response (IPR), 244
Indices, 225
Individual components, 268t
Individuals, 32
Inferred mandatory classes, 229
Inferred mandatory entities, 230
Information
 active, 389
 exchange, 3-12
 exchange framework, 9f, 129f, 184f
 losslessness, 249-250
 systems interoperability, 363t
Inheritance, 72, 84-86
 of aspects, 85-86, 86f
 clashes, 87
 specializations/variables, 84-85
Initialization, 374
Instance template model tree structure, 415f
Integrated architectures, 372-382
Intelligence, Surveillance and Reconnaissance (ISR), 357
Intended world structures, 5
Intent, 276
Intentional interpretation, 6
Interaction diagrams, 29
International Standards Organization (ISO), 41
Interoperability
 conceptual, 362-363, 364t
 end-to-end, 372
 joint, 373f
 LCIM, 362
 LISI, 362

Index

Interoperation of services, 368t
Interpulse periods, 257t
I/O representation, 345f
I/O specification, 340-343
Island social relations ontology, 19f
Isomorphism-checking techniques, 229
ISR. *See* Intelligence, Surveillance and Reconnaissance

J

Java object, 17
Java package, 94
Joint interoperability, 373f
Joint Single Link Implementation Requirements Specification (JSLIRS), 326
JSLIRS. *See* Joint Single Link Implementation Requirements Specification

K

KIF. *See* Knowledge interchange format
Knowledge interchange format (KIF), 18
Knowledge model, 28f

L

Language analogy example, 293t
Layer to layer transmission, 284f
LayerDeficit, 283
Layered decomposition, 282f
Layered processing networks, 277-281
LCIM. *See* Levels of Conceptual Interoperability Model
Legacy transition, 304
Levels of Conceptual Interoperability Model (LCIM), 362
Levels of Information Systems Interoperability (LISI), 362
Linear algebra, 290
Linguistic levels, 363-368, 367f
Link-16 Specification, 197f
Link-16 Standard, 196
LISI. *See* Levels of Information Systems Interoperability
Logical material implication, 165
Lowest possible level, 126

M

Managing pragmatic equivalence, 301-303
Mandatory entities, 229f

Manner, 46
Mappings, 113-127
MatchSequence, 359
Math model ontology, 24
Maxims, 45-48
Maximum depth, 393f
MDA. *See* Model driven architecture
Measure of effectiveness (MOE), 339
Measures of performance (MOP), 339, 370
Merging, 214, 220f
Metadata, 36
Metadata deficit, 281t
Metadata harmonization, 294
Metadata processing, 284f
Metadata profiles, 246, 252
MIL-STD-6016c document, 327f
Minimal testable input/output pair, 343-344
Minimal testable pairs, 351, 352f
Mixer, 260t
Model driven architecture (MDA), 403
Model realizations, 203-206
Modeling, 365
Modeling/defense acquisition, 356-357
Modeling/simulation, 367f
Modeling/simulation-based data engineering, 388-417
 data engineering, 294
 further research/development, 412
 SES framework, 404-412
 strict hierarchy axiom, 392
 UML/ontology development, 403-404
 vertical/horizontal relationships, 390-391
 web ontology language, 400-402
MOE. *See* Measure of effectiveness
MOP. *See* Measures of performance
Multiaspect, 83, 110, 115, 118f
 completion, 141
 expansion restructuring, 164
 illustrating expansion of, 120
 inheritance, 87f
 interaction, 162-164
 paired, 185
 pruning, 136f
 restructuring, 117-120
Multilevel testing, 195
Multiple inheritance, 84-85, 396t, 397f
 illustrating, 85
 of specializations/variables, 84-85
Multiple levels, 371f

Multiple restriction relations, 171-173
Multiple specializations, 58f
Multiple system views, 372
Multiplicities, 19
Multi-specializations, 53
Multivalued logic, 290

N

Names, 225
National Imagery Transmission Format (NITF), 38-39
Natural hierarchy, 196f
Natural language
  input, 104
  specification, 107, 108-109, 117
  tool, 57, 102
  translations, 291
Nature evolutionary, 338
NECC. *See* Net-Enabled Command Capability
Net-centric environment, 323-360
  automation, 324-337
  conceptual interoperability model, 362-363
  employing agent-based technology, 337-354
  integrated architectures, 372-382
  linguistic levels, 363-368
  simultaneous testing at multiple levels, 368-371
  testing in, 361-387
Net-centric systems, 195, 208
Net-Enabled Command Capability (NECC), 377
Network evolution, 286
Network layer, 365f
NITF. *See* National Imager Transmission Format

O

Object Constraint Language (OCL), 406
Object Modeling Group (OMG), 403
Object-Orientation (OO), 414
Obscurity, 48
Occurrence, 79
OCL. *See* Object Constraint Language
ODM. *See* Ontology definition metamodel
OKBC. *See* Open Knowledge Base Connectivity
OMG. *See* Object Modeling Group
On-demand test development, 338
Ontologies, 3-12
  for affection in community, 21
  common, 221
  common reference, 221
  connective, 25-27
  definition of, 5-6
  design, 12
  development, 18, 30f
  environments, 13-34
  exercise for, 7-8
  geospatial sensor data, 35-49
  integration, 220, 228
  island social relations, 19f
  management, 237
  math model, 24
  payment, 23f
  pragmatic equivalence of, 223f
  prelude to framework, 8
  registration, 15
  semantic web, 28-29
  semantics, 17
  simulation based, 251
  for social relations, 27f
  for social relations on Island, 4
  wine, 402f
Ontology definition metamodel (ODM), 403
OO. *See* Object-Orientation
Open Knowledge Base Connectivity (OKBC), 25, 31-32
Operational-view (OV), 372, 376
Opportunistic message tester, 357-360
Opportunistic test, 347f
OV. *See* Operational-view
Overhead requirements, 39
OWL. *See* Web Ontology Language
Own facets, 31-32
Own slots, 31-32

P

Paired aspects, 185
Paired entities, 185
Paired multiaspects, 185
Paired specializations, 185
Parallel lines, 53
Parent labels, 188
Parent/Grandparent, 115f
Partial pruning, 170f, 307
Partial specification, 393f
Passive information, 389
Passive state, 349

Index

Path identifiers
  using context to shorten, 179-180
  using top ancestors, 182f
Payload state parameter, 252
Payment ontology, 23f
PersonTypes, 21
PES. *See* Pruned entity structures
Physical decomposition, 55f, 121, 133
PhysicalDec, 99
Polynomial representations, 251
Pragmatic equivalence, 222-224, 302f
Pragmatic frame, 3, 8, 9-12, 35-49, 36f, 223, 273-287, 304, 363
  analogy, 293t
  associated with node, 282
  attributes, 275t
  based filtering, 232-233
  changes, 305f
  conjecture, 310-319
  data nature and, 10
  for downstream image processing, 41-44
  elements, 296t
  intent values, 297f
  principles, 45-48
  semantics enhanced by, 306-308
  static, 295-301
  targeting, 300f
Primary data, 36
printTree, 93
Processing function, 283
Processing messages, 342f
Processing net model, 280f
Processing network pragmatics, 274-277
Processing networks, 273-287
  layered, 277-281
  repository of discoverable data, 284-285
Processing nodes, 278, 279f, 281
Processing order, 159
Processing subnet, 282f
Producer, 3
Product definition data, 252
Profile conformance, 229-230
Profile entities, 230f
PROMPT, 228
Propositional logic, 166
Prototypical DEVS test federation, 383f
Prunable entity structures, 135-138, 142f
  validation/completion state of, 144
  XML attribute choice, 151f, 152f

Pruned entity structures (PES), 116, 128-145, 176-191, 215f
  change-based information exchange, 177-178
  example of, 191f
  extract, 182-183
  extracting data, 177-178
  family of, 145
  recursive pattern applied to, 184f
  synchronizing, 238f
PrunedState, 169, 172
Pruning, 120
  comparison, 187f
  completion, 139-140
  constrained, 146-175
  opening up multiAspect, 136f
  operator, 158
  partial, 170f, 307
  process, 82
  relation-based, 155, 156f, 164f, 167-170
  restrictive relations to, 157f
  root-based, 159-160
  rule satisfaction, 385-387
  SES/JAVA tools used by, 141-144
  state, 174
  synchronized, 237-239
  two prunings in same SES, 183-186
  unique complete, 307

Q

Quality, 46, 47
Quantity, 46, 47
Quantized event sets, 202f
Quantrized form, 203

R

Range restrictions, 64-66
Range sets, 225
Range specialization, 159
Range specifications, 110
Raw sensor data, 43t
Reception, 259-260t
Recursive depth, 268f
Recursive pattern, 97
Registration ontology, 15
Relation-based approach, 154-157
Relation-based pruning, 155, 164f
  overall process of, 156f
  theory support, 167-170
Relations
  activity-IE, 380t
  class, 91-93
  component-activity, 380t

430 Index

conformsTo, 138-139
restrictive, 156-160
UML binary, 397f
vertical, 390-391
Relationships, 54, 295
Renaming, 233
Replace method, 115
Representation boundaries, 400f
Representation completing, 392-394
Representing Input/Output pair, 341f
Restrictive relations, 156-160
Restructurings, 113-127
   aggregation, 234-236, 396t
   effects, 235t
   multiaspect, 117-120
   source, 235t
   tracking, 236-237
   transformations, 114f
   variables, 121, 122, 234
Root-based pruning, 159-160
Rule capture component, 330
Rule formalization component, 333
Rule set analyzer component, 332
Rule-based approach, 147-148

S

SAR. *See* Synthetic Aperture Radar
SAR phase history, 244
SAR phase history generation, 243f
SAR sensing, 243-245
Satellite-transmitted weather data, 37t
Scalable entity structure modeler (SESM), 413-417
Schema representations, 100, 102-105
Schema validation, 106
SchemaWithOptions, 117
Second order, 254
Selection rules, 165-167
Semantic equivalence, 224
Semantic inclusion, 224
Semantic integration, 239
Semantic web ontologies, 28-29
Semantical expressive power, 17
Semantics, 13-34, 363
   definition of, 13
   ontology and, 17
   testing web services, 14
Semi-automation, 335-337
Sensor, 36f
   accumulated use, 277f
   data, 43-44t
   data types, 43t

description, 39f
model, 44-45, 45
Sentence order, 109
Sequence removing, 117
Service Oriented Architecture (SOA), 273, 288, 375
Service-oriented computing, 338
SES. *See* System Entity Structure
SES framework, 404-412
SES items, 54
SES natural language description, 254
SES representation, 397f
SES specialization, 405f
Shared substructures, 232f
Simple processing network, 274f
Simulation based ontology, 251
Simulation based testing, 325-326
Simulation systems, 364
Simulation-based data engineering methodology, 389f
Simulator transparency, 382
Simultaneous testing, 371f
Single choice, 117
Single inheritance, 396t
Singleton equivalence classes, 188
Slot frames, 34
Small Group, 166, 173
SOA. *See* Service Oriented Architecture
Source restructuring, 235t
Spatial data, 252, 256-262
Specialization, 59, 72, 83, 84-85, 109, 114-117, 121
   completion ratio, 140
   element, 116
   HasEntity, 82, 132, 139, 140
   labels, 116f
   partitioning continuous ranges using, 123
   restructuring a variable into, 122
   specificity, 125-126
Standardization, 40
Standards conformance testing, 337
State of Completion, 140
State transition diagram, 15f
Static description, 251
Static pragmatic frames, 295-301, 303
Static world state, 128
Stereotype, 394n
Stimulation model elements, 198t
Stove-piped chains, 48
Stove-piped solutions, 289

Strict hierarchy, 71
 axiom, 392
 requirement, 79-80
Structured representation, 227-228
StyleSpec, 166, 172, 173
Subnet, 281
Subspecializations, 162
Substitution groups, 397f
Substitution types, 396
Substructures, 217f, 234
SUT. *See* System Under Test
SV. *See* Systems view
SV leaf entities, 262
Synchronized prunings, 237-239
Syntax, 13-34, 100, 364
 definition of, 13
 XML and, 15-17
Synthetic Aperture Radar (SAR), 242
 generic formulation of, 251-252
 illustrating timing considerations, 245f
 sensors, 243-245
 structure/descriptors, 252f
System Entity Structure, 25, 53-70
 for bicycle with attached variables, 63f
 with multiAspects, 82-84
 problems for, 88
System Entity Structure (SES), 38, 195-208, 375
 alternative, 153
 architecture/pruning, 379f
 attribute choice, 150f, 153f
 axioms, 71-88
 basic concept summary, 68t
 black box multiplication representation, 67f
 for book with chapters/sections and conforming, 181f
 for book with core composed of pages, 61
 class relation of, 91-93
 components of, 216f, 252-253
 composition, 212-213, 215f, 216f
 computational representations, 90-112
 computed values and formulas, 66-68
 constructing, 92f
 directed graph from, 74
 extracting testable behaviors from, 378f
 forms, 109-111

harmonization supported by, 228-231
harmonized, 237-239, 238f
illustrating, 57
inheritance, 84-86
interaction between aspects/specializations, 57-58
interrelation, 91f
managing, 210-218
mapping, 198t
merging, 210-213
multiAspects, 135
natural language specification syntax, 107-109
ontology/implantation levels, 129f
prunings comparison, 183-186
range restrictions, 64-66
relating, 407t, 408t
relational specification of, 72-76
Relation-based pruning, 164f
representing, 94-96, 376
restating information exchange framework in, 184f
restrictive relations, 158f
revisiting, 390-391
schema generated from, 267t
schema representations for, 102-105, 103t
schema validation of, 106
as semantical concept, 100
strict hierarchy for, 79-80
substructures comparison, 231f
summarizing, 409t-411t
uniformity application, 76-78
unique path labeling, 81
for validity, 100
as XML file, 96-98
System modeling, 251
System specification, 201t
System Under Test (SUT), 329, 357
Systems of Systems, 325f
Systems view (SV), 372

T

Target geometric parameters, 252
Targeted access, 217f
Targeting, 299-301
TCDE. *See* Technical Committee on Data Engineering
Technical Committee on Data Engineering (TCDE), 294
Technical interoperation, 367
Technical view (TV), 372
Template facets, 32

Template model tree structure, 415f
Template slots, 32
Tense, 276, 296
Test agent construction, 349-352
Test agents, 352-354
  deploying, 380
  federation, 381f
Test control, 353
Test deployer, 348
Test derivation process, 378f
Test detector, 347
Test development approach, 372-375
Test driver deployment, 375
Test federations, 380-382
Test generation component, 333
Testable behaviors, 339, 374, 377
Testable I/O behaviors, 340
Testing
  multilevel, 195
  in net-centric environment, 361-387
  semantics, 14
  service synthesis, 339
  simulation based, 325-326
  simultaneous, 371f
  standards conformance, 337
  web services, 14
Top ancestor, 187-190
  illustrating, 188f
  path identifiers, 182f
  tree of, 190-191
Total function, 131
Tracking restructurings, 236-237
Transformation description, 119
Transformations, 113-127
Translation, 204
  natural language, 291
  operator, 205f
  relation, 239t
Tree depictions, 76
Tree operations, 114-117
TV. *See* Technical view
TV set Standard, 248f

U

UML binary relations, 397f
UML generalizationSet, 405f
UML/ontology development, 403-404
Unified Modeling Language (UML), 22-24, 44, 389, 394-400
Uniform Resource Locators (URLs), 15
Uniformity, 71
  application, 76-78, 111
  axiom, 77f

Unique complete pruning, 307
Unique identifiers, 131
Unique path labeling property, 81
Unique specification, 307f
Units/representation, 250
Universal Phase History Data (UPHD), 242-272
  abbreviations, 269-270
  global range definitions, 262-264
  goals for, standard development, 245-247
  managing, master/components, 264-265
  many-to-many producers to consumers, 246f
  SAR sensing, 243-245
  spatial data, 256-262
  syntactic/semantic/pragmatic, 250-251
  system modeling, 251
  table of all paths, 270-272
  top level components, 253f
  universality/scope of, 247-250
  VH, 256
  WF, 253-256
Universe of discourse, 5
Untranslated format, 288
UPHD. *See* Universal Phase History Data
Urgency, 275, 276, 295
URLs. *See* Uniform Resource Locators
Usage, 226

V

Valid brothers axiom, 78-79
ValidatePruning, 144
Validation, 144f
Variable replacement, 125f
Variables, 54, 84-85, 123, 331
Vehicle identification number (VIN), 14
Vehicle position, 263f
Verification, 144
Vertical relationships, 390-391
VH. *See* Wideband vector header data
VIN. *See* Vehicle identification number

W

waitNotReceive, 348
waitReceive, 348
WaitReceive primitive, 349f

Index

Wave form parameters (WF), 252, 253-256
Weather Generator Simulation model, 214-215
Web Ontology Language (OWL), 28-29
WF. *See* Wave form parameters
Wideband vector header data (VH), 252, 256
Wine ontology, 402f
World state change, 10

X

XML. *See* eXtensible Markup Language
XML Schema document (XSD), 395
XML-S specification, 416f
XSD. *See* XML Schema document